Maths

Teacher's Guide

2

Ray Huntley
Tony Cotton

Caroline Clissold
Linda Glithro
Cherri Moseley
Janet Rees

OXFORD
UNIVERSITY PRESS

OXFORD
UNIVERSITY PRESS

Great Clarendon Street, Oxford, OX2 6DP, United Kingdom

Oxford University Press is a department of the University of Oxford. It furthers the University's objective of excellence in research, scholarship, and education by publishing worldwide. Oxford is a registered trade mark of Oxford University Press in the UK and in certain other countries.

British Library Cataloguing in Publication Data

Data available

ISBN 9781382017275

12

Paper used in the production of this book is a natural, recyclable product made from wood grown in sustainable forests. The manufacturing process conforms to the environmental regulations of the country of origin.

Printed and bound by CPI Group (UK) Ltd, Croydon, CR0 4YY

Acknowledgements

The publisher and authors would like to thank the following for permission to use photographs and other copyright material:

Cover: Artwork by Peskimo. **Photos: pv(t):** foto-bee/Alamy Stock Photo; **pv(c):** Alistair McDonald/Shutterstock; **pv(b):** 2R fotografia/Shutterstock; **px:** Monkey Business Images/Shutterstock; **pxviii:** monkeybusinessimages/iStockphoto.

Artwork by Q2A Media Services Pvt. Ltd.

Every effort has been made to contact copyright holders of material reproduced in this book. Any omissions will be rectified in subsequent printings if notice is given to the publisher.

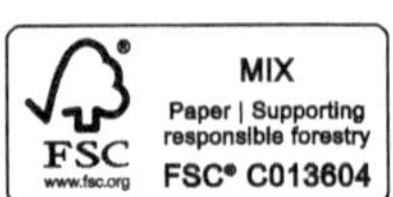

The manufacturer's authorised representative in the EU for product safety is Oxford University Press España S.A. of El Parque Empresarial San Fernando de Henares, Avenida de Castilla, 2 – 28830 Madrid (www.oup.es/en or product.safety@oup.com). OUP España S.A. also acts as importer into Spain of products made by the manufacturer.

Contents

Introduction

The joy of learning maths

We are living in an ever-changing world, where the way we work, live, learn, communicate and relate to one another is constantly shifting. In this climate, we need to instil in our learners the skills to equip them for every eventuality so they are able to overcome challenges, adapt to change and have the best chance of success. To do this, we need to evolve beyond traditional teaching approaches and foster an environment where students can start to build lifelong learning skills for success. Students need to learn how to learn, how to problem solve, be agile and work flexibly. Going hand-in-hand with this is the development of self-awareness and mindfulness through the promotion of wellbeing to ensure students learn the socio-emotional skills to succeed.

With *Oxford International Primary Maths*, students develop lifelong learning skills as well as mathematical skills. The course promotes the development of real-world skills including financial literacy. The activities in the Student Books and Practice Books offer numerous opportunities to think creatively and develop interpersonal skills. Fundamentally, *Oxford International Primary Maths* promotes students' self-development as critical thinking and motivation are at the heart of the problem-solving approach in the course.

This series is based on the English National Curriculum Programme of Study for Primary Maths. The books for each stage meet all the learning objectives. Each lesson includes the learning objectives from the curriculum and summary of the key teaching points. A full mapping grid identifying the unit and lesson where each objective can be found is available online at https://www.oxfordowl.co.uk/

Oxford International Primary Maths: a problem-solving approach

In this second edition of *Oxford International Primary Maths*, there is a strong focus on using a problem-solving approach. Whilst mathematical facts are important, it is unlikely that simply giving students the information they need will result in them understanding the mathematics and being able to apply their learning in new problem-solving situations. This is often described as a move from 'surface learning' to 'deep learning'.

Many people remember mathematics lessons as places where the teacher stood at the front of the class writing on the board. The students copied the information down, maybe worked through a couple of examples with the teacher and then proceeded to complete a series of exercises to practise the skill that they had been taught. This can be described as a *didactic* approach and it relies on the idea that direct instruction is the appropriate strategy to adopt. The authors of this series would argue that *heuristic* strategies encourage students to explore the mathematics for themselves supported by the teacher. 'Heuristic' derives from the Greek word meaning to discover, and in mathematics learning, heuristic strategies are ones where the student engages in exploration and discovery to solve a problem. Heuristic strategies include making a visual representation of a problem, making a calculated guess or estimate, simplifying a problem or following a known method. This results in a deeper understanding for the student.

When faced with any problem in mathematics, there are recognised stages to go through in order to solve the problem, and these have been developed and agreed by many researchers. One version that summaries the problem-solving process comes from Georg Polya:

1. Understand the problem

2. Devise a plan

3. Carry out the plan

4. Check the reasoning

In following these stages, students will engage in a number of skills which support problem solving such as trial and improvement, working systematically, pattern spotting, visualising, conjecturing and generalising.

Embedding a mastery approach

In recent years, the term 'mastery' has been used in conjunction with mathematics learning. It has been drawn from teaching approaches in countries where mathematics performance is deemed to be very high. The essence of mastery is to produce students who have deep conceptual understanding and procedural fluency through learning in a collaborative and problem-solving context. Mastery learning incorporates use of manipulatives, exposure to different methods of solving a problem, dialogue and explanation.

Following a Concrete Pictorial Abstract (CPA) approach

One of the more successful approaches to learning was provided by Jerome Bruner in his model of enactive, iconic and symbolic modes. This has been developed in recent years to form the CPA approach, which stands for concrete, pictorial and abstract, each of which aligns with Bruner's modes. The concrete phase is about making use of physical manipulatives to help understand the learning, before moving to record the learning in pictorial form as an individual student. As the learning develops, students will begin to recognise how to record their learning in a more general and abstract way. The CPA approach is not necessarily sequential, and students might move between the different modes as they work through a problem.

Oxford International Primary Maths **and the use of manipulatives**

Throughout the series, students are encouraged to use manipulatives, or concrete objects, to model addition, subtraction, multiplication and division. These manipulatives include:

- base-ten equipment (ones-cubes, tens-rods, hundreds-flats and thousands-cubes)

- place-value counters

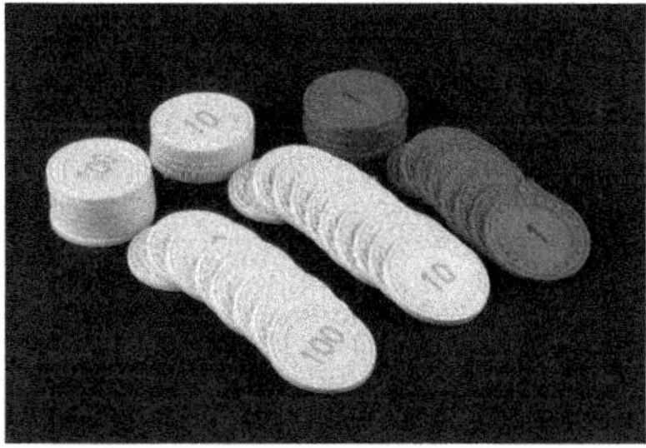

- number rods

Such manipulatives are used to explain to students how the written methods 'work', for example, by modelling exchanging 10 ones cubes for 1 tens rods in an addition.

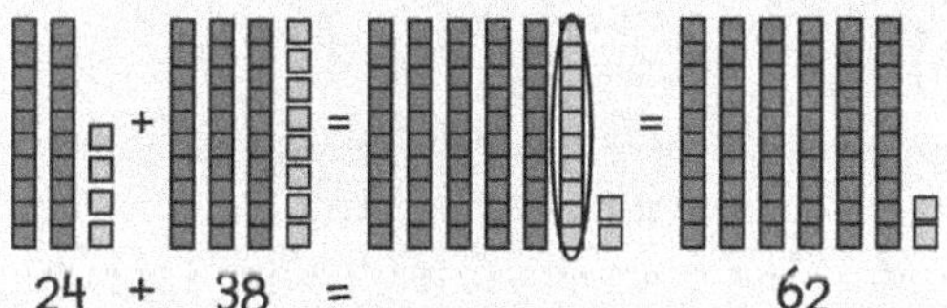

$$24 + 38 = 62$$

Differentiation

There are several ways that you can differentiate learning in the classroom. These include:

- Differentiation by task
- Differentiation by outcome
- Differentiation by support
- Differentiation by grouping.

It has been traditional in some schools to offer up to three different levels of tasks for each lesson. This is differentiation by task. It is important that all students are exploring the same area of mathematics as they

can collaborate and discuss their mathematics in a way that is not possible if students are engaged on different activities. This approach has been extensively researched and published by Jo Boaler of Stanford University, California. In her book 'The Elephant in the Classroom' (2nd Edition, 2015, Souvenir Press) she outlines projects which gave students in different schools either a differentiated approach in lessons, or lessons where everyone worked on the same task. Where all abilities worked on the same task, every student made and sustained 'better than expected' progress, and performed better on statutory tests and exams. The Education Endowment Foundation teacher's toolkit suggests that collaborative learning can result in a five-month acceleration in student learning. (See https://educationendowmentfoundation.org.uk/resources/teaching-learning-toolkit .)

The expectation in this series is that all students will be offered the same starting point. The activities are carefully designed to be accessible to all students in your class and the teacher's notes for the activity offer differentiated outcomes for students. It is also important that you offer differentiated support to different students. You will mainly do this through the sort of questioning that you engage in and support you offer. You will ask challenging questions and supporting questions to help all students access the task. For example, when engaging in a simple counting activity with some students you may model the action of counting by placing a finger on each object as you count and emphasise the last number you say to model that the last number you say gives the number of objects in the set. For other students engaged in the same activity you may ask them to compare two sets, or to find one more or one less than the set they are counting.

Grouping students to promote a growth mindset

When engaging in learning mathematics, it is expected that you will use a variety of student groupings. This may be a change for some teachers who have previously grouped students by prior attainment in their classroom. Research has shown that grouping students 'by ability' which usually means grouping students using test results, can have a negative impact on their future attainment. It is more effective to use a range of ways of grouping students. You will decide on the most appropriate way of grouping students depending on the activity. You are also given advice in the teacher's notes. It is important that the teacher is active in deciding which form of grouping is appropriate. It is also important that students learn how to operate in a range of different groups and with a range of different students so that they get used to working in a variety of ways and with different people.

There are three main ways of grouping students:

- Friendship
- Ability/Prior experience
- Mixed attainment.

Friendship groups, are most appropriate for activities in which the students have been given some element of choice. Perhaps they are carrying out some research for a data handling project or exploring data on animals to develop their understanding of measurement. This grouping is the default if teachers do not actively group students.

Ability groups, or groups based on the prior experience of students, may be helpful if the lesson requires a very specific prior knowledge. You can group students who you know have this knowledge together as they can then work with minimal teacher guidance which then allows you to focus on groups who need additional support.

Mixed-attainment groups are the grouping that is encouraged for the majority of the activities. This is also the form of grouping favoured by those following a mastery approach. Working in collaborative, all-attainment groups also supports students' wellbeing and promotes a growth mindset, as described in research by Carol Dweck. She found that students who were put in ability groupings tended to stay in those groupings throughout their school life, and regard themselves as having a fixed ability that could not be changed. This has dire consequences for students in middle or lower sets. By using mixed ability groupings, all students can develop a growth mindset which enables them to believe they can learn and improve, whatever their starting point. (Dweck, C., 2007, The Perils and Promise of Praise, Educational Leadership, October 2007, 65(2), 34-39). A growth mindset is promoted when students do not feel that their future success is predicated on prior achievement. This kind of grouping is particularly helpful for students new to English. Mixed-attainment groups allow students who are less confident in English to hear their more confident peers using mathematical vocabulary. Research has shown that mixed-attainment groups benefit both high attainers, who become more secure in their mathematics knowledge through explaining their thinking to peers, and to those less secure in their mathematical knowledge as peer teaching has been shown to be effective.

Whatever form of grouping you choose, it is helpful to assign roles to individuals in the group. Some teachers use 'role cards' to remind members of the group of the role they should play. Examples of these roles are:

- Leader: You should make sure everyone has a chance to speak and focus the discussion around the task.

- Time keeper: You should encourage the group to stay on task. Announce when the time is half way through and when time is nearly up.

- Recorder: You should write down the group members ideas or draw a collective graphic. You will write on the board during the presentation.

- Presenter: You will present the group's findings to the whole class at the end of the session.

- Resource organiser: You will make sure the group has all the resources they need during the task.

Assessment is the process of establishing how each student is progressing and what they have achieved, or a means of measuring their learning. Assessment is usually carried out in two main ways – assessment of learning and assessment for learning.

Assessment of learning is sometimes called summative assessment, and takes place at the end of a lesson, a unit, a term or even a year. It measures what students know at that point as a summary of their learning to that point. In *Oxford International Primary Maths*, summative assessment opportunities are provided in the Review lesson at the end of each unit in the Student Book, whilst half-termly summative assessment opportunities are provided through printable resources, available online.

Assessment for learning is an approach brought to prominence by Paul Black and Dylan Wiliam and is based on the notion that students have a full, clear sense of what they are learning, where they have reached in their learning and what they need to do to improve further. It is carried out during lessons and gives teachers continuous data on each student's learning, as well as allowing students to track their own learning, which provides greater motivation. (Black, P., Harrison, C., Lee, C., Marshall, B., and Wiliam, D. (2004) Inside the Black Box: Assessment for learning in the classroom. Phi Delta Kappan, Vol. 86 No. 1 pp8-21)

It is suggested that there are five key strategies for assessment for learning. These are outlined below with suggestions of how you can do this in your classroom.

1 Being clear about learning objectives and success criteria with the students.

Each activity has at least one learning objective. At the beginning of a lesson, share the activity's learning objective with the students. This should be more than simply stating the objective. You should make sure that students understand the objective and how you will measure success. For example, you might say: *I know that you can all count 10 objects* and all count to 10 as a class. Then you point to '20' on a number line and ask: *Does anyone know what this number is?* If a student knows it is 20 praise them, if no-one knows, tell them it is twenty and say: *By the end of the lesson I will be able to listen to you count to twenty.*

2 Planning student discussions that give you evidence of their learning.

Every activity plan in the Teacher's Guide offers the opportunity for small-group or whole-class discussion. There are also examples of probing questions that you can ask to assess students' current understanding. For example, if a group has been counting two sets of objects you can ask: *Were there more or less in the second group? How do you know?*

3 Giving students feedback that helps them move forward.

This allows students to know whether or not they are meeting the success criteria and what they can do next to move their learning on. Developing the example above, if a group has been comparing two sets and understands the concept of 'more' and 'less' you could ask them to make sets that are 'one more' and 'one less' or even 'two more, and, two less'.

4   Activating students to act as instructional resources for each other.

Collaborative group work in mixed-attainment groups, as described by Jo Boaler in her research (see under Differentiation earlier), gives students the opportunity to operate both as learners and teachers, with peer learning being highly effective. Not only is understanding of the mathematics enhanced, but students can support each other in assessing their progress.

5   Activating students as owners of their own learning.

The key point here is to listen carefully to the students and adapt your questioning to support individual development and to follow individual interests.

Questioning is key

The most skilled mathematics teachers can ask open questions to elicit students' current understandings. Skilful open questioning also allows students to articulate their current understanding carefully and though this process either consolidate their understanding or come to realise where they have made a mistake. The list below offers a series of open questions that can be used whatever mathematics you are teaching:

- *How are these the same/different?*
- *About how many/how long/many more …. do you think there will be?*
- *What would happen if …?*
- *How else could you have done that?*
- *Why did you ….?*
- *How did you …?*
- *How do you know that is correct?*

If you want students to check their solutions and consolidate their learning it is helpful to ask them to explain how they reached their solution to a friend. Similarly, to support students in reflecting on their learning you might ask:

- *What mathematics did you use to solve the problem?*
- *What new mathematics did you learn?*
- *What key words did you use?*
- *What was the most challenging part of the activity?*
- *What did you do when you got stuck?*
- *What other questions could you ask?*
- *Did this remind you of any other areas of mathematics?*

In *Oxford International Primary Maths*, there is an opportunity to ask these reflective questions, and for students to reflect on their learning, at the end of each unit in the Review lesson of the Practice Book.

Word problems

Word problems are useful as an assessment of children's understanding of the correct mathematics to use in any given situation. In *Oxford International Primary Maths* word problems are included throughout the units and on every Student Book Review page as part of the end-of-unit assessment. Many teachers find teaching word problems a challenge. This area is particularly challenging for students with a limited English vocabulary as word problems are tightly bound to linguistic ability. We have to decode and understand what the problem is asking us to do before we can begin to apply our mathematical knowledge. Some teachers have found the following acronym helpful when working with students on solving word problems.

R: Read the problem carefully.

U: Understand what the problem is asking you to do.

C: Choose the mathematics or arithmetical operations that you need to use to solve the problem.

S: Solve the problem.

A: Answer the problem.

C: Check the answer is accurate and reasonable.

It is often helpful for students to underline key facts and write down the operations they are going to use before they solve the problem. For example:

> Tony rode his bicycle 7 miles to school with his friend. On his way home he took a short cut which was only 5 miles. How far did he cycle altogether?
>
> *This will be an addition calculation.*

It is a useful activity for students to annotate word problems and write down the operation(s) they will use without carrying out the calculation as this focuses on the skill of understanding the problem and choosing the operations appropriately.

Another activity which helps students becomes skilled at solving word problems is asking them to write their own word problems based on a picture or a set of objects. For example:

- How many black cubes are there? (3)
- Two friends took three cubes each. How many were left? (2)
- If I take out the black cubes, how many are left? (5)
- If I share the cubes equally between two people, how many do they each get? (4)

Wellbeing and *Oxford International Primary Maths*

It is thought that children learn more and feel more connected to their learning when they are active in their lessons. OIPM has active learning at its heart. Most lessons start with a whole-class session which usually includes a range of physical or active activites. You will see this signified by a star jump icon in the Teacher Guide.

Many adults and children have felt anxious about their learning of mathematics at some stage. This anxiety is reduced by working collaboratively in all-attainment groups. There is also a reflective session at the end of each lesson and the formative assessment activity in the Practice Book asks students to reflect on their learning across the unit.

Wellbeing is also supported by effective questioning to support and stretch students and by planning group work carefully. These areas have already been discussed above.

Language Support

The challenges

Ministries of Education at both local and national level are increasingly adopting the policy of English Medium Instruction (EMI), for either one or two subjects or across the whole curriculum. The rationale for doing so varies according to the local context, but improving the levels of achievement in English is an important factor.

In international schools an additional reason is likely to be that students do not share a mother tongue with each other or perhaps the teacher. English is, therefore, chosen as the medium for instruction so that all students are in the same position and to provide the opportunity to develop proficiency in an international language.

This does not mean that the mathematics teacher is now being asked to replace the English teacher, or to have the same skills or knowledge of English (though in many primary schools one teacher may indeed teach both). What it does mean, however, is that the mathematics teacher has to view his/her role differently: he/she has to become much more language aware. It is this recognition of the need to ensure that the delivery of the content is not negatively impacted by the use of the second language that informs the planning and methodology of EMI.

This raises significant challenges, including:

- the teacher's knowledge of English
- students' level of English (which may vary considerably in international schools)
- resources which provide appropriate language support
- assessment tools which ensure that it is the content and not the language which is being tested
- differentiation which acknowledges different levels of proficiency in both language and content.

Meeting the challenges positively

Perhaps lack of confidence in their own English proficiency is one of the most common concerns among teachers. However, while it is a factor, success in EMI is not necessarily linked to the teachers' proficiency in English. Teachers who have English as their mother tongue may well lack the sensitivity to, or awareness of,

the language that a non-native speaker has acquired through learning and studying the second language. Developing this awareness and demonstrating it in both materials and method is the key to effective EMI.

Classroom language/Teacher Talk

Often non-native-speaker teachers are more concerned about their ability to run and manage the whole class in English than they are about the actual teaching of the mathematics concepts, as the resources or textbook should help them with the latter. However, this use of English in the class is very important as it provides exposure to the second language, which plays a valuable role in language acquisition. It is also true that the teacher talk for purposes such as checking attendance and collecting homework does not have to be totally accurate or accessible to the students. When teaching the mathematics concepts, however, it is essential that the Teacher Talk is comprehensible. Some basic strategies to ensure this include:

- simplification of your language
- use short, simple sentences and project your voice
- paraphrase (say in a different way) as necessary
- use visuals, write or draw on the board, gestures and body language to clarify meaning
- repeat as necessary
- plan before the lesson
- prepare clear, simple instructions and check understanding.

Creating a language-rich environment

Primary teachers often excel at providing a colourful and engaging physical environment for students. In the EMI classroom, this becomes even more important. Posters, 'Word walls', lists of key structures, students' work, English signs and notices all provide a backdrop which provides the opportunity for language exposure and language acquisition.

Planning

When planning, look carefully at each stage of the unit and identify what the language demands are. This means thinking about what language students will need to understand or produce, and deciding how best to scaffold the learning to ensure that language does not become an obstacle to understanding the concept. This involves providing language support and goes beyond the familiar strategy of identifying key vocabulary.

Support for listening and reading

Listening and reading are receptive skills, requiring understanding rather than production of language. If you are asking your students to listen to or read texts in English, ask yourself the following questions when you are planning the unit:

- Do I need to teach any vocabulary before they listen/ read?
- How can I prepare them for the content of the text so that they are not listening 'cold'?

- Can I provide visual support to help them understand the key content?

- How many times should I ask them to read/listen?

- What simple question can I set before they listen/read for the first time to focus their attention?

- How can I check more detailed understanding of the text? Can I use a graphic organiser (e.g. tables, charts and diagrams) or gap-fill task to reduce the language demands?

- Do I need to differentiate the task for those students who find reading/listening difficult?

- Could I make the tasks interactive (e.g. jigsaw reading, when students access different information before coming together, and information share)?

- How am I going to check their answers and give feedback?

Support for speaking and writing

Speaking and writing are productive skills because students doing these need to produce language. They are different to the receptive skills of listening and reading where students receive language from other sources. These skills may require more input from the teacher.

When you plan to use a task which requires students to *produce* English (speak or write), you need to think about how to help them do this.

This means that you have to think in detail about what language the task requires (Language Demands, LD) and what strategies you will use to help them use English to perform the task (Language Support, LS).

You need to ask yourself the following questions:

- What *vocabulary* does the task require? (LD)

- Do I need to teach this before they start? How? (LS)

- What *phrases/sentences* will they need? Think about the language for learning mathematics, e.g. predicting and comparing. What structures do they need for these language functions? (LD)

- Will they be able to produce these sentences or should I provide some *scaffolding* [e.g. sentence starters/sentence frames/gapped sentences (see below)]? (LS)

    A square has ＿＿＿ sides.

    A triangle has ＿＿＿ sides.

    A quadrilateral has ＿＿＿ sides.

    A pentagon has ＿＿＿ sides.

- While I am *monitoring* this task is there any way I can provide further support for their use of English (especially for the less-confident students)? (LS)

- What language will students need to use at the *feedback* stage (e.g. when they present their task)? Do I need to scaffold this? (LD, LS)

Teaching vocabulary and structures

Vocabulary

Learning the key mathematics vocabulary is central to EMI and 'learning' means more than simply understanding the meaning. Knowing a word also involves being able to *pronounce* it accurately and *use* it appropriately. Below is a list of strategies which could be useful:

- Avoid writing the list of vocabulary on the board at the start of the unit and 'explaining' it. The vocabulary should be introduced as and when it arises in the unit. Word boxes are provided on each page of the Student Books and Practice Books with the key words for the lesson. This helps students associate the word or phrase with the concept and context.

- Record the vocabulary clearly on the board when you first introduce it in the lesson, and check that you are confident with the pronunciation and spelling. before the lesson. If you think students may struggle to pronounce words, decide how best to model this pronunciation.

- Give students a chance to say the word once they have understood it. The most efficient way to do this is through repetition drilling.

- Use visuals whenever possible to reinforce students' understanding of the word.

- Ensure students are recording the vocabulary systematically in their glossaries, at the back of the Student Books, and, if possible, use a 'Word wall' which lists the vocabulary under unit/topic headings.

- Remember to use and revise the vocabulary.

Structures

In order for students to talk or write about their mathematics, they will need to go beyond vocabulary: they will also need to use those phrases and sentence frames which a particular task requires.

For example, they may need the following expressions in mathematics:

> *X is the same as Y.*
>
> *The sides are the same length.*
>
> *The next number in the sequence.*
>
> *I predict that X will happen.*
>
> *If X happens, then Y happens.*
>
> *The next step is …*

You need to build up these banks of common mathematics phrases and encourage students to record them. This is an important part of identifying the language demands and providing the necessary support. The teacher does not have to focus on grammar here as the language can be taught as phrases rather than specific grammatical structures.

Every unit of the Teacher's Guide begins with some useful background information. This includes:

The Big idea: The main mathematical concept covered in the unit.

Look out for: Tricky concepts that may need explaining prior to any learning taking place.

Common misconceptions: Common errors that students make or misunderstandings that students have. This section offers advice on how to deal with these misconceptions.

Key vocabulary: The key mathematical words used in the unit.

Coverage in lessons: The English National Curriculum objectives covered in the unit.

Every lesson in the Student Book and Practice Book has corresponding lesson notes in the Teacher's Guide. Each comprehensive set of lesson notes includes:

A mini reproduction: The relevant pages from the Student Book.

Global skills: These are the skills that aim to foster a classroom environment where students develop the skills for success. The skills are: *Creative skills* where students are problem solving, investigating or exploring new maths content; *Real-world skills* where students are taking part in research, or presenting and interpreting information, or if they are dealing with money and developing their financial literacy; *Interpersonal skills* where students are practising their teamwork and communication, often through working in pairs or larger groups; and *Self-development skills* where students have the opportunity to reflect on their learning and talk about what went well and what they are still uncertain about.

The key vocabulary and resources: A list of key vocabulary used in the lesson and the concrete resources required for the activity.

Language support: A range of strategies, including card sorts and card games, word walls, team games to define or explain words, use of similar words to explain meaning and exploration of the origins of words.

The key principles underpinning the language support are:
Words should be introduced and explained carefully.
The word should be explained in context.
Repetition is vital.
Words should be linked to pictures or actions.
Students should develop their own glossaries.
The learning of mathematics vocabulary should be fun.
Language should not be a barrier to effective learning of mathematics.

Detailed lesson notes: Comprehensive lesson notes, including an introduction activity and main activity. These notes refer to the Student Book and Practice Book,

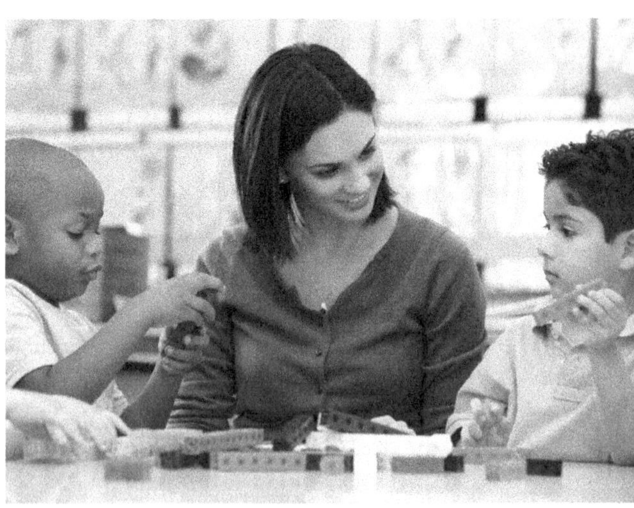

where relevant. The notes include probing questions for formative assessment, which are italicised. Icons are used to suggest the groupings that should be used at each point of the activity (whole class, small group, pairs, individual). A separate 'star-jump' icon indicates that the activities give students an opportunity for physical movement (standing up, jumping, moving around) rather than doing activities sitting down.

Differentiation: The Teacher's Guide offers strategies for you to *support* those students who may have difficulty accessing the task; to *consolidate* the learning for those students who need a little more practice; and to *extend* the learning for those who need more challenge.

The Teacher's Guide also offers differentiated outcomes. These outcomes are listed in the form of:

All students

Most students

Some students

Stretch zone: Each activity in both the Student Book and the Practice Book has a stretch zone question to support deeper learning. The Teacher's Guide provides additional notes on these activities.

Reflection time: Suggestions on how to bring the class back together to reflect on the learning and share ideas.

Answers: Answers to all the Student Book and Practice Book activities are provided.

Review pages: The Teacher's Guide provides notes on the Review pages of the Student Book (summative assessment), including answers to the assessment questions, and the Practice Book (a formative, reflective review).

Digital resources: Where it is appropriate to use digital resources in a lesson, such as sharing the interactive Student eBook page on an interactive whiteboard (IWB), suggestions are embedded in the lesson plan.

Resources sheets: these photocopiable resources can be used with some of the main activities. They are referenced in the resources section of the lesson plan and are available from the *Oxford International Primary Maths* page on the Oxford Owl website (www.oxfordowl.co.uk).

Tour of a typical unit

The 'Big question' provides a discussion stimulus about the key idea of the unit.

1 Numbers and counting

How do we use numbers?

In this unit you will:

- count, read and write numbers to 100
- count in twos, fives and tens
- know and make numbers using objects and pictures
- use words such as equal to, more than, less than (fewer), most, least
- read and write numbers from 1 to 20 in words.

Learning objectives are stated clearly at the beginning of every unit.

Engage

Which numbers can you see in the classroom?

Which numbers can you see on your way to school?

What is the biggest number you have ever seen?

Further questions allow students to develop communication skills.

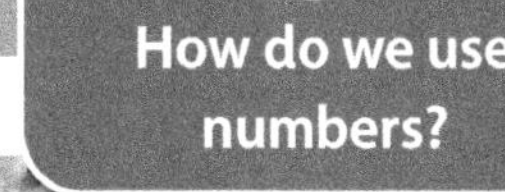

6

The Engage spread is bright and colourful, with artwork or photos to spark interest in young students and provide discussion points.

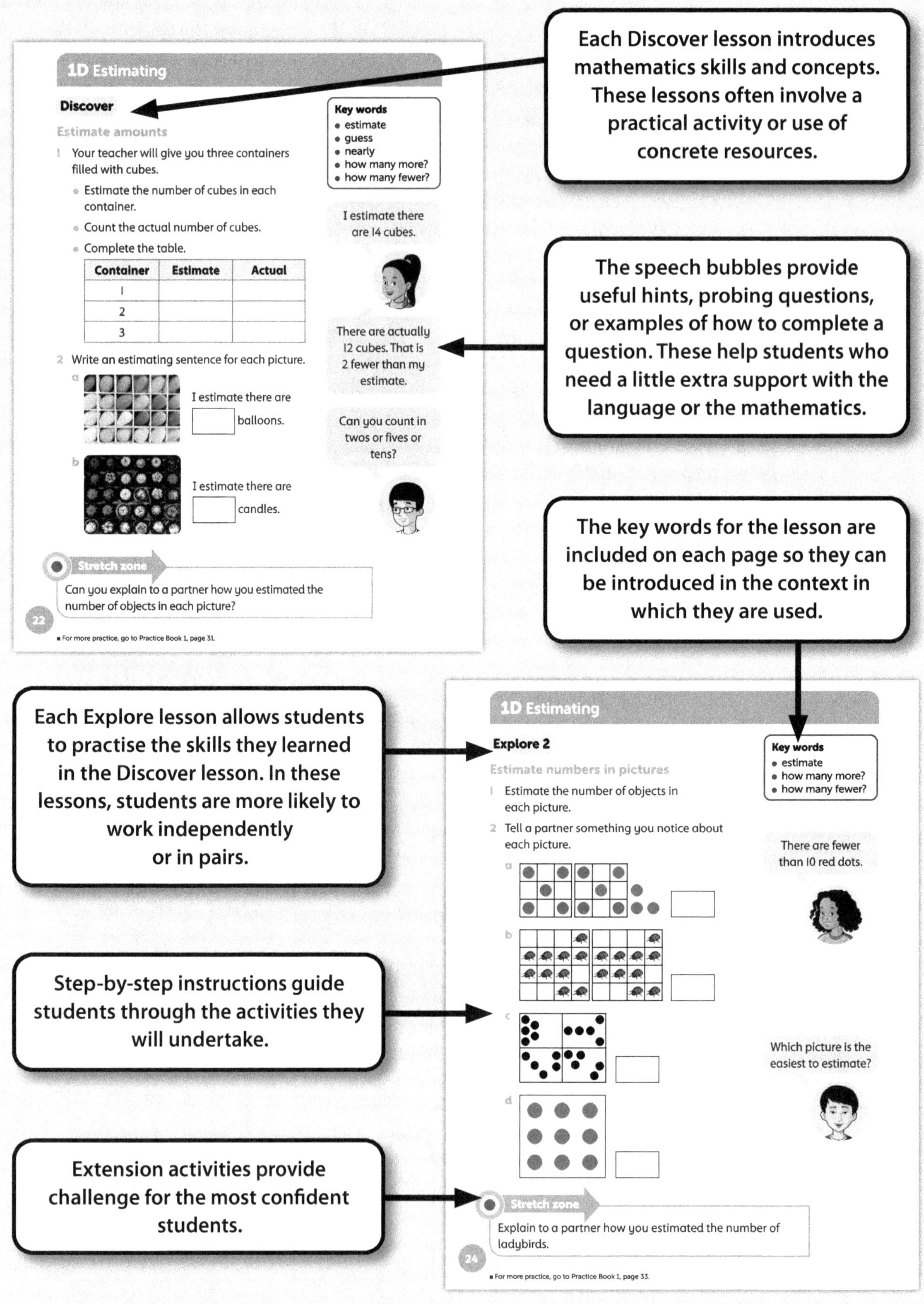

Each Discover lesson introduces mathematics skills and concepts. These lessons often involve a practical activity or use of concrete resources.

The speech bubbles provide useful hints, probing questions, or examples of how to complete a question. These help students who need a little extra support with the language or the mathematics.

The key words for the lesson are included on each page so they can be introduced in the context in which they are used.

Each Explore lesson allows students to practise the skills they learned in the Discover lesson. In these lessons, students are more likely to work independently or in pairs.

Step-by-step instructions guide students through the activities they will undertake.

Extension activities provide challenge for the most confident students.

The Connect lesson makes [links] between the [diff]erent areas of [ma]thematics in the unit.

1 Numbers and counting

Connect

Make a number poster

Work as a group.

1. Collect some magazines. Talk about which magazines might have numbers in them. What do the numbers tell us?

[Con]nect activities [are] often set in [real]-life contexts [to] make the [li]nk between [mat]hematics and [th]e real world.

2. Cut out pictures that have numbers.

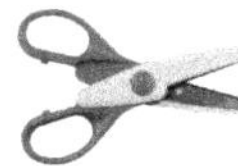

3. Make a poster to display in class.

4. Talk in your group about the numbers you have found.

Stretch zone

Take photographs of numbers on the way home from school. What job are the numbers doing? Explain your ideas to a partner.

We use numbers to count or to say how many of something there are.

The 'Big idea' sums up what students have discovered in the unit. It answers the Big question on the Engage page.

What is the biggest number on your poster?

What is the smallest number on your poster?

A further extension activity provides a challenge for the most confident students.

Students' progress is assessed through the questions and tasks at the end of each unit. In Student Books 2 and 6, these questions reflect the style of the SATs (national Standard Assessment Tests).

1 Numbers and counting

Review

1 Draw the beads and write the numbers in the spaces.

Beads	Numbers	Words
	5	
		sixteen
		three
	12	
		nineteen
	1	
		four
	14	
		twenty

2 Samir has a bracelet with 19 beads. Lina's bracelet has one more bead than Samir's. How many beads are on Lina's bracelet?

Celine's bracelet has 10 more beads than Lina's. How many beads are on Celine's bracelet?

26

A word problem is always included on the Review page.

Practice Book Discover and Explore

Practice Book activities can be completed in the school lesson or **as** homework.

Most of the Discover and Explore lessons in the Student Book have a corresponding page in the Practice Book. These activities provide opportunities for students to consolidate and deepen their learning.

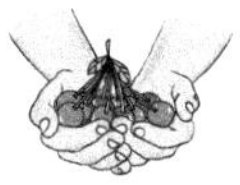

Step-by-step instructions guide students through the activities they will undertake.

If students require concrete resources, these are listed in a box at the top of the page. This is particularly useful if students are completing the activities at home.

Extension activities provide challenge for the most confident students.

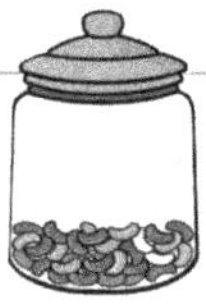

Each Review page in the Practice Book includes a reminder of all the topics learned in the unit.

1 Numbers and counting

Review

 I Draw a face next to each bubble to show how you feel about your learning.

counting objects

reading and writing numbers

counting in twos, fives and tens

estimating quantities

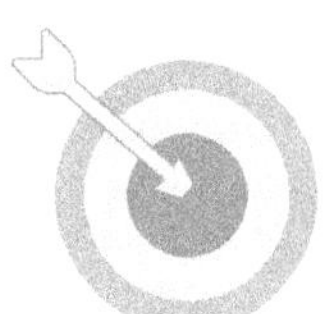 2 Tell a partner about one thing you did really well in this unit.

3 Draw or write about things you found easy, challenging or really hard.

Self-assessment activities help students to reflect on their learning

What work did you feel confident doing?

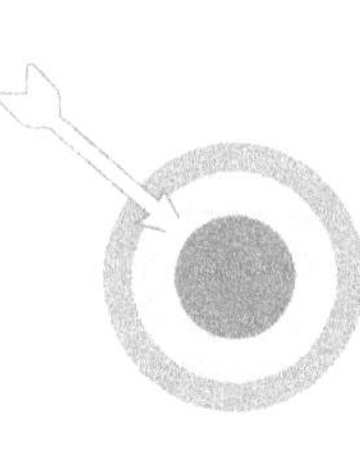

What work was challenging?

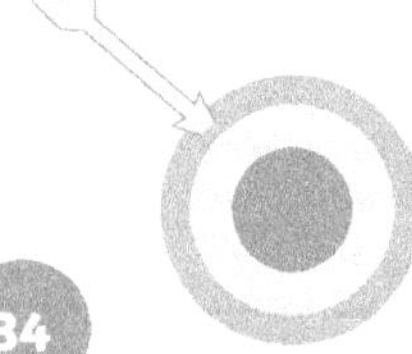

Is there any work you might need some extra help with?

34

The Student Books

The Student Books are write-in textbooks for students to read and use. There are six Student Books: one for each school year at primary school. The Student Books introduce learning through a mixture of practical, discussion and independent activities.

Student Book	Typical student age range
Student Book 1	Age 5–6
Student Book 2	Age 6–7
Student Book 3	Age 7–8
Student Book 4	Age 8–9
Student Book 5	Age 9–10
Student Book 6	Age 10–11

  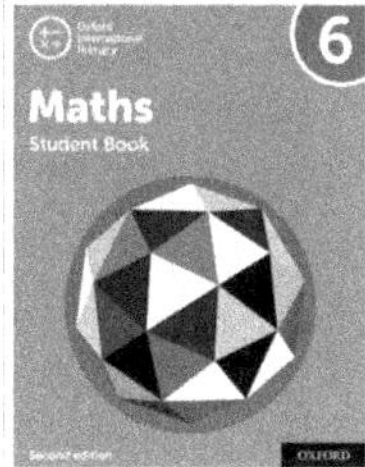

The Practice Books

The Practice Books are write-in workbooks for students to read and use. There are six Practice Books: one for each school year at primary school. The Practice Books provide deeper learning opportunities through a range of independent activities, which can be completed in school or at home.

Practice Book	Typical student age range
Practice Book 1	Age 5–6
Practice Book 2	Age 6–7
Practice Book 3	Age 7–8
Practice Book 4	Age 8–9
Practice Book 5	Age 9–10
Practice Book 6	Age 10–11

 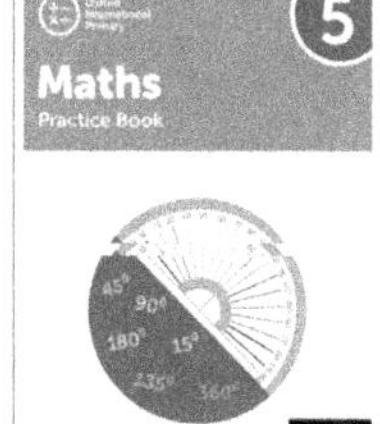 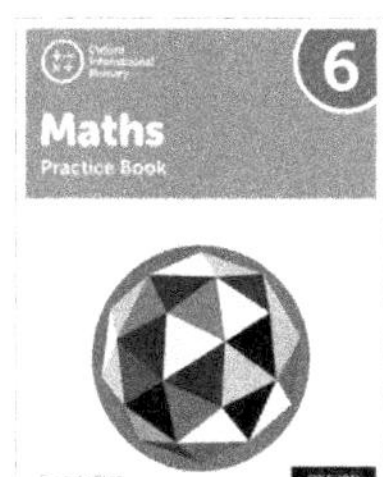

The Teacher's Guides

There are six Teacher's Guides: one for each school year at primary school. Each Teacher's Guide includes:

- An introduction with advice about delivering mathematics in primary schools using *Oxford International Primary Mathematics*.
- A unit overview, giving advice on teaching each unit, including common misconceptions and how to deal with them.
- A lesson plan for every lesson in the Student Book and corresponding pages in the Practice Book.
- Model answers to each question in the Student Book and Practice Book.

  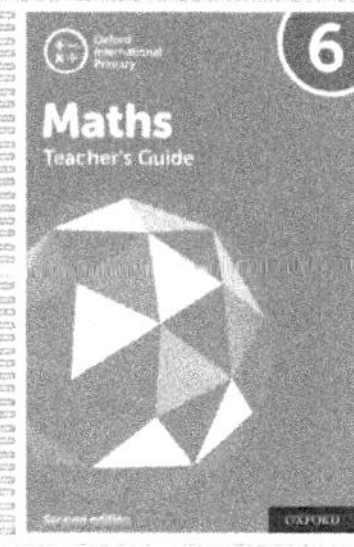

Digital resources

Interactive eBooks

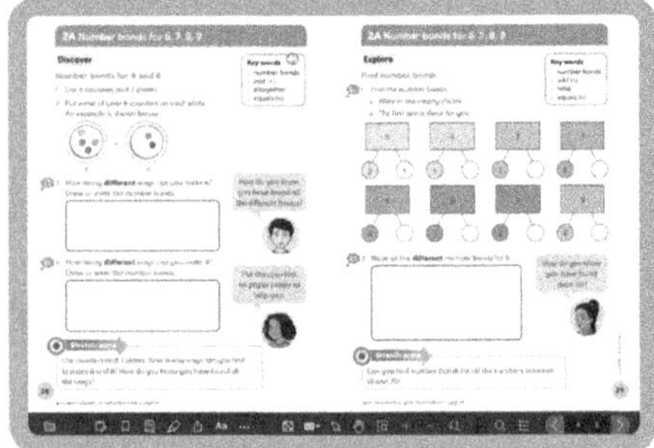

For the teacher

Teachers can access the Student Books, Practice Books and Teacher's Guides online in eBook format, on the Oxford Owl website (www.oxfordowl.co.uk).

The enhanced eBooks show the course content on screen, making it easier for teachers to deliver engaging lessons.

For the students

Teachers can allocate an eBook version of the Students Books to the students for use at home. The Student eBooks include interactive activities, worksheets and audio of all the key vocabulary,

Assessment resources

The downloadable assessment materials offer you additional opportunities to assess students' progress. The materials include:

- end-of-unit summative assessment
- end-of-year summative assessment.

Every test comes with everything you need to assess and record progress including:

- Answers

- Mark schemes and guidance on assessment

Oxford Primary Illustrated Maths Dictionary

The *Oxford Primary Illustrated Maths Dictionary* gives comprehensive coverage of the key maths terminology children use in the course. Entries are in alphabetical order, and each includes a clear and straightforward definition along with a fun and informative colour illustration or diagram to help explain the meaning. The dictionary is suitable for Students with English as an Additional Language.

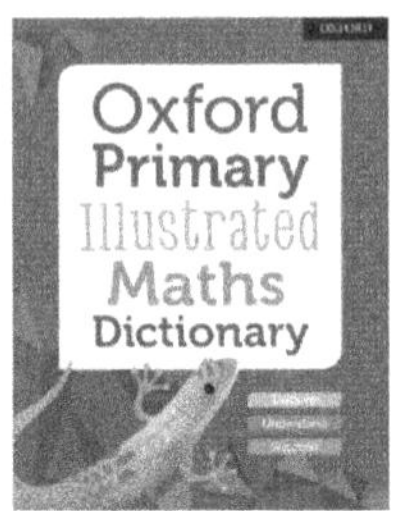

The curriculum

The Oxford International Curriculum offers a new approach to teaching and learning focused on wellbeing, which places joy at the heart of the curriculum and develops the global skills students need for their future academic, personal and career success.

Through six subjects – English, Maths, Science, Computing, Wellbeing and Global Skills Projects – the Oxford International Curriculum offers a coherent and holistic approach to ensure continuity and progression across every student's educational journey, equipping them with the skills to shape their own future. Through this approach, we can help your students discover the joy of learning and develop the global skills they need to thrive in a changing world.

1 Numbers and counting

Big idea

Two of the Big Ideas for this unit are place value and recognising and using the patterns within the decimal number system. The third Big Idea is that we count to find out how many, or how much.

In the decimal number system, the position of a digit gives its value – ones, tens, hundreds, thousands and so on. The digits 0 to 9 are repeated again and again. This pattern helps students to develop a clear image of the number system, supported by the 100-square and number line.

There are many patterns in the way we count, and students need to explore and describe the number patterns to enhance their understanding of number: for example, the way that every number when we count in tens has a 0 in the ones place, or how every number is either even or odd.

Awareness of the place-value system is crucial to being able to count and to calculate, and mastering the features of the system with small numbers equips students to be able to calculate with much larger numbers later.

Look out for

- **Students who confuse the teen numbers and decade numbers.** The names we give to numbers, particularly teen numbers and decade numbers, are very similar. For example, students may say 'fifty' for 'fifteen'. This confusion is very common and may extend beyond the words to the written numerals, with 15 written as 51. The main way to distinguish between the words is by stressing the parts: fifTEEN and FIFty.
- **Students who, when finding a number between two others, include the end numbers.** By crossing out the end numbers, we can ensure that students identify the correct 'between numbers'. For example, to find a number between 20 and 30 on a number line, they should cross out 20 and 30 and choose a number from the ones remaining.

- **Students who confuse the greater than (>) and less than (<) signs.** Looking at the sign as an arrow on the number line may help to avoid this confusion. Alternatively, viewing the sign as a greedy mouth, always open towards the biggest meal, that is, the highest number, is another useful image to help clarify understanding. You could display copies of the signs turned into animals or monsters with large mouths to reinforce understanding.

Possible misconceptions

- **Students may, when counting on from a number, include the starting number.** For example, counting on 3 from 5 they may say 5, 6, 7 and get 7 as their answer. Model how to count correctly along a number line and count together.
- **Students may not, when estimating numbers of objects, take into account the relative sizes of the objects.** For example, 15 large items may appear to be 'more' than 20 smaller items.

Key vocabulary

- count, digit, number, zero, one, two, three, twenty, thirty, forty, fifty, ninety, hundred
- row, column, forward, backward, place value, multiple, tens, ones, twos, fives, threes, fours, value, diagonal lines
- count on, count back, even, odd
- number pattern, sequence, number line, steps
- multiple of 10, multiple of 5, multiple of 2
- estimate, more than, less than, next to, between, smaller, larger
- round up, round down, nearest multiple of 10
- greater than >, less than <, smallest, largest
- ordinal numbers, first, second, third, …

Learning focus	Learning outcomes (the ENC objectives)
Counting in tens and ones	Count in tens from any number, forward and backward. Recognise the place value of each digit in a 2-digit number (tens, ones).
Counting in tens	Count in tens from any given number, forward and backward. Read and write numbers to at least 100 in numerals and in words.
Counting in twos	Count in steps of 2 from any number, forward and backward. Read and write numbers to at least 100 in numerals and in words.
Counting in fives	Count in steps of 5 from any number, forward and backward. Read and write numbers to at least 100 in numerals and in words.
Counting in threes and fours	Count in steps of 3 [and 4] from any number, forward and backward.
Estimating and counting	Identify, represent and estimate numbers using different representations, including the number line.
Numbers in between	Identify numbers that lie between given numbers, or missing numbers when counting in steps of 2, 3, 4, 5 or 10. [Optional non-ENC objective]
Rounding to the nearest 10	Round any number up to 100 to the nearest 10, recognising that halfway numbers round upwards. [Optional non-ENC objective]
Less than and greater than	Compare and order numbers from 0 up to 100; use <, > and = signs.
Ordinal numbers	Compare and order numbers from 0 up to 100; use <, > and = signs. Use ordinal number names, for example, first, second, third and so on. [Optional non-ENC objective]

Engage Student Book page 6

Big question

- How can I estimate numbers? How can I order numbers?

Global skills

- **Creative skills:** problem solving/exploring
- **Interpersonal skills:** communication/teamwork

Key vocabulary

- estimate, tens, ones, twos, fives

Resources

- base-ten apparatus (tens-rods and ones-cubes)
- cubes or counters

Language support

Listen to the range of estimates students are giving. These will help you to check whether students have an idea of the quantity value of the numbers they use. Model counting in tens as often as you can – ask students to fill in the gaps when you count in tens and miss some numbers out.

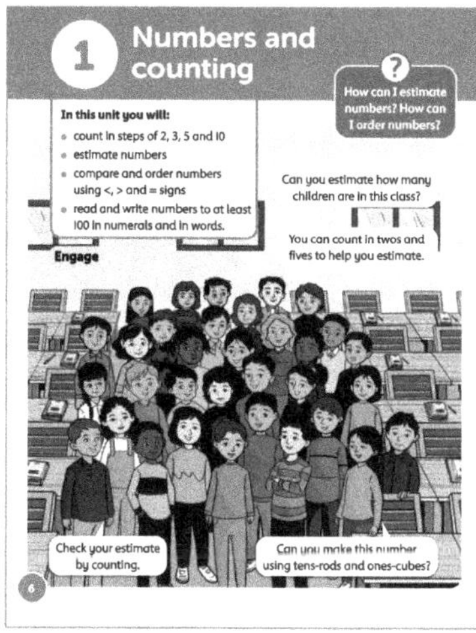

Introductory activity

Ask students to look around the room and guess, without counting, how many students are in the class. Give some guidance by asking, for example, *Are there more than 5 students? Are there more than 10 students? Are there fewer than 50 students? Are there 100 students?* These questions should help students estimate a realistic total, perhaps 25, 30 or 40.

Can they estimate how many students there might be in the whole school? 20? 50? 100? More than 100?

Main activity

Ask students whether they can recall what the word 'estimate' means (to make a sensible guess). Ask them to look at the picture on page 6 of the Student Book. Display it on the IWB if possible. Ask: *Can you estimate how many students are in this class?*

Ask whether 1 would be a good estimate – allow students to explain why not. Ask whether 1 million would be a good estimate – allow students to explain why this would not be a good estimate.

Arrange students in small groups and ask each group to talk to each other and agree an estimate for the number of students in the picture. Share the estimates, drawing up a list on the board under the heading 'First estimate'.

Discuss how students reached their estimates and talk about how they might be able to find a more accurate estimate. This will lead to students discussing counting in twos or fives instead of ones or tens, for example. Students can then use base-10 equipment or other concrete resources, such as cubes and counters, to count and represent how many. Agree how many (33), and deal with any differences in counting by having groups model their counting on the board.

Differentiation

Supporting: Ask students to work in mixed-ability groups so that less-confident students can hear the numbers modelled by their peers.

Consolidating: Ask students why they think that their estimate is sensible.

Extending: Ask students how confident they are that their estimate is accurate. How close do they think their estimate will be?

Reflection time

Look at the spread of results. Are they close together? Decide whether the estimates were good ones and discuss what makes a good estimate, for example being within 5 of the right answer (or being within 20 if it's a very large number).

Which estimates were close to this total?

1A Counting in tens and ones

Discover
Student Book page 7 • Practice Book page 16

Specific learning foci
- Count, read and write numbers to 100 and back again.
- Count on in ones and tens from single- and 2-digit numbers and back again.

Global skills
- **Interpersonal skills:** communication/teamwork

Key vocabulary
- count on, count back, tens, ones, rows, columns, sequence, number pattern

Resources
- large 100-square for the front of the class
- timer

Language support
Listen carefully to how each individual says the numbers out loud. Listen out for confusion between '-teen' and '-ty' and correct, explaining that '-teen' means one ten, which is why the numbers 13, 14, 15, 16, 17, 18 and 19 are known as teen numbers. Model how to say the words: fifTEEN, FIFty.

Introductory activity

Display a large 100-square. **Count on** together from 1 to 10 and **count back** again, pointing to the numbers on the top row of the 100-square. Ask students what we call the line of numbers from 1 to 10 on a 100-square (a **row**) and ask, *How many rows of numbers in this 100-square?* (10)

Choose a row. Count along that row together, forward and backward. Ask students what happens as they count forward and backward along any row. (The **ones** digit changes but the **tens** digit stays the same until the final number.) Practise counting along other rows.

Now point to the numbers 10, 20, 30, 40, … , 100. Ask what we call a line of numbers like this on the 100-square (a **column**). Ask how many columns of numbers there are in a 100-square. (10)

Choose a column. Count down that column together, then count up the column together. *What happens when you count down and up any column?* (The tens digit changes but the ones digit stays the same, apart from the final number.) Practise counting down/up other columns.

Choose a number on the 100-square, say 4, and count on along a few places: *4, 5, 6, 7, 8.* Explain that this is called a **sequence** of numbers. *How much does this sequence change from each number to the next?* (1) Say that sequences have a **number pattern** that we can find and describe. This helps us find more numbers in the sequence.

Point to the right-hand column of the 100-square. Count together in tens (10, 20, 30, …), then ask students to say how many tens for each number and then count together this way (1 ten, 2 tens, 3 tens, …). Remind students that these numbers have names that tell you how many tens they contain.

Repeat the count from 10 to 100, forward and backward, using the correct number names. Count up and down a different column and then ask students *What do you notice about the numbers? Can you see a pattern?* Can they see for themselves that the tens change but that ones stay the same? Choose another column. Count down/up that column together.

Organise students into small groups of up to six to play a counting game. Explain that the first person chooses a number and counts on (forward) or counts back (backward) from that number in ones or tens. After saying about 10 numbers, the next student in the group repeats the last number said, then chooses whether to count in ones or tens, on or back. Counting continues until everyone has had a turn. The aim is to continue counting without any pauses.

Ask students to complete the missing numbers in the counts on page 7 of the Student Book individually. They also complete the sentences to describe each count. Once they have completed the questions, ask them to take it in turns with a partner to describe each sequence. Do they notice any patterns? Can they describe them?

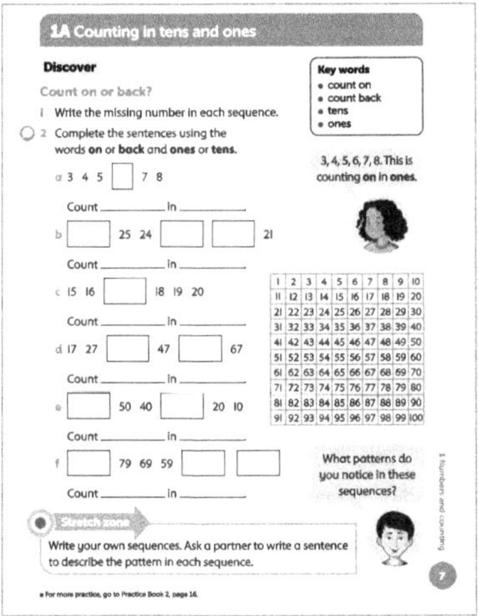

Differentiation
Supporting: Model counting out loud to students. Leave gaps in your counts for them to fill. Emphasise fifTEEN, FIFty so that students can hear the language patterns.

Consolidating: Encourage students to develop their own counting patterns in ones and in tens.

Extending: Ask students whether they can continue their patterns beyond 100.

Stretch zone: *Write your own sequences. Ask a partner to write a sentence to describe the pattern in each sequence.*

Check that students have correctly described the pattern using language such as counting on and back in tens and ones.

 Reflection time

Count with the whole class acting as one large group. Once the count is started, anyone, including any adults in the room, can be named to continue the count. Again, the aim is to continue counting without any long pauses. Set a timer for 5 minutes. Can the class continue the count for that long? Try extending the time for longer than this.

Practice Book: Students can complete page 16 of the Practice Book. They can do this directly after the main activity, as homework, or as the focus of a separate mathematics session to help consolidate their learning and build fluency. Encourage students to say their sequences aloud so they have additional practice in saying the numbers.

Differentiated outcomes	
All students	should count to 100 in ones with support from peers and the teacher.
Most students	will understand the difference between counting in ones and counting in tens.
Some students	may continue their counts beyond 100.

Answers

Student Book page 7

a 3, 4, 5, 6, 7, 8 Count on in ones

b 26, 25, 24, 23, 22, 21 Count back in ones

c 15, 16, 17, 18, 19, 20 Count on in ones

d 17, 27, 37, 47, 57, 67 Count on in tens

e 60, 50, 40, 30, 20, 10 Count back in tens

f 89, 79, 69, 59, 49, 39 Count back in tens

Practice Book page 16

1 18, 28, 38, 48, 58, 68, 78   **4** 49, 59, 69, 79, 89, 99

2 82, 72, 62, 52, 42, 32, 22   **5** 97, 87, 77, 67, 57, 47, 37

3 65, 55, 45, 35, 25, 15, 5

Stretch zone: Students are likely to answer that they can see the tens all in one column and can count down and up it.

1A Counting in tens and ones

Explore   Student Book page 8 · Practice Book page 17

Specific learning foci

- Count, read and write numbers to at least 100 and back again.
- Count on in ones and tens from single- and 2-digit numbers and back again.

Global skills

- **Creative skills:** investigating
- **Interpersonal skills:** communication/teamwork

Key vocabulary

- count on, count back, tens, ones, rows, columns, digits, value, place value

Resources

- large 100-square for the front of the class, small 100-squares for each table
- set of place-value cards for each pair of students cut out from Resource Sheet 1, copied onto card

Language support

Display a large 100-square to support students' counting. Then listen carefully to how each individual says the numbers out loud. Correct any confusions between '-teen' and '-ty'.

Introductory activity

Give each pair of students a set of **place-value** cards. Explain that they are called place-value cards because they help us to see which place each **digit** is in, so we know its **value**. Ask students to sort their cards into two rows, a row with the ones in order (01, 02, 03, …, 09) and a row with the tens in order (10, 20, 30, …, 90). Ask students to take the cards showing 20 and 4. *Place them together so they show 24. How many tens and ones are there in this number?* (2 tens, 4 ones) Repeat for other 2-digit numbers, for example, 25, 57 and 82. Ask each time: *How many tens are there in this number? How many ones?*

 Main activity

Display a 100-square, on the whiteboard if possible. Ask two students to each choose a number on the 100-square. Start from one of the numbers and count forward or backward to the other in steps of 1 or 10.

For example, to move from 23 to 57, the count might be like this: 23, 24, 25, 26, 27, 37, 47, 57. Now work together to find another way, for example, 23, 33, 43, 53, 54, 55, 56, 57. Talk about how many moves it takes. *Is it always the same number of moves?*

Ask students to work in pairs, one choosing two numbers, the other counting in ones and tens to get from one number to the other, forward and backward, then swap roles.

Ask students to complete page 8 in the Student Book. They should complete the first two questions individually and the latter part of question 3 and question 4 in pairs. Pairs may join with another pair to share the sequence of the fewest number of moves they found to get all three numbers.

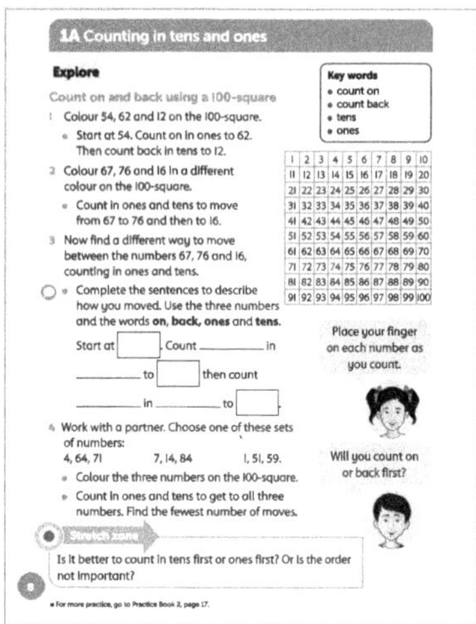

Differentiation

Supporting: Pick numbers in the 100-square for students to name. Model counting from one number to another on the 100-square by touching each number and saying the number name.

Consolidating: Ask students to describe how they moved from one number to the next.

Extending: Ask students to count how many tens and ones they moved when counting from one number to the next. For example, from 31 to 54 is a move of 2 tens and 3 ones, or 23.

Stretch zone: *Is it better to count in tens first or ones first? Or is the order not important?*

Ask students to explain which they think is better and why, for example, counting in tens first is doing bigger jumps, or they might say that it doesn't make a difference which way, it's the same number of steps. Encourage them to support their answer with examples.

 **Reflection time**

Ask students to share some of their journeys between numbers on the 100-square. Did they always do tens first then ones, or vice versa? Did they mix their steps? For example, from 15 to 38, they might have moved: 15, 25, 26, 27, 37, 38.

Ask whether anyone chose numbers that needed backward counting. Ask them to describe how they did it and whether they were counting on or back in ones or tens.

Practice Book: Students can now complete the activity on page of 17 in the Practice Book. They can do this directly after the main activity, as homework, or as the focus of a separate mathematics session to help consolidate their learning and build fluency. You are aiming for students to be able to efficiently count on or back between numbers in steps of 1 or 10, as this will help them when they come to addition and subtraction.

Differentiated outcomes	
All students	should count in ones and tens on a 100-square with support.
Most students	will count on and back in ones and tens between numbers on the 100-square.
Some students	may find different ways to count from one number to another in ones and tens on the 100-square.

Answers

Student Book page 8

You should try to mark this activity as you walk around the classroom. Check that students have correctly coloured the numbers on the 100-square, and that they have written accurate sentences to describe how they moved between the numbers on the square.

Practice Book page 17

1 83, 84, 85, 86, 87 88, 89, 90   83, 73, 63, 53, 43

2 65, 66, 67, 68, 69, 70, 71, 72, 73   65, 55, 45, 35

3 48, 49, 50, 51, 52, 53, 54, 55, 56   48, 38, 28, 18

4 52, 53, 54, 55, 56, 57, 58, 59, 60   52, 42, 32, 22

5 66, 67, 68, 69, 70, 71   66, 56, 46, 36, 26, 16, 6

Stretch zone: To arrive at 52 in tens, you could have started on 2, 12, 22, 32 or 42.

1B Counting in tens

Specific learning focus
- Count in tens on a 100-square.

Global skills
- **Creative skills:** investigating
- **Interpersonal skills:** communication

Key vocabulary
- number, zero, one, two, three, …, twenty, thirty, forty, …, ninety, hundred, multiple of 10, count on/back

Resources
- large 100-square for the front of the class, small 100-squares for each table
- set of place-value cards for each pair of students cut out from Resource Sheet 1, copied onto card
- coloured pencils

Language support
Use the large 100-square to support students' counting. Listen to how each individual says the numbers out loud when counting up or back in tens. Correct any confusions between '-teen' and '-ty'.

Introductory activity

Count together from 0 (**zero**) to 100 in tens and then back. Ask students what they notice about the pattern in how the **numbers** sound (all the numbers after 10 end in '-ty'). *What do you notice about how the numbers are written?* (all end in 0) Explain that numbers ending in 0 are called **multiples of 10**, because they come from counting in tens.

Ask a series of questions about 10 more and 10 less, for example: *What number is 10 more than 50? What number is 10 less than 80?*

Main activity

Ask a student to come to the front of the class and point to a single-digit number in the top row of the large 100-square, for example 4. Now point down the column as the class recite the numbers from 4 to 94 going up in tens and then back again. Ask students what they notice about the pattern in how the numbers sound (all the numbers after 14 end in '-ty four'). *What do you notice about how the numbers are written?* (all end in 4) Students should notice that, as they count in tens, the tens change but not the ones.

Now repeat twice more with other starting numbers, asking students about how the numbers sound each time and how they are written. Give them a number and ask them to tell you the number that is 10 more and 10 less than that number, for example if you say 47, they should be able to tell you that 37 is 10 less than 47 and that 57 is 10 more than 47.

Ask students to complete the activity on page 9 of the Student Book individually. Ask students to share their answers for question 4 in pairs. Did they notice a similar or different pattern? Can they explain why?

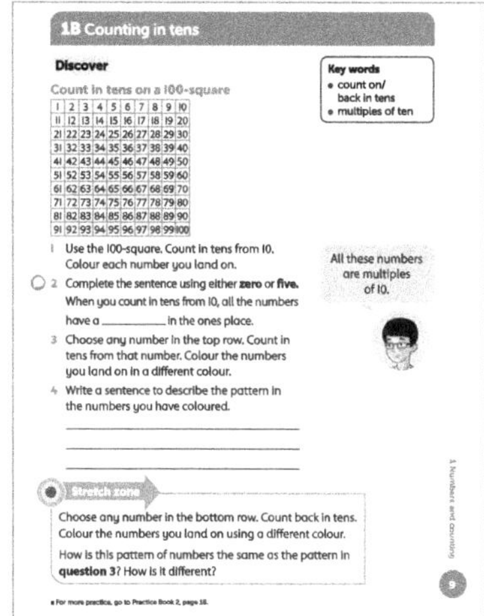

Differentiation
Supporting: Model counting out loud to students. Leave gaps in your counts for them to fill and have a small 100-square available for support. Emphasise the repeated endings so that students can hear the language patterns, for example, twenTY-SEVEN, thirTY-SEVEN, forTY-SEVEN. and so on.

Consolidating: Ask students to continue your counting patterns in tens up and back from any starting number.

Extending: Ask students whether they can continue their patterns beyond 100.

Stretch zone: *Choose any number in the bottom row. Count back in tens. Colour the numbers you land on using a different colour. How is this pattern of numbers the same as the pattern in question 3? How is it different?*

Ask students to share their observations with another student. Can they see the pattern?

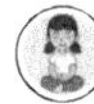 Reflection time
Choose a student to pick a number on the large 100-square. Ask the class to tell you what number is 10 more than the number and 10 less than the number. For example, if the student chooses 48, then 10 more is 58 and 10 less is 38. Repeat with two more examples. Now say, for example, *I am thinking of a number that is 10 less than 49. What number am I thinking of?* and *When I count back 2 tens from this number, the digit in the ones place is 5 and the digit in the tens place is 4. What was my start number?* Students should notice that the ones digits stay the same when counting back in tens. Select students to ask similar questions for the rest of the class to answer.

Practice Book: Students can now complete the activity on page of 18 in the Practice Book. They can do this directly after the main activity, as homework, or as the focus of a separate mathematics session to help consolidate their learning and build fluency. You may want to complete this activity after the next lesson when students will have had more practice counting in tens along a number line. Go through the worked example, counting 3 jumps on the number line together and asking, *How many jumps is this? Three jumps of 10 make what?* (30)

Differentiated outcomes	
All students	should count to 100 in tens with support from peers and the teacher.
Most students	will understand the difference between counting in tens from 0 and counting in tens from other numbers.
Some students	may continue their counts beyond 100.

1B Counting in tens

Explore   Student Book page 10 • Practice Book page 19

Specific learning focus

- Count in tens on a number line.

Global Skills

- **Interpersonal skills:** communication/teamwork

Key vocabulary

- number, zero, one, two, three, …, twenty, thirty, forty, …, ninety, hundred, number line

Resources

- large 0–100 number line

Language support

Use the 0–100 number line to support students' counting. Listen to how students say the numbers when counting up or back in tens. Correct any confusions between '-teen' and '-ty' and make sure that the jumps on the number line match the counting.

 **Introductory activity**

Count together as a class from 0 to 100 in multiples of ten. Use a large 0–100 **number line** to help. Then start from another number ending in 0 and count on in tens to 100 from that number. Choose a different number (e.g. 50) and repeat. Invite a student to choose a multiple of 10 and lead the class in counting on in tens.

Student Book page 9

1   Check that the numbers 0, 10, 20, 30, 40, 50, 60, 70, 80, 90 and 100 are coloured.

2   'When you count in tens from 10, all the numbers have a zero in the ones place.'

3, 4   Check that the coloured numbers are in a vertical column and that an accurate sentence is written to describe them.

Practice Book page 18

Stretch zone: Students should draw 7 tens-rods and 3 ones-cubes and write 73.

Using the number line to help, count back together from 100 to 0, and then from different numbers ending in 0, for example, count back in tens on the number line from 60.

 **Main activity**

Show students a 0–100 number line marked in tens on the board or have them look at the one at the top of page 10 in their Student Books. Start at one of the marked numbers, say, 30. Say *We are at 30. Count on 1 jump of 10. Which number do we land on?* (40) *I make 2 more jumps of 10. What numbers will I land on?* (50, 60) *What number did I finish on?* (60)

Now ask, for example: *I start at 80 and finish on 50. How many jumps of 10 is this? In which direction?* Ask students to make a series of jumps of 10, forward or backward, from different numbers. For example, 4 jumps of 10 forward starting from 20 (60), 3 jumps of 10 back from 80 (50), and so on.

Ask students to complete the activity on page 10 of the Student Book individually. Encourage them to form their numbers carefully as well as say them out loud as they count.

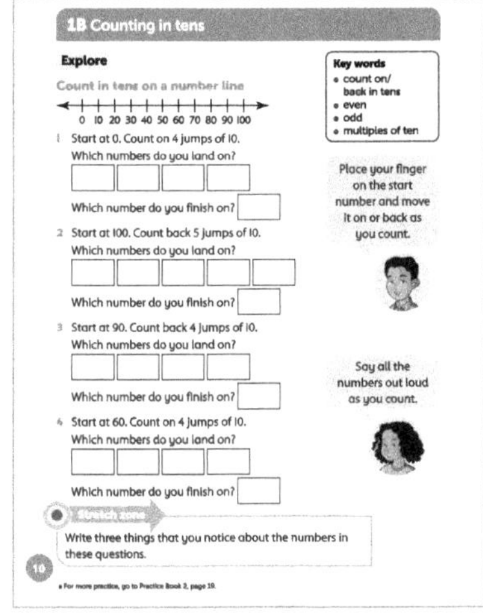

Differentiation

Supporting: Assist students with keeping their place on the number line as they count on or back in tens – you can suggest they use their fingers.

Consolidating: Choose a multiple of 10 (e.g. 30) and ask, *What is next multiple of 10 after 30? What was the multiple of 10 before 30?*

Extending: Ask questions to get students counting backward as well as forward in several tens, e.g. *I count back in 4 steps of 10 from 70 and then 2 steps of 10 forward. What number do I land on?*

Stretch zone: *Write three things that you notice about the numbers in these questions.*

Students should notice that the numbers all end in 0, that, for example, 3 forward jumps of 10 is the same as finding 30 more and that 5 backward jumps of 10 is the same as finding 50 less.

 **Reflection time**

Choose a student to pick a number and mark it on a 0–100 number line, all numbers labelled. Ask the class to tell you what number is 10 more than the number and 10 less than the number. For example, if the student chooses 23, then 10 more is 33 and 10 less is 13. Draw the jumps to the new numbers on the number line. Ask students: *What do you notice about a jump of 10 and finding 10 more or 10 less?* For example, they may notice that a forward jump of 10 is the same as finding 10 more. Repeat with three more examples.

Select students to ask similar questions for the rest of the class to answer.

Practice Book: Students can now complete the activity on page of 19 in the Practice Book. They can do this directly after the main activity, as homework, or as the focus of a separate mathematics session to help consolidate their learning and build fluency.

Differentiated outcomes	
All students	should count forward in multiples of 10 from along a number line.
Most students	will count on and back from different multiples of 10, for example, count on in tens from 20 or back in tens from 90.
Some students	may be able to work out what number they will land on without having to count each jump.

Answers

Student Book page 10

1 10, 20, 30, 40; 40

2 90, 80, 70, 60, 50; 50

3 80, 70, 60, 50; 50

4 70, 80, 90, 100; 100.

Practice Book page 19

1 20, 30, 40, 50, 60, 70, 80, 90, 100

2 90, 80, 70, 60, 50, 40, 30, 20

3 90, 80, 70, 60, 50, 40, 30

4 90, 80, 70, 60, 50, 40

Stretch zone: Answers will vary.

1C Counting in twos

Discover Student Book page 11 • Practice Book page 20

Specific learning focus

- Count in twos and use grouping in twos to count larger groups of objects.

Global Skills

- **Creative skills:** investigating
- **Interpersonal skills:** communication

Key vocabulary

- count on/back, twos, multiples of 2, even, odd

Resources

- large 0–50 number line for display
- 100-squares: large one for the front of class and small one for each pair

Language support

Display a 100-square with the even numbers coloured in. Label these 'multiples of 2'. Alongside the 100-square, display the meaning of the word 'multiple' as 'lots of' or 'groups', for example, multiples of 2 are groups of twos.

 **Introductory activity**

Display the large 100-square and count in **twos** with students, to 50, pointing to each **multiple of 2** in the first instance. Count again, encouraging students to use the 100-square for support as necessary, but to try to do this without looking at it. Ask students how they knew which number to say next, and to explain using the pattern that they see on the 100-square (e.g. it is 2 more than the number before, it is every other number, I am counting in twos and so on).

As a whole class, count to 100 and then back in twos. As students count, highlight each number on the 100-square.

Main activity

Cover or remove the 100-square or the numbers to 20 and challenge students to count from 2 to 20 in twos as before. Reveal the numbers as you go along.

Ask students to complete the activities on page 11 in the Student Book individually. Ask them to then discuss in pairs the questions in the speech bubbles using their answers to support their thinking. Students should be able to tell you that the red highlighted numbers are all **even** numbers, because they all end in 0, 2, 4, 6 or 8.

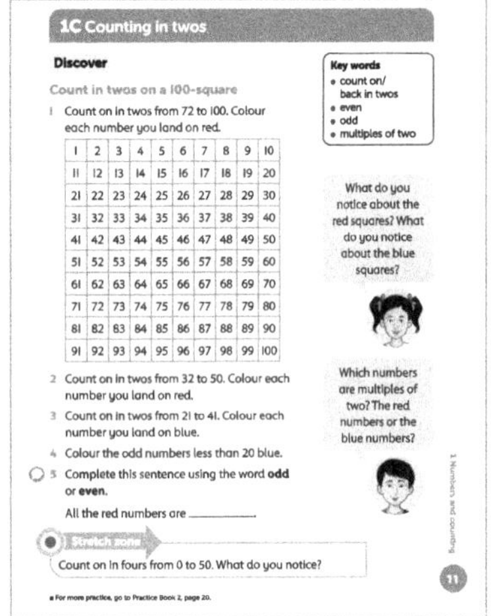

Differentiation

Supporting: Give students 100-squares with multiples of 2 up to 20 highlighted to support counting in twos.

Consolidating: Ask students to predict whether or not a number will be included when counting on in twos from a range of numbers.

Extending: Ask students to predict what happens to even numbers beyond 100 and explain how they know this.

Stretch zone: *Count on in fours from 0 to 50. What do you notice?*

Help students to recognise that counting on in fours lands on the same numbers as counting on in twos, missing every other number. You can do this by counting in twos, alternately whispering one number and saying the next aloud, while the class count in 4s using the 100-square. They should notice that the numbers you say aloud are the same as the numbers they are saying.

Reflection time

Ask students what they notice about the pattern they have counted on the 100-square. (The numbers counted have 0, 2, 4, 6 or 8 in the ones place.) Ask students to imagine counting in twos to much higher numbers. Would they say 134? 157? 171? 224? How do they know? Explain that large numbers such as 224 and 134 are also multiples of 2.

Practice Book: Students can complete Practice Book page 20. They can do this directly after the main activity, as homework, or as the focus of a separate mathematics session to help consolidate their learning and build fluency. Remind students of their work with counting jumps of 10 on page 18 of the Practice Book and say that they will do the same here, but with jumps of 2.

Differentiated outcomes	
All students	should count to 100 and back in twos with support.
Most students	will count in twos confidently and recognise even numbers.
Some students	may know how the pattern of even numbers continues beyond 100.

Answers

Student Book page 11

1 72, 74, 76, 78, 80, 82, 84, 86, 88, 90, 92, 94, 96, 98, 100

2 32, 34, 36, 38, 40, 42, 44, 46, 48, 50

3 21, 23, 25, 27, 29, 31, 33, 35, 37, 39,

4 1, 3, 5, 7, 9, 11, 13, 15, 17, 19

5 All the red numbers are even.

Practice Book page 20

Check that students have counted on correctly in jumps of 2, depending on which card they choose.

Stretch zone: Students may draw two tens-rods and four ones-cubes combined to represent 24, or two groups of one tens-rod and two ones-cubes.

1C Counting in twos

Explore Student Book page 12 · Practice Book page 21

Specific learning focus

- Count in twos on a number line.

Global Skills

- **Interpersonal skills:** communication

Key vocabulary

- count on/back, twos, even, odd, multiples of 2

Resources

- large 0–30 number line for front of class, marked in intervals of 2
- small 0–30 number line, marked in intervals of 2, for each pair; or mini whiteboards and markers

 **Introductory activity**

Count together as a class from 0 to 30 in twos. Use the number line to help. Then start from another number and jump on in twos to 30 from that number. Choose a different number and repeat.

Using the number line to help, count back from 30 to 0, and then from different numbers, for example, count back in twos on the number line from 24.

 **Main activity**

Show students a number line from 0 to 30 marked in intervals of 2. Ask questions to get students talking about the number line, such as *What is the start/end number? What do you notice about the numbers in between?*

Start at one of the marked numbers, say 12, and count on one jump of 2. *Which number do I land on?* (14) Now jump a further two jumps of 2 from 14. *Which numbers do I land on?* (16, 18) *Which number do I finish on?* (18)

Start at 20 and jump back two jumps of 2. *Which number do I finish on?* (16) Now jump back a further three jumps of 2. *Which number do I finish on?* (10)

Ask students to work in pairs. Give them a 0–30 number line marked in intervals of 2 or ask them to draw one on a mini whiteboard. Ask them to make a series of jumps of 2, forward or backward, from different numbers on the number line. For example, one student chooses a starting number (14), makes four jumps of 2 forward, landing on 22, then three jumps of 2 back, landing on 16, and so on.

Ask students to complete the activity on page 12 of the Student Book individually. Once they have completed this, they can return to their partners and compare answers. They should also discuss the question in the first speech bubble and practise counting back in twos together from different starting numbers.

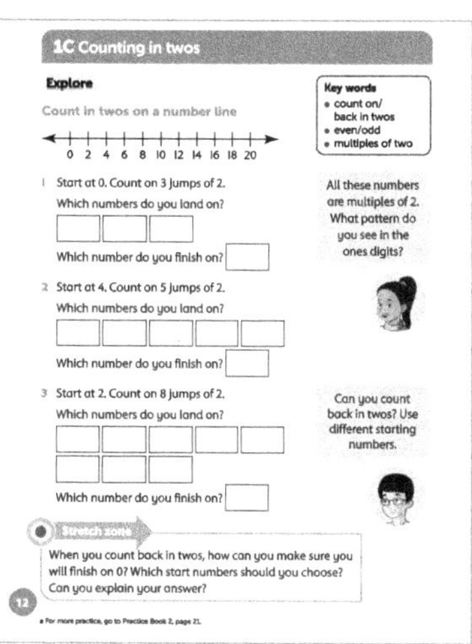

Differentiation

Supporting: Help students with keeping their place on the number line as they count on or back in twos. Model pointing to the numbers with a finger.

Consolidating: Where students are counting on from different numbers (e.g. counting on in twos from 8), ask them to predict which numbers they will land on.

Extending: Ask students to work out the finishing number or number landed on without looking at the number line or using it only to check.

Stretch zone: *When you count back in twos, how can you make sure you will finish on 0? Which start numbers should you choose? Can you explain your answer?*

Students should notice that starting from numbers ending in 0, 2, 4, 6, or 8, will mean finishing on 0, because the count will be on even numbers.

 **Reflection time**

Choose a student to pick a number and mark it on a large 0–30 number line. Ask the class to tell you which number is 2 more than the number and 2 less than the number. For example, if the student chooses 16, then 2 more is 18 and 2 less is 14. Draw the jumps to the new numbers on the number line. Ask students: *What do you notice about a jump of 2 and finding 2 more or 2 less?* For example, they might notice that a forward jump of 2 is the same as finding 2 more. Repeat with three more examples.

Ask students to ask similar questions for the rest of the class to answer.

Practice Book: Students can complete page 21 of the Practice Book. They can do this directly after the main activity, as homework, or as the focus of a separate mathematics session to help consolidate their learning and build fluency. Encourage students to read the numbers they land on, for additional practice in saying the names of each number.

Differentiated outcomes	
All students	should count on in twos from 0.
Most students	will count on from any number in twos on a number line.
Some students	may count back confidently in twos without the number line from any number.

Answers

Student Book page 12

1 2, 4, 6; 6

2 6, 8, 10, 12, 14; 14

3 4, 6, 8, 10, 12, 14, 16, 18; 18

Practice Book page 21

1 12, 14, 16, 18, 20

2 18, 16, 14, 12, 10, 8

3 18, 16, 14, 12, 10, 8, 6

4 17, 15, 13, 11, 9, 7, 5

Stretch zone: Answers will vary.

1D Counting in fives

Discover Student Book page 13 • Practice Book page 22

Specific learning focus

- Learn and recognise multiples of 5.

Global Skills

- **Creative skills:** exploring/investigating
- **Interpersonal skills:** communication

Key vocabulary

- multiple of 5, fives, count on, count back

Resources

- counting stick
- 100-squares and 0–100 number lines marked in intervals of 10

Language support

Display 100-squares and number lines showing multiples of 5 and multiples of 10. Label and add speech bubbles.

 **Introductory activity**

Show students a counting stick with 10 intervals marked.

Tell them that one end of the counting stick is zero and that you are going to count up in intervals of 2. Ask: *How many intervals can you see?* (10) *What would the end number be then?* (20) *What would the middle number be?* (10) *What about this number?* Point to 18. Students should discuss the questions in pairs first. Repeat with other divisions.

Repeat the activity but this time starting at 0 and counting in tens. Then repeat, counting in fives.

 **Main activity**

Count together in tens from 0 to 100 and back again. Look at the pattern of tens on the 100-square and draw a 0–100 number line, marking each interval of 10 only. Now count in **fives** to 100 and back on the 100-square.

Ask students to look at the 100-square on page 13 of the Student Book and colour in each **multiple of 5**, starting from 5. Ask students how the pattern is the same and how is it different to counting in tens. (When you count in fives, you land on all the multiples of 10. When you count in tens, you land on every second multiple of 5.) They complete the remaining questions and discuss questions 2–4 in pairs.

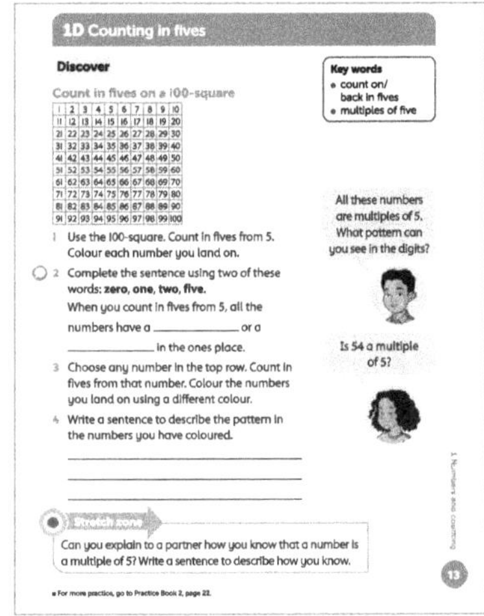

Differentiation

Supporting: Provide students with 100-squares and number lines to support counting in fives and tens.

Consolidating: Ask students to predict whether or not a number is a multiple of 5 or 10 and ask them to explain why.

Extending: Ask students to tell you multiples of 5 and 10 up to 100. Ask them to explain how they know.

Stretch zone: *Can you explain to a partner how you know that a number is a multiple of 5? Write a sentence to describe how you know.*

Students should describe multiples of 5 correctly as numbers that end in a 0 or 5.

 **Reflection time**

Ask students to look again at the patterns on the 100-square. Check that students recognise that all the numbers have 0 or 5 in the ones place if we are counting in fives, but only 0 in the ones place if we are counting in tens. Ask students to imagine counting in fives or tens to much higher numbers. Would they say 135? 150? 171? 220? How do they know? Remind students that we call numbers multiples of 5 if we would say them when counting in fives and multiples of 10 if we would say them when counting in tens.

Practice Book: Students can complete page 22 of the Practice Book. They can do this directly after the main activity, as homework, or as the focus of a separate mathematics session to help consolidate their learning and build fluency.

Differentiated outcomes	
All students	should count in fives and tens from zero using a 100-square.
Most students	will recognise and count in multiples of 5 from any number less than 100 using a 100-square.
Some students	may see patterns between multiples of 5 and 10 and use these to identify multiples of 5 and 10 greater than 100.

1D Counting in fives

Explore Student Book page 14 • Practice Book page 23

Specific learning focus

- Count on and back in fives on a number line.

Global skills

- **Interpersonal skills:** communication

Key vocabulary

- multiple of 5, fives, count on, count back

Resources

- large 0–50 number line for front of class

Language support

Use the 0–50 number line to support students' counting. Listen to how each individual says the numbers out loud when counting up or back in fives.

Introductory activity

Count together as a class from 0 to 50 in tens. Use a large 0–50 number line to help. Then start from another multiple of 5 and count on in tens from that number: for example, choose 15 and say 35, 45. Choose a different number and repeat. For numbers that end in 5, the counting will always be numbers that end in 5.

Using the number line to help, count back from 50 to 0, and then from different numbers ending in 0, for example, count back in fives on the number line from 40.

Student Book page 13

1 5, 10, 15, 20, 25, 30, 35, 40, 45, 50, 55, 60, 65, 70, 75, 80, 85, 90, 95, 100

2 When you count in fives from 5, all the numbers have a **zero** or a **five** in the ones place.

3 Check that students have correctly counted on in fives from their chosen number and coloured the correct numbers.

4 Students may write, for example, 'The numbers go up by 5 each time and make a pattern of two columns on the 100-square.'

Practice Book page 22

Check that students have counted on correctly in jumps of 5, depending on which card they chose, and filled in the missing numbers correctly for each question.

Stretch zone: Answers will vary.

Main activity

Show students a large 0–50 number line marked in intervals of 5. Start at one of the marked numbers, say 35, and count on one jump of 5. *What number do we land on?* (40) Now jump a further two jumps of 5 from 40. *What number do we finish on?* (50)

Start at 30 and jump back two jumps of 5. *What number do we land on?* (20) Now jump back a further three jumps of 5. *What number do we finish on?* (5)

Ask students for the answers to a series of jumps of 5, forward or backward, from different numbers. For example, four jumps of 5 forward starting from 20 (40), three jumps of 5 back from 35 (20), and so on.

Ask students to complete the activity on page 14 of the Student Book individually. Once they have completed the questions, ask them to discuss in pairs the questions in the speech bubbles. Encourage them to use examples from their work to support their answers.

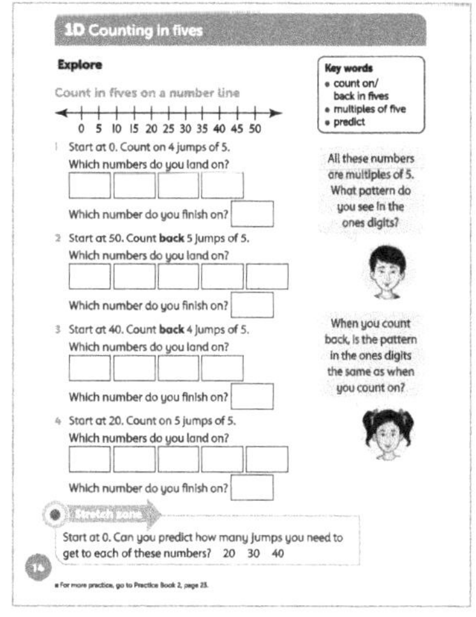

Differentiation

Supporting: Help students to keep their place on the number line as they count on or back in fives, by modelling using a finger to keep the place.

Consolidating: Ask students to predict which numbers they will land on when counting on in fives from different starting numbers.

Extending: Suggest that students try to count back from 50 without the number line.

Stretch zone: *Start at 0. Can you predict how many jumps you need to get to each of these numbers? 20 30 40*

Students might predict that it takes two jumps of 5 to get to 10, so four jumps of 5 will get to 20, and so on.

 **Reflection time**

Ask students to look at the multiples of 5. Ask *What pattern do you see in the ones digits? When you count back, is the pattern in the ones digits the same as when you count on?'*

Choose a student to pick a number and mark it on a 0–50 number line with all numbers marked and labelled. Ask the class to tell you what number is 5 more than the number and 5 less than the number. Draw the jumps to the new numbers on the number line. Repeat with three more examples.

Ask students, *What do you notice about a jump of 5 and finding 5 more or 5 less?* They may notice, for example, that a forward jump of 5 is the same as finding 5 more.

Select students to ask similar questions for the rest of the class to answer.

Practice Book: Students can complete page 23 of the Practice Book. They can do this directly after the main activity, as homework, or as the focus of a separate mathematics session to help consolidate their learning and build fluency.

Differentiated outcomes	
All students	should count on in fives from 0.
Most students	will count on from any number in fives on a number line.
Some students	may count back in fives on the number line from any number.

Answers

Student Book page 14

1 5, 10, 15, 20; 20

2 45, 40, 35, 30, 25; 25

3 35, 30, 25, 20; 20

4 25, 30, 35, 40, 45; 45

Practice Book page 23

1 15, 20, 25, 30, 35

2 5, 10, 15, 20, 25, 30, 35, 40, 45, 50.

3 45, 40, 35, 30, 25

4 45, 40, 35, 30, 25, 20, 15

Stretch zone: Answers will vary.

1E Counting in threes and fours

Discover Student Book page 15 • Practice Book page 24

Specific learning focus

- Begin to count on in small constant steps such as threes and fours.

Global skills

- **Interpersonal skills:** communication

Key vocabulary

- number pattern, count on/back, diagonal lines, threes, fours

Resources

- counting stick
- 100-squares
- poster with two large 100-squares, one with multiples of 3 coloured and one with multiples of 4 coloured

Language support

Display a poster with two large 100-squares, one with the multiples of 3 coloured in and one with the multiples of 4 coloured in. Label these 'multiples of 3' and 'multiples of 4'. Add a speech bubble asking, *How is the pattern of multiples of 4 the same as the pattern for multiples of 2? How is it different?* Include some blank speech bubbles for students to record their answers. Introduce formally the vocabulary 'horizontal', 'vertical' and 'diagonal'. Model how to pronounce them correctly.

 **Introductory activity**

Use a counting stick to count with students in twos, fives and tens, forward and back. After each separate count, point to lines on the stick and ask students which number that would represent for that particular count.

 **Main activity**

Ask students to raise three fingers on their left hand (or their thumb and two fingers) and count along them in ones repeatedly to 30. They should whisper or think the numbers for the first two fingers and only say the third one out loud, so that they are counting in **threes**, for example whisper '1' (move thumb), whisper '2' (move index finger), say '3' (move middle finger), whisper '4' (move thumb), whisper '5' (move index finger), say '6' (move middle finger) and so on. Ask students to do this in pairs to keep a check on each other. Repeat.

Now ask students to work in pairs to count on in threes and colour the correct numbers on the 100-square on page 15 of their Student Books.

Ask students what they notice about the patterns. Unlike the number patterns for 2, 5 and 10, the lines are not straight, so explain to students that these lines are called **diagonal lines**.

Next, ask them to repeat the activity to count in **fours**, using four fingers this time, and then colour each multiple of 4 on the 100-square in their Student Book.

Students then answer questions 3 and 4.

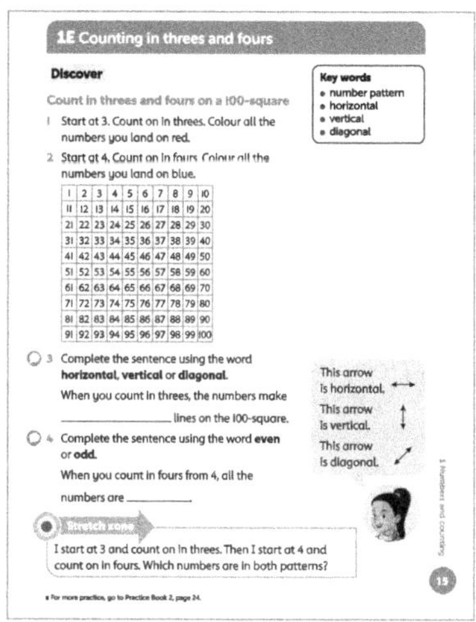

Differentiation

Supporting: Show students how to count on in threes and fours on the 100-square, using a finger to keep the place after each count.

Consolidating: Ask students to use the 100-square to count back in threes and fours.

Extending: Ask students to think about the link between multiples of 2 and multiples of 4 and be able to explain this.

Stretch zone: *I start at 3 and count on in threes. Then I start at 4 and count on in fours. Which numbers are in both patterns?*

Check that students can correctly identify the numbers common to both counting patterns. The numbers are 12, 24, 36, 48, 60, 72, 84, 96.

 **Reflection time**

Ask students to highlight the even numbers (multiples of 2) on a 100-square. Ask students how this is the same and how it is different from the pattern they found for multiples of 4. Agree that every other even number is missed out when counting in fours. As an extension question ask, *Can you explain why?* Some students may say that this is because two lots of 2 are needed to make each 4.

Practice Book: Students can complete page 24 of the Practice Book. They can do this directly after the main activity, as homework, or as the focus of a separate mathematics session to help consolidate their learning and build fluency.

Differentiated outcomes	
All students	should count up in threes and fours using a 100-square.
Most students	will count on and back in threes and fours using a 100-square.
Some students	may predict what the next multiple of 3 and 4 will be on a 100-square after identifying those up to, for example, 24.

Answers

Student Book page 15

1   3, 6, 9, 12, 15, 18, 21, 24, 27, 30, 33, 36, 39, 42, 45, 48, 51, 54, 57, 60, 63, 66, 69, 72, 75, 78, 81, 84, 87, 90, 93, 96, 99

2   4, 8, 12, 16, 20, 24, 28, 32, 36, 40, 44, 48, 52, 56, 60, 64, 68, 72, 76, 80, 84, 88, 92, 96, 100

3   When you count in threes, the numbers make diagonal lines on the 100-square.

4   When you count in fours from 0, all the numbers are even.

Practice Book page 24

Answers will vary as students choose their own number of jumps each time. Check that students have counted on correctly in jumps of 3 and 4, depending on which card they chose, and have filled in the missing numbers in the boxes correctly for each question.

Stretch zone: Students may say that when you count in threes and fours you land on 12, 24 and 36. When you count in fours all the numbers are even, but when you count in threes every second number is even.

1E Counting in threes and fours

Explore Student Book page 16 • Practice Book page 25

Specific learning focus

- Begin to count on in small constant steps of threes and fours along a number line.

Global Skills

- **Interpersonal skills:** communication/teamwork

Key vocabulary

- number pattern, number line, count on/back, threes, fours, steps

Resources

- counting stick
- calculators, counters or cubes
- 100-squares
- poster showing a large 0–30 number line with intervals of 3 marked and large 0–40 number line with intervals of 4 marked

Language support

Display a poster with a large 0–30 number line with intervals of 3 marked and labelled and a large 0–40 number line with intervals of 4 marked and labelled. Label these 'multiples of 3' and 'multiples of 4'. Add a speech bubble asking, *How is the pattern of multiples of 4 the same as the pattern for multiples of 2? How is it different?* Include some blank speech bubbles for students to record their answers.

 **Introductory activity**

Use the counting stick to count with students in threes and fours, forward and back. After each separate count, point to lines on the stick and ask students which number that would represent for that particular count.

 **Main activity**

Arrange students in groups of three in a circle. Count in threes around the circle, starting from zero, until all groups have said a number. Repeat from a different starting point. Any students not in a group can be checkers, making sure the correct numbers are said. Repeat the same activity in groups of four.

Ask students to look at page 16 in the Student Book. *What do you notice about the first number line?* (It goes up in threes.) *What about the second one?* (It goes up in

fours.) Practise counting jumps on each number line as a class before students work through each question on the page independently. *On the first number line I count on 5 jumps of 3 from zero. What numbers do I land on?* (3, 6, 9, 12, 15) *What number do I finish on?* (15)

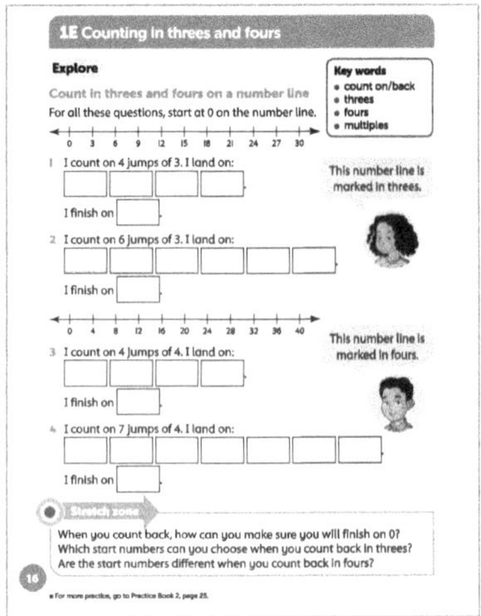

Differentiation

Supporting: Ask students to group cubes or counters and count them to check their answers.

Consolidating: Encourage students to write down the numbers landed on during each set of jumps and describe the pattern in the numbers.

Extending: Ask students to write multiplication sentences for the answers to the questions in the Student Book, for example, 4 jumps of 3 is written as $4 \times 3 = 12$.

Stretch zone: *When you count back, how can you make sure you will finish on 0?*
Which start numbers can you choose when you count back in threes?
Are the start numbers different when you count back in fours?

Check that students understand how to start on a multiple so that when they count back they end on 0. For example, can they start counting back in fours from a multiple of 4 to reach 0?

 **Reflection time**

Show students how to use the repeat function or key on a calculator. Ask them to press 3 + 3 = and ask '*What number does it show?*' (6) Then tell them to repeatedly press = to make the calculator count in threes. Ask, *How many times do you need to press = to get to 30?* Change the number to count in other **steps**. Students may enjoy watching the numbers increase rapidly.

Practice Book: Students can complete page 25 of the Practice Book. They can do this directly after the main activity, as homework, or as the focus of a separate mathematics session to help consolidate their learning and build fluency. For additional challenge, encourage students to count up in their chosen number of jumps in their head, recording each number they land on, and then use the number line to check.

<table>
<tr><td colspan="2">Differentiated outcomes</td></tr>
<tr><td>All students</td><td>should count on in threes and fours from 0.</td></tr>
<tr><td>Most students</td><td>will count on in threes or fours from any number on a number line.</td></tr>
<tr><td>Some students</td><td>may count back in threes or fours from any number on a number line.</td></tr>
</table>

Student Book page 16

1 3, 6, 9, 12; 12   **3** 4, 8, 12, 16; 16

2 3, 6, 9, 12, 15, 18; 18   **4** 4, 8, 12, 16, 20, 24, 28; 28

Practice Book page 25

1 3, 6, 9, 12, 15, 18

2 3, 6, 9, 12, 15, 18, 21, 24

3 4, 8, 12, 16, 20, 24

4 4, 8, 12, 16, 20, 24, 28, 32

Stretch zone: Answers will vary.

1F Estimating and counting

Discover   Student Book page 17 · Practice Book page 26

Specific learning foci

- Give a sensible estimate of up to 100 objects (e.g. choosing from 10, 20, 50 or 100).
- Count up to 100 objects.

Global Skills

- **Creative skills:** investigating
- **Real-world skills:** presenting information
- **Interpersonal skills:** communication/teamwork

Key vocabulary

- estimate, count, tens, ones, more than, less than

Resources

- bags of small classroom objects (e.g. counters, cubes, pencils), with 20–100 of one type of object in each bag (at least one bag for each pair)
- bowl and dried beans or counters

Language support

When students are counting in tens, listen in and support pronunciation where necessary. Students may not know the actual name of some of the objects they are counting and how to write them. Have these labelled so that you can share and say the names with students who need this support.

Introductory activity

Show students a bag of counters or cubes. Tell them that you are going to ask them to **estimate** how many cubes might be in the bag. *Who can tell me what 'estimate' means?*

Pass the bag around the class for them to look at and feel. Ask students to get into pairs to discuss and then to decide on an estimate. Record the estimates on the board.

Ask one student to come to the front of the class and count the objects. Ask another student to come and check the total by counting in a different way. If necessary, model grouping the cubes in tens and then counting in tens and ones to find out how many there are. Compare the total with the estimates. Ask the pair that had the closest estimate: *How did you decide on that estimate? Was your total* **more than** *or* **less than** *the estimate? Was your estimate close to the total?*

Main activity

Tell students that they will work in pairs to estimate and count the contents of six bags.

For each bag, they must first decide with their partner a reasonable estimate, then count the objects. They record their estimate and their count in the Student Book on page 17. Explain that they should draw the object that they count (e.g. red cube, green counter, coloured pencil, paperclip) on the picture of the bag in the Student Book as well as write the name of the object.

Distribute one bag to each pair. When students finish their count, tell them to swap bags with another pair or take a spare bag.

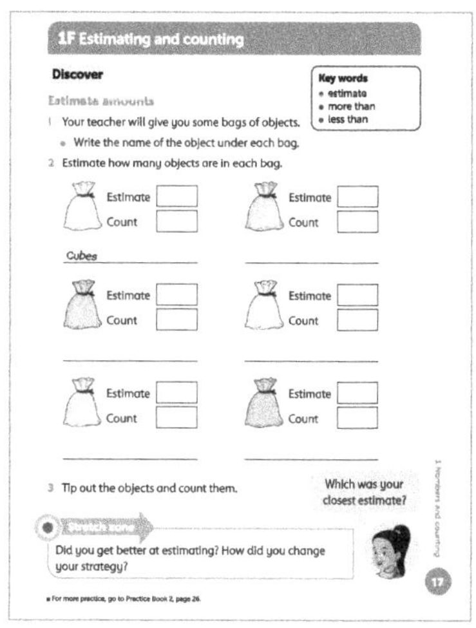

Differentiation

Supporting: Support students to count in twos or organise items into piles of 10 to count.

Consolidating: Ask students to show you how many by counting in twos and then tens.

Extending: Invite students to put bags together to estimate larger numbers.

Stretch zone: *Did you get better at estimating? How did you change your strategy?*

Check students' explanations of how they estimated and how they improved their estimating by adopting a different strategy. Students can also share their strategy with others to consolidate this thinking.

 **Reflection time**

Choose two bags and compare students' estimates with their actual counts. Challenge students to find the difference between the estimate and the count to find out whose estimate was closest.

Practice Book: Students can complete page 26 of the Practice Book. They can do this directly after the main activity, as homework, or as the focus of a separate mathematics session to help consolidate their learning and build fluency. Students will need a bowl and dried beans or similar small objects to complete this activity. Explain that you would like them to make an estimate that is a multiple of ten. Give an example: *I think there are about 26 beans in the bowl. 26 is between two multiples of 10: 20 and 30. If I think about 26 on a number line, 26 is closer to 30 than 20 so I estimate 30 beans.*

Differentiated outcomes	
All students	should make sensible estimates and count to check the total.
Most students	will group objects in twos, fives or tens to make counting quicker.
Some students	will be able to put bags together to estimate and count larger numbers.

Answers

Student Book page 17

Check that students have estimated and counted the objects in each bag.

Check that students have written the correct number of objects to match the number they have estimated and counted.

Practice Book page 26

Check that students have estimated and counted the beans in the bowl.

Check that students have written the correct number of beans to match the number they have estimated and counted.

Stretch zone: Answers will vary.

1F Estimating and counting

Explore · Student Book page 18 · Practice Book page 27

Specific learning foci

- Give a sensible estimate of up to 100 objects (e.g. choosing from 10, 20, 50 or 100).
- Count up to 100 objects.

Global Skills

- **Creative skills:** investigating
- **Real-world skills:** research/presenting information

Key vocabulary

- estimate, count, tens, ones, less than, more than

Resources

- a bag of 62 counters (or cubes) and a bag of 33 counters (or cubes)
- three shallow trays with 20, 50 and 100 counters or cubes on them
- labels 20, 50 and 100 to label the trays
- images of different numbers (e.g. 20–80) of the same thing, such as flocks of birds

Language support

Display a list of number words and numerals for 0 to 9 and 10 to 100. Underline the stressed part of the word or write it in capitals to support pronunciation.

 #### Introductory activity

Explain to students that, to help with estimating, it is useful to look at known numbers of that object.

Put the three trays of counters and the two bags of counters on a table for students to look at. Tell students that they are going to use these trays to help them make estimates of how many cubes there are in each bag. Show them the three labels (20, 50 and 100). *What do you notice about these amounts?* (They are all multiples of 10.)

Which label do you think matches which tray? Why? Agree on the appropriate label for each tray and add the labels to the trays.

Pass the two bags around the class, asking students to feel each bag to get a sense of how many counters may be inside and to compare to each of the trays. Collect estimates for each bag and write a selection on the board.

Ask a student to come to the front of the class and count the number in each bag. Agree 20 and 50 as good estimates.

Ask, *What do we mean by a good estimate? Is it when the guess is quite close to the actual number? Can we make our estimates better by thinking about whether the number is less than or more than another number?*

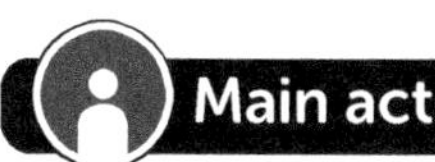 **Main activity**

Ask students to complete page 18 in the Student Book individually. Explain that they should look at each box of smiley faces in the questions and compare with the pictures of 20, 50 and 100 smiley faces at the top of the page to decide whether 20, 50 or 100 would be the closest estimate. After drawing a circle around their chosen estimate, they should count the smiley faces. Ask, *How can we count them easily? Can we arrange them to make it easier to count?* Students may suggest drawing a circle round groups of 10. *Why are some arrangements easier to count than others?*

As they work on the activity, ask individuals how they decided on their estimate.

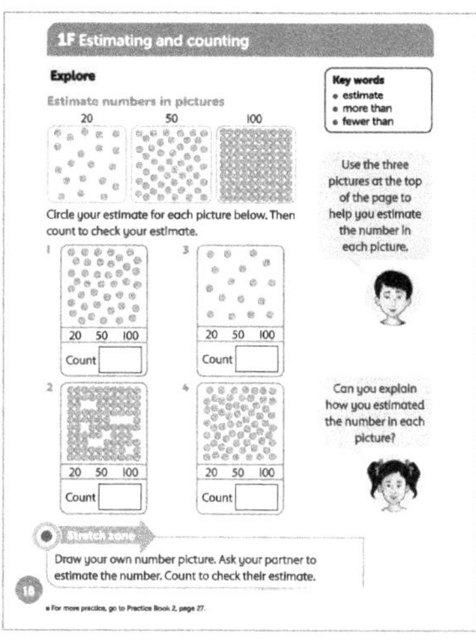

Differentiation

Supporting: Model grouping objects to count more quickly. Count alongside students to support them.

Consolidating: Ask students to explain how they estimated the number in each picture.

Extending: Ask students to make bags of counters for a partner to estimate.

Stretch zone: *Draw your own number picture. Ask your partner to estimate the number. Count to check their estimate.*

Encourage students to draw a number than is more than 20 but less than 100. Also, encourage them to explain their estimate once they have made it. (The estimate should be to the nearest multiple of 10.) Check that students are estimating the numbers drawn by their partner and ask what strategy they are using.

 ## Reflection time

Ask students: *How did you decide on an estimate? How did you use the pictures at the top of the page to help you? What was the actual count? Were there fewer or more smiley faces than your estimate?*

Display photographs showing, say, 20–80 of the same item and ask students to estimate the number to the nearest multiple of 10. For example, you could use images of flocks of birds.

Practice Book: Students can complete page 27 of the Practice Book. They can do this directly after the main activity, as homework, or as the focus of a separate mathematics session to help consolidate their learning and build fluency. Students will need access to four containers of varying sizes and enough dried beans (or similar small objects) to fill the containers in order to complete this activity.

Differentiated outcomes	
All students	should make sensible estimates and count accurately.
Most students	will group objects to count more quickly.
Some students	may group objects in more efficient ways to enable accurate counting.

Answers

Student Book page 18

1 50   42

2 100   89

3 20   17

4 50   65

Practice Book page 27

Check that students have estimated the number of beans needed to fill each container and have counted the actual number. Check whether their estimates are accurate.

Stretch zone: Answers will vary.

1G Numbers in between

Specific learning focus

- Say a number between any given neighbouring pairs of multiples of 10 (e.g. 44 is between 40 and 50).

Global skills

- **Creative skills:** exploring/investigating
- **Interpersonal skills:** communication/teamwork

Key vocabulary

- next to, between, larger, smaller, multiple

Resources

- set of large number cards showing multiples of 10, for 0 to 100
- 100-squares
- washing line and pegs

Language support

Display a washing line and number cards for students to order on the line. Make sure that the line is only long enough for about 10 cards, then label the line with the task, such as 'Choose any 10 numbers to order on the washing line'. Use captions such as 'Remember to look at how many tens, then how many ones.'

 **Introductory activity**

Ask three students to come to the front of the class. Give each student one of the large number cards showing multiples of 10 (e.g. 20, 50, 90) and ask them to arrange themselves in a line, in ascending order (smallest to biggest).

Call out eight more students, one at a time. Give them one of the cards and ask the rest of the class where to position each student in the line. Encourage the use of a range of positional words such as **next to** and **between**. You could introduce these terms first by describing a range of classroom objects, for example, *the window is next to the door, the board is between those two desks*.

Once the line is complete, ask everyone to count along it, forward and back. Write 'multiple' on the board and tell students that they were counting in multiples of 10. Make sure that the students understand that they must start at zero and count in steps of equal size to find multiples.

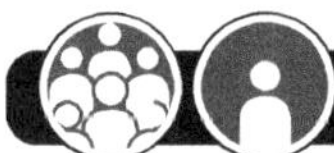 **Main activity**

Give out the large multiple-of-10 number cards to ten students and ask them to arrange themselves in ascending order again but this time leaving some space between them.

Stand between any two multiples of 10, for example, 20 and 30. Say, *I am not a multiple of ten, what number could I be?* (any number from 21 to 29) Model students' suggestions on an empty number line drawn on the board: a 20–30 number line, for example, with 25 marked as accurately as possible.

Once students have some understanding of the word 'between', ask them to complete page 19 in the Student Book individually. Have 100-squares available for support. As you walk around the classroom, observing students, ask questions such as: *How do you know that 46 is between 40 and 50?* (by counting on from the starting number)

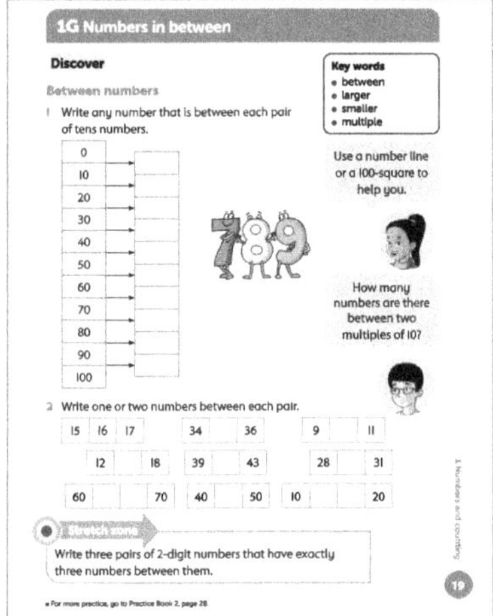

Differentiation

Supporting: Give students a 100-square to use to check their answers; you can support them to use it initially.

Consolidating: Ask students to work in pairs using a 100-square, with one student picking any two numbers, the other saying a number between them.

Extending: Ask students to create their own questions, (e.g. finding a number that is between 40 and 50 and a multiple of 2, or an odd number)

Stretch zone: *Write three pairs of 2-digit numbers that have exactly three numbers between them.*

Check that students notice that they can choose any 2-digit number as their start number and then count on 4 to find their end number and that there will always be three whole numbers between them.

 **Reflection time**

Ask students how many numbers there are between two multiples of 10, say 50 and 60. *Is this the same for any two multiples of 10?*

Explore the 'between' questions from the Student Book such as:

12   18

Ask students to discuss in pairs how many possible answers there are. (There are five possible numbers between 12 and 18: 13, 14, 15, 16, 17.) Repeat for the other questions.

Practice Book: Students can complete page 28 of the Practice Book. They can do this directly after the main activity, as homework, or as the focus of a separate mathematics session to help consolidate their learning and build fluency. Provide a 100-square for students who may need additional support.

Differentiated outcomes	
All students	should count on in tens accurately and use a 100-square to find numbers in between multiples of 10.
Most students	will know more than one number between two given multiples of 10.
Some students	will know how many numbers are between two given numbers.

Student Book page 19

1   Check that the students have chosen a number between each of the tens.

2   Check that the students have chosen numbers within the ranges given.

Practice Book page 28

Check that the students have chosen numbers within the ranges given.

Stretch zone: Answers will vary as students choose their own numbers. Check that their numbers fit the question.

1G Numbers in between

Explore   Student Book page 20 • Practice Book page 29

Specific learning focus

- Say a number between any given neighbouring pairs of numbers on a 100-square.

Global Skills

- **Creative skills:** investigating
- **Real-world skills:** presenting information

Key vocabulary

- next to, between, larger, smaller

Resources

- large 100-square
- place-value cards cut out from Resource Sheet 1, copied onto card
- two sets of large digit cards 0–9
- digit cards 0–9 (one set per student)

Language support

Reinforce the vocabulary of 'before', 'after' for numbers. Help students to understand that a number 'between' is one that comes after one number and before another number. For example, 24 is between 23 and 25.

Introductory activity

Ask three students to come to the front of the class. Give each student a pair of large digit cards from two 0–9 sets and ask them to make a 2-digit number with their cards. Then ask them to arrange themselves in a line, in ascending order.

Call out two more students, one at a time. Give them a pair of cards to make a 2-digit number, then ask the rest of the class where to position each student in the line. Encourage the use of a range of positional words such as 'next to' and 'between'.

Ask students to make up statements about the ordered line, e.g. 54 is between 18 and 73; 54 is **larger** than 18, and **smaller** than 73.

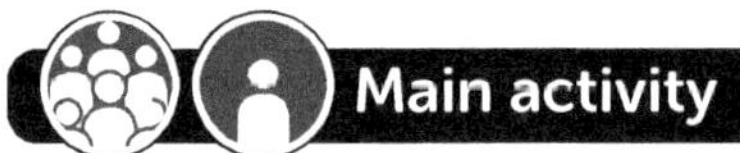

Using the same line of five students holding number cards, stand between any two numbers, for example, 18 and 31. Say, *I am not a multiple of two, what number could I be?* (any number from 19, 21, 23, 25, 27, 29). Model students' suggestions on an empty number line drawn on the board: a 15–35 number line, for example, with 25 marked as accurately as possible in the middle.

Ask students to complete page 20 in the Student Book individually with the support of place-value cards. As you walk around the classroom, observing the students, ask questions such as: *How do you know that 34 is between 20 and 35? What other numbers are between 20 and 35?*

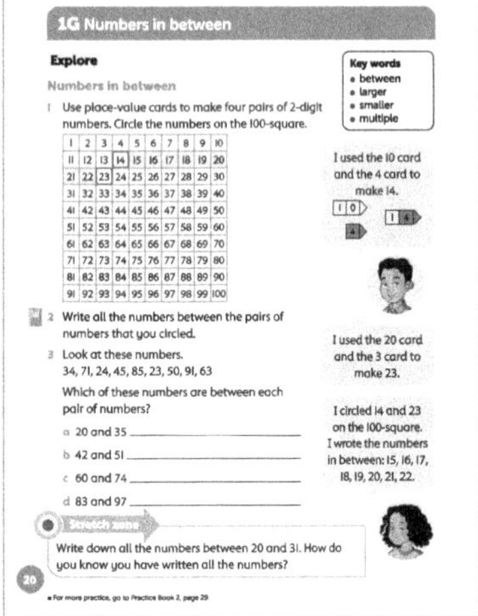

Differentiation

Supporting: Revise how to use place-value cards to build 2-digit numbers before students begin working on the Student Book activity.

Consolidating: Ask students to describe some of the numbers in between one of their pairs, e.g '25 is between 19 and 26. It is a multiple of 5. It has 2 tens and 5 ones. It is 1 less than 26'.

Extending: Ask students to create their own questions (e.g. finding a number that is between 58 and 74 and a multiple of 3 and an odd number).

Stretch zone: *Write down all the numbers between 20 and 31. How do you know you have written all the numbers?*

Check that students have written 21, 22, 23, 24, 25, 26, 27, 28, 29, 30. Ask students to describe their strategy, perhaps counting on from 20 to 31 in ones.

 **Reflection time**

Explore one of the between questions from the Student Book, for example, between 42 and 51.

After finding which of the listed numbers is between 42 and 51, ask students to discuss in pairs how many other possible numbers there are. (There are nine possible numbers between 42 and 51: 43, 44, 45, 46, 47, 48, 49, 50, 51.) Ask students to describe the numbers in other ways: *Which of these numbers are odd? Even? Which is larger, 48 or 44? How do you know?* Repeat for the other questions.

Practice Book: Students can complete page 29 of the Practice Book. They can do this directly after the main activity, as homework, or as the focus of a separate mathematics session to help consolidate their learning and build fluency. Provide children with digit cards to make their 2-digit numbers. To extend the activity, ask students to describe the numbers they have made in at least three ways.

Differentiated outcomes	
All students	should use a 100-square to find numbers in between any two given numbers.
Most students	will be able to find numbers between two given numbers and check with a 100-square.
Some students	may know how many numbers are between two given numbers, suggesting more than one and describing the numbers in different ways – odd, even, multiple of….

Student Book page 20

1 Check that the numbers have been correctly circled.

2 Check that the in-between numbers have all been included.

3 **a** 34, 24, 23

   b 45, 50

   c 71, 63

   d 85, 91

Practice Book page 29

Check that students' numbers made from their digit cards have been correctly sorted as smallest and largest and two examples of numbers in between have been recorded.

Stretch zone: Students' answers will vary, depending on which numbers they chose.

Discover Student Book page 21 • Practice Book page 30

Specific learning focus

- Use a number line to round to the nearest 10.

Global skills

- **Interpersonal skills:** communication/teamwork

Key vocabulary

- rounded, round up, round down, nearest multiple of 10

Resources

- 0–100 number lines labelled in intervals of 10
- sets of digit cards 0–9 (one set for each pair)

Language support

Model the key language such as 'round up', 'round down'. Create a poster showing these phrases as sentence starters and displaying numbers with an arrow to indicate whether they round up or down.

 **Introductory activity**

Draw a large empty 0–100 number line on the board. Mark 30 on the line on the board. Ask students: *Which numbers are 1 more than 30 and 1 less than 30?* (31 and 29). Ask *What is the **nearest multiple of 10** to 29?* (30) *What is the nearest nearest multiple of 10 to 31?* (30) Now ask *Which numbers are 2 more and 2 less than 30?'* (32 and 28) Ask, *What is the nearest multiple of 10 to 28 and 32?* Agree that 30 is the nearest multiple of 10 to both 28 and 32. Repeat with other numbers, leaving out any multiples of 5 and mention of rounding until the main activity.

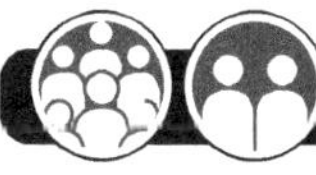 **Main activity**

Draw a section of a number line on the board showing 20–30, labelled at its ends only. Mark 23 on the line and ask, *Which multiples of 10 is 23 between?* (20 and 30) *Which multiple of 10 is 23 nearest to?* (20) Now ask, *Which other numbers between 20 and 30 are nearer to 20 than to 30?* Students should see that 21, 22, 23 and 24 are closer to 20 than 30.

Now mark 26 on the line and ask: *Which multiples of 10 is 26 between?* (20 and 30) *Which multiple of 10 is 26 nearest to?* (30) Now ask, *Which other numbers between 20 and 30 are nearer to 30 than 20?* Students should see that 26, 27, 28 and 29 are closer to 30 than 20.

Is 25 closer to 20 or 30? Agree that 25 is the same distance from 20 as it is from 30.

Explain that numbers between two multiples of 10 are sometimes **rounded** to the nearest multiple of 10. Ask, *Which multiple of 10 is 18 nearest to?* (20) Say, *18 **rounds up** to 20. What about 34? What is the nearest multiple of 10?* (30) Say, *34 **rounds down** to 30.*

Discuss when rounding might be used, for example, to give an estimate for a number of objects. Ask students other examples of how numbers round to the nearest multiple of 10. Then ask students to round 25. Ask, *Is it closer to 20 or 30?* When students say that 25 is halfway between 20 and 30, explain that, in this case, we always round up, so 25 rounds to 30.

Students can now work on Student Book page 21 in pairs, working together to round their four numbers, but each recording their own two numbers in their Student Book. Encourage students to discuss with their partners whether they should round up or down and how they know.

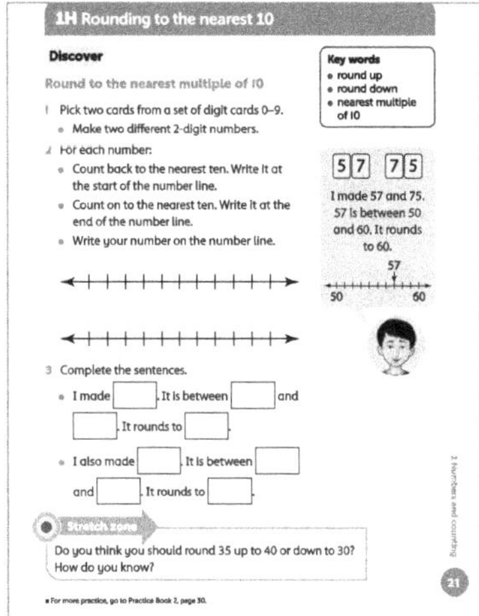

Differentiation

Supporting: Using a 0–100 number line labelled in intervals of 10, support students to copy the relevant sections of the number line to their Student Book.

Consolidating: Ask students to explain how they decide which way to round their numbers.

Extending: Challenge students to work with increasingly larger numbers, rounding between 70 and 80, 80 and 90, 90 and 100.

Stretch zone: *Do you think you should round 35 up to 40 or down to 30? How do you know?*

Students should explain how they know which way to round 35.

 **Reflection time**

Ask some students to share the sentences they completed in the Student Book activity for the numbers they rounded from the digit cards. They should tell the class the number they made from the digit cards, which multiples of 10 it falls between and which it rounds to.

Practice Book: Students can complete page 30 of the Practice Book. They can do this directly after the main activity, as homework, or as the focus of a separate mathematics session to help consolidate their learning and build fluency. Once students have rounded all numbers and placed them on the number line, ask them to explain to a partner or an adult how they decided to round each number.

Differentiated outcomes	
All students	should name the multiples of 10 either side of a number.
Most students	will know which multiple of 10 is nearer to the given number.
Some students	may round any number to the nearest multiple of 10 including rounding up 'halfway' numbers.

1H Rounding to the nearest 10

Explore   Student Book page 22 · Practice Book page 31

Specific learning focus

- Use a 100-square to round up or down to the nearest 10.

Global skills

- **Interpersonal skills:** communication/teamwork

Key vocabulary

- round up, round down, nearest multiple of 10, more than, less than

Resources

- large 100-square for front of class

Language support

Use the language of estimation (nearly, about, close to, about the same, just over, just under, too many, too few) while rounding, relating the words to what is seen on the 100-square. Repetition of these words allows students to become more familiar with the vocabulary.

 Introductory activity

Ask students to recall from the previous lesson how they know when to round down or when to round up between two multiples of 10. Ask, *Which digits in the ones tell you to round down to the lower multiple of 10?* As students recall that numbers ending in 1, 2, 3 or 4 round down, highlight those numbers on a large 100-square to show they all lie in the same four columns.

Answers

Student Book page 21

Observe students while they work and check that they have placed their numbers correctly on the number lines. Check they have completed the sentences for question 3 correctly for the numbers they chose.

Practice Book page 30

1 50   **2** 80   **3** 20   **4** 90   **5** 10

Stretch zone: Students' answers will vary.

Now ask which ones digits tell you to round up. As students recall that numbers ending in 5, 6, 7, 8 or 9 round up, highlight those numbers on a 100-square to show that they all lie in the same 5 columns.

 Main activity

Display the large 100-square. Write on the board the following numbers: 23, 41, 74, 92. Ask students: *Do these numbers round up or down to the nearest multiples of 10? How do you know? What do the numbers all have in common?* Students should be encouraged to look at the ones digits if they can't see what the numbers have in common. All the ones digits are less than 5, so all the numbers round down.

Repeat using the numbers 87, 28, 36, 95, 19. Students should see that all the numbers round up as the ones digits are all 5 or above.

Students should then complete page 22 in the Student Book in pairs. Once they have completed both questions, ask them to discuss why they think we round up a number that has 5 in the ones place. They can then share their ideas later in reflection time.

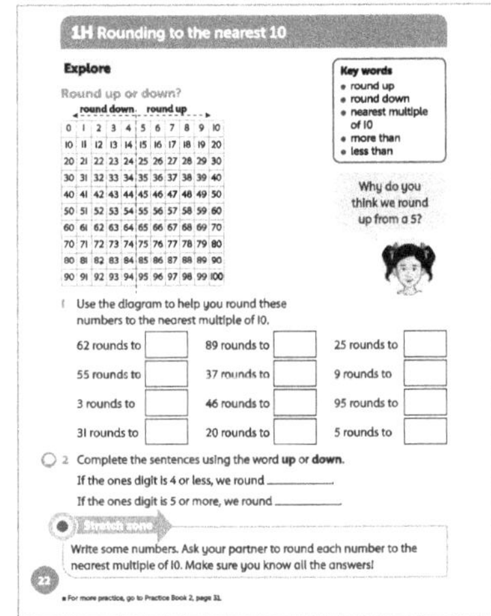

Differentiation

Supporting: Use smaller numbers for rounding practice, for example, up to 30. Help students to use the 100-square or draw number lines to help them visualise how to round up or down.

Consolidating: Ask students to explain how they are deciding which way to round.

Extending: Ask students to consider how they might round 3-digit numbers to the nearest multiple of 10.

Stretch zone: *Write some numbers. Ask your partner to round each number to the nearest multiple of 10. Make sure that you know all the answers!*

Check that students are rounding correctly.

 **Reflection time**

Ask students to stand up and fold their arms. As you call out a number, they should decide whether the number rounds up or down, and either raise their hands or lower them accordingly.

Include some numbers that end in 5, to make sure students understand that numbers ending in 5 round up. You can also say multiples of 10 and see whether students keep their arms folded as there is no need to round up or down for multiples of 10.

Practice Book: Students can complete page 31 of the Practice Book. They can do this directly after the main activity, as homework, or as the focus of a separate mathematics session to help consolidate their learning and build fluency. Once students have rounded all numbers, ask them to explain to a partner or an adult how they decided to round each number either up or down.

Differentiated outcomes	
All students	should name the multiples of 10 either side of a number.
Most students	will know which multiple of 10 is nearer the given number and will round to this number.
Some students	may extend their understanding of rounding to the nearest 10 to numbers over 100.

Answers

Student Book page 22

1 60   90   30

   60   40   10

   0   50   100

   30   20   10

2 If the ones digit is 4 or less, we round down.

   If the ones digit is 5 or more, we round up.

Practice Book page 31

1 60 **3** 50 **5** 20 **7** 40 **9** 90 **11** 100

2 60 **4** 20 **6** 10 **8** 20 **10** 100 **12** 90

Stretch zone: 40.

1I Less than and greater than

Discover Student Book page 23 • Practice Book page 32

Specific learning focus

- Compare two numbers up to 100 saying 'is greater than' or 'is less than' using the > and < inequality symbols.

Global skills

- **Real-world skills:** presenting
- **Interpersonal skills:** communication/teamwork

Key vocabulary

- between, larger, smaller, greater than >, less than <

Resources

- large number cards showing multiples of 10, from 0 to 100
- mini whiteboards and markers, or A4 paper
- set of large digit cards 1–9
- large < and > symbols on sheets of A4 paper

Language support

For students who find it difficult to remember which inequality sign is which, show them that the pointed end of the sign always points to the smaller number. So 4 < 8 is correct because 4 is less than 8, but 4 > 8 is incorrect because 8 is not less than 4. Use a visual demonstration such as:

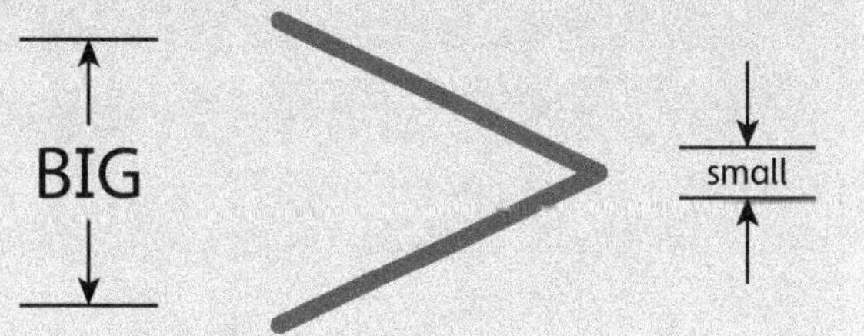

 **Introductory activity**

Call two students to the front of the class and give them each a multiple-of-10 number card, such as 20 and 30.

Ask, *Which is the smaller number?* (20) Say, *20 is **less than** 30* and write it on the board, underlining 'less than'. *Which number is larger or greater?* (30) Say, *30 is **greater than** 20* and write it on the board, underlining 'greater than'. Invite all other students to write a 2-digit number on a piece of paper or a mini whiteboard. Ask a student to show their number to the class and say it aloud. Each student at the front of the class says whether the number is greater than or less their number, in a complete sentence, for example, '19 is less than 20'. Repeat with another student sharing their number.

Choose a new pair of students to come to the front of the class to choose cards and compare their numbers to those of the rest of the class.

Main activity

Using one of the examples from the introductory activity, for example 20 and 30, ask two students to come to the front and hold these number cards. Show students how mathematicians write the sign for 'is less than', <, drawn on a sheet of paper. Ask another student to stand between the students with 20 and 30, holding the sheet of paper with the sign <. Read the number sentence 20 < 30 together, saying *20 is less than 30*. Repeat for the 'is greater than' sign (>). Ask how you could change the number sentence to use this sign. Agree the sentence 30 > 20, swapping < for > and rearranging the students. Repeat using different students and numbers.

Explain that many people confuse the two signs, but if you look at them with a number line it is easy to see which is which. Show students page 23 of Student Book. Ask students to look at number 8 and notice the sign 'is less than' just above it. Point out that the sign looks a bit like an arrow, pointing back along the line to numbers smaller than 8. Draw their attention to the number sentence 4 < 8 and ask them to suggest other 'is less than 8' sentences. Move on to explore 'is greater than 8' in a similar way. The 'is greater than' sign opens towards numbers that are bigger than 8, and we can choose any of those numbers (including those beyond 20) and write, for example, 16 > 8.

Ask students to work in pairs to complete the number sentences on page 23 of the Student Book, talking to each other about the numbers they are choosing and why.

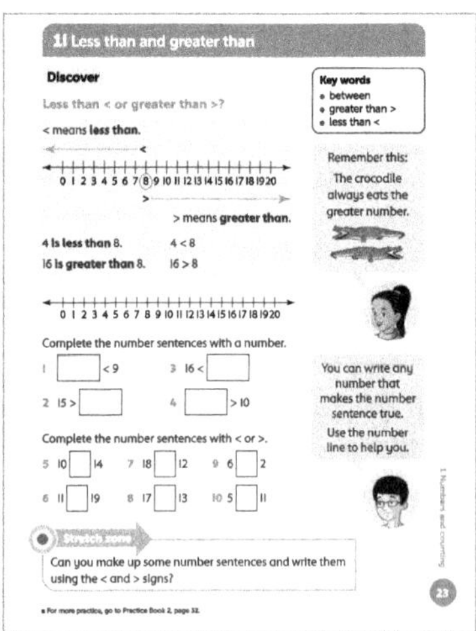

Differentiation

Supporting: Use digit cards and symbols cards to make number sentences and read them to students. For example, make 13 < 17 and say *13 is less than 17*.

Consolidating: Ask students to read their number sentences aloud.

Extending: Use digit cards 1–9 and < and > symbols. Ask questions such as: *How many different number sentences can you make using 3 digits picked randomly from the pack? One of your numbers should be a 2-digit number.*

Stretch zone: *Can you make up some number sentences and write them using the < and > signs?*

Encourage students to describe how they decided which symbol to use in each sentence.

Reflection time

Ask four students to come to the front. Each should pick two digit cards and make a 2-digit number. They should quickly form two accurate number sentences using the < or > symbol. Ask them to rearrange themselves to make a different pair of accurate number sentences.

Practice Book: Students can complete page 32 of the Practice Book. They can do this directly after the main activity, as homework, or as the focus of a separate mathematics session to help consolidate their learning and build fluency. Work through the steps of the activity once together so that students understand how to complete the activity.

Differentiated outcomes	
All students	should understand the use of < and > symbols with support.
Most students	will confidently use < and > symbols.
Some students	may make a range of number sentences using < and > symbols.

Answers

Student Book page 23

Observe students while they work and ask them to explain how they worked out the missing numbers. For example, they might use the number line to search for numbers less than the given one.

5	<	**8**	>
6	<	**9**	>
7	>	**10**	<

Practice Book page 32

Check that students have written the correct numerals and words for their numbers and inserted the correct symbol, either < or >.

Stretch zone: Listen for answers that refer to comparing the tens first and then the ones to decide which number is bigger or smaller.

1I Less than and greater than

Explore 1   Student Book page 24 · Practice Book page 33

Specific learning focus

- Order numbers to 100; compare two numbers using the > and < symbols.

Global skills

- **Creative skills:** exploring
- **Real-world skills:** presenting information
- **Interpersonal skills:** communication/teamwork

Key vocabulary

- between, larger, smaller, greater than, >, less than, <

Resources

- place-value cards 10–90 and 1–9, from Resource sheet 1 (one set per pair)
- base-10 equipment: tens-rods and ones-cubes
- number cards 0–100

Language support

Some students will find it helpful to be reminded that 'greater than' means the same as 'more than'. So, when comparing two numbers, they can order them with the smallest first, then put in the sign pointing to the smaller one and read it correctly.

 **Introductory activity**

Ask each student to write a 2-digit number on a small piece of paper. They should then walk around the room and, on your signal, form a pair with another student. They then say their numbers and compare them using 'greater than' and 'less than'. For example, if a pair have the numbers 36 and 82, they would say:

I am 36 and I am less than 82.

I am 82 and I am greater than 36.

Repeat several times.

 **Main activity**

Give each pair of students a set of place-value cards and ask them to create four 2-digit numbers. Explain that they need to make each number using base-10 equipment and then order the numbers from smallest to largest. They then write them in the grid on page 24 of the Student Book and draw how they represented each number using base-10 equipment.

Pairs then use the numbers they have both made to write a range of sentences using < and >.

Explain that you want them to talk in pairs about their sentences so that you can check that they can use the two signs correctly.

You could write some sentences on the board to demonstrate what is required, for example, 'My two numbers are 53 and 14 and 53 > 14.

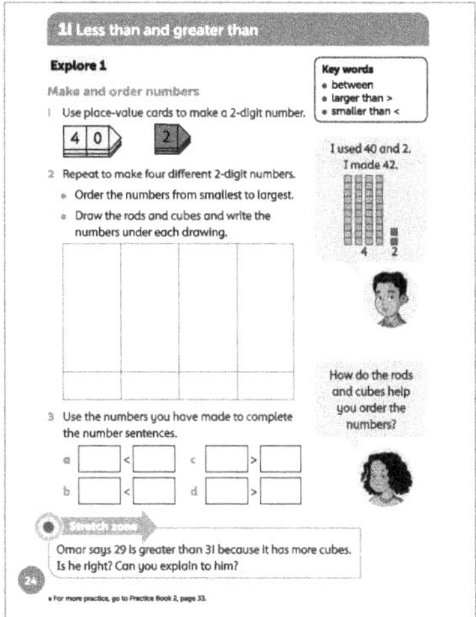

Differentiation

Supporting: Use digit cards and symbols cards to make number sentences. Help students build the numbers with base-10 equipment.

Consolidating: Ask students to read their number sentences aloud, describe how they know they are correct and how the base-10 equipment helped them to compare the numbers.

Extending: Ask students to make a new set of 2-digit numbers that are all different to the previous set and form number sentences comparing pairs of the numbers using < and >.

Stretch zone: *Omar says 29 is greater than 31 because it has more cubes. Is he right? Can you explain to him?*

Students might explain that 9 cubes represent 9 ones and that this is more than 1 cube, which represents 1 one. 29 has only two rods (tens) but 31 has three rods (tens) so it is greater.

 **Reflection time**

Ask students to read one of their number sentences to the whole class. Ask questions such as: *Is that true? How do you know? Could you write another number sentence using the same numbers but with the other sign? Is your new sentence also true?*

Practice Book: Students can complete page 33 of the Practice Book. They can do this directly after the main activity, as homework, or as the focus of a separate mathematics session to help consolidate their learning and build fluency. Some students will benefit from having ones-cubes and tens-rods to make their numbers.

Differentiated outcomes	
All students	should represent and compare numbers using base-10 equipment.
Most students	will use the vocabulary 'greater than' and 'less than' and the symbols < and > with increasing confidence and be able to explain them.
Some students	may confidently write a range of number sentences using < and > symbols.

1I Less than and greater than

Explore 2 Student Book page 25 • Practice Book page 34

Specific learning focus

- Order 2-digit numbers using place value.

Global skills

- **Real-world skills:** presenting information
- **Interpersonal skills:** communication/teamwork

Key vocabulary

- between, greater than, less than, <, >, smallest, largest

Resources

- place-value cards 10–90 and 1–9, from Resource sheet 1 (one set per pair)
- 100-squares
- base-10 equipment

Language support

Reinforce the idea of the tens being worth more than the ones and help students with the language of '-teen' and the '-ty' for comparing.

 **Introductory activity**

Write two 2-digit numbers on the board, for example, 27 and 61. *Which number do you think is larger? Why?* If students suggest 27 because the 7 is more than the 1, remind them to compare tens first because tens are worth more than ones. Agree that 61 has more tens than 27 so 61 is larger than 27. Repeat with other pairs,

Answers

Student Book page 24

Observe students while they work and ask them to explain how they completed their number sentences. For example, they might compare the tens first to see which number has more tens and, where this is the same for two or more numbers, compare the ones.

Practice Book page 33

Check that students have completed the right-hand side of the table with appropriate numbers and used the correct symbol between them, either < or >.

Stretch zone: Students' answers will vary but will probably mention that the greater the number of rods, the larger the number.

for example, 39 and 62. Ask, *Which digits should we compare first?* (3 and 6) *Why?* (Because they are in the tens place.) *So, which is larger, 3 tens or 6 tens?* (6 tens) *Can we say now which number is larger?* (Yes, 62) Repeat with an example where the tens digits are the same, for example, 23 and 27.

 **Main activity**

Give each pair a set of place-value cards and ask them to work together to make a set of nine 2-digit numbers. Ask them to look at the tens in their numbers and decide which number is the smallest. Ask, *How do you know which number is the smallest?* Students should be able to answer by saying, for example, it's the number with the fewest tens, which means it will be in the range 11–19.

Now ask them to work together to find the order for all their numbers, recording them from smallest to largest in the grid on page 25 of the Student Book. Once they have sorted the numbers, they write some number sentences (e.g. 17 < 35, or 83 > 39) to compare pairs of their numbers. Have base-10 equipment available for support for any students who find it helpful.

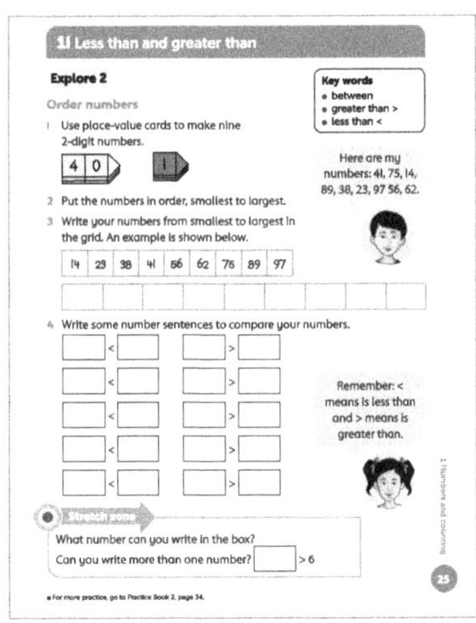

Differentiation

Supporting: Build the numbers with base-10 equipment to help students order them.

Consolidating: Ask students to explain how they are sorting their numbers into order, encouraging them to compare the tens first and then the ones.

Extending: Challenge students to find the pairs of numbers they made with the smallest and greatest difference. How do they know?

Stretch zone: *What number can you write in the box [to make a number > 6]? Can you write more than one number?*

Students should be encouraged to find several possible numbers that are greater than 6, and they could also find the smallest and greatest 2-digit odd or even number possible, for example.

 **Reflection time**

Write on the board the numbers 27 and 83. *Which is larger? Why?* Students should say 83 because the number of tens is greater. Now swap the tens over and ask students what numbers you will have. (87 and 23) Ask, *Which is the larger now?* Students should see that it is still the one with the most tens, so 87.

Ask students to write some pairs of 2-digit numbers and then swap the tens. Which number is bigger in each pair? Agree that it will always be the one starting with the larger tens digit.

Practice Book: Students complete page 34 of the Practice Book. They can do this directly after the main activity, as homework, or as the focus of a separate mathematics session to help consolidate their learning and build fluency. Some students may struggle to come up with numbers. Provide them with digit cards or dice to help them generate numbers. Continue to provide base-10 equipment for students to represent their numbers if they choose.

Differentiated outcomes	
All students	should compare two 2-digit numbers using base-10 equipment by looking at the tens first.
Most students	will compare two 2-digit numbers by looking at the tens first.
Some students	may find the smallest and greatest difference between pairs of numbers.

Answers

Student Book page 25

Observe students while they work and ask them to explain how they worked out their order and how they chose their number sentences. Check that they have written correct sentences and can read their sentences aloud correctly.

Practice Book page 34

Check that students have written their numbers correctly in figures and words, and that they have written accurate sentences comparing their numbers.

Stretch zone: eighty-six > twenty-three

1J Ordinal numbers

Discover Student Book page 26 • Practice Book page 35

Specific learning focus

- Recognise and use ordinal numbers up to at least the 10th number and then to the 20th number.

Global Skills

- **Creative skills:** exploring
- **Interpersonal skills:** communication/teamwork

Key vocabulary

- ordinal numbers, first, second, third, fourth, fifth, sixth, seventh, eighth, ninth, tenth, …, twentieth

Resources

- large number cards 1–10
- 20-bead string, beads, cubes

Language support

Some students find it hard to pronounce words with a -th sound at the end, such as fourth, fifth and so on. Make some double-sided cards with a picture of a thumb and 'th' on one side, and a picture of fingers and 'f' on the other side. When saying -th words, students should have the thumb side of the card in front of them as a reminder.

 **Introductory activity**

Ask five students to come to the front of the class and stand in a line facing the rest of the students. Indicate one end of the line and tell students that this is the front of the line. Ask questions such as: *Who is **first** in the line? Who is **second**? Who is **third**?* and so on.

Move to the other end of the line, explaining that this is now the front of the line and repeat the questioning. Give these students large number cards 1 to 5 and read along the line saying: *first, second, third, **fourth**, **fifth**.*

Ask students if they noticed that the student who was first held number 1, the student who was second held number 2, the student who was third held number 3 and so on.

Ask another student to come out and be sixth in the line. Ask the rest of the students which number they need. Continue to the tenth student, linking the position to the number.

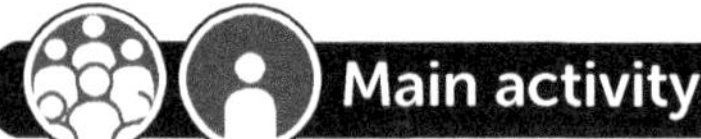

Main activity

Ask students to look at the table in the Student Book page 26 and compare the numbers and words on the same line. *What do you notice about the words in the second and third columns?* Check that students recognise that after 1, 2 and 3 the numbers up to 10 have 'th' after them and that for each it is the digit for the number followed by the last two letters of the **ordinal** word that you write: 1st, 2nd, 3rd, 4th and so on.

Explain that we use 'first', 'second' and 'third' so often that they have special names to show us the order of something. 'Third' is a bit of a mixture, partly a new word but with the beginning of three in it.

Hold up a 20-bead string and show students, say, 13 beads. Then ask them, *Which bead is this?* See whether they can come up with 'thirteenth' following the linguistic pattern and repeat for all others to 20. Then record each on the board and model the correct pronunciation (putting the tongue behind the top row of teeth or between both sets of teeth) as 'th' can be a challenging sound for some students to make.

Ask students to complete the questions on page 26 in the Student Book individually. As they work, circulate and ask questions such as: *How did you know which colour to use for that bead? What colour would the fifteenth (or eighteenth or twentieth) bead be?*

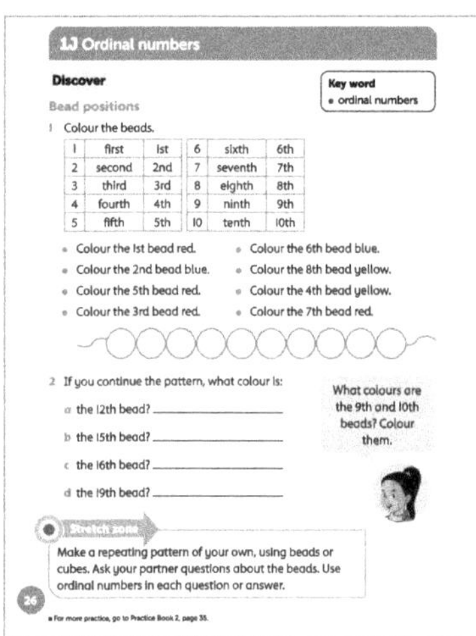

Differentiation

Supporting: Count alongside students to model the vocabulary.

Consolidating: Ask students to describe patterns that they can see in the classroom using ordinal numbers.

Extending: Ask students to use ordinal numbers to explore and create repeating patterns of shapes and colours.

Stretch zone: *Make a repeating pattern of your own, using beads or cubes. Ask your partner questions about the beads. Use ordinal numbers in each question or answer.*

Students might describe their pattern compared with that of a partner and be able to explain the repeating pattern involved.

Reflection time

Give students a colour or letter pattern such as 'blue, green, white, red' or B, G, W, R. In pairs, they should sketch the repeating patterns. Ask them to predict what the tenth colour or letter will be. *What about the twelfth? The fifteenth?* and so on.

Students work in pairs, taking turns to ask each other an order question about the alphabet, numbers, words and so on.

Practice Book: Students can complete page 35 of the Practice Book. They can do this directly after the main activity, as homework, or as the focus of a separate mathematics session to help consolidate their learning and build fluency. Students will need colouring pencils to complete this activity.

Differentiated outcomes	
All students	should use ordinal numbers up to the tenth place.
Most students	will use ordinal numbers up to the twentieth place.
Some students	will understand how ordinal numbers are formed up to 100.

Answers

Student Book page 26

Observe students while they work and check they have coloured the beads correctly.

1 bead string: red-blue-red-yellow-red-blue-red-yellow

2 a yellow   **b** red   **c** yellow   **d** red

Practice Book page 35

Check students have coloured the beads correctly.

1	Y	**6**	Y	**11**	Y	**16**	Y
2	B	**7**	B	**12**	B	**17**	B
3	G	**8**	G	**13**	G	**18**	G
4	R	**9**	R	**14**	R	**19**	R
5	P	**10**	P	**15**	P	**20**	P

Stretch zone: The 21st bead would be yellow. The 30th bead would be purple.

1J Ordinal numbers

Explore Student Book page 27 • Practice Book page 36

Specific learning focus

- Recognise and use ordinal numbers up to the thirty-first number.

Global Skills

- **Creative skills:** exploring
- **Real-world skills:** interpreting information
- **Interpersonal skills:** communication/teamwork

Key vocabulary

- ordinal numbers, first, second, third, fourth, fifth, sixth, seventh, eighth, ninth, tenth

Resources

- large copy of the current month calendar

Language support

Display the date every day. Change it each morning with the students. Ask a student to tell you yesterday's date and what must be changed to turn it into today's date. Focus on correct pronunciation of ordinal numbers, for example, yesterday the date was the 15th April, today it is the 16th April.

 **Introductory activity**

Ask students whether they know what today's date is. Look on a calendar and explain how the calendar works. Ask questions to check understanding: *What do all the words in this row tell us?* (They are the days of the week.) *What do the dates in this column all have in common?* (e.g. They are all Thursdays.) Check that students realise that, although there are ordinary numbers on the calendar, we say the ordinal version of that number (first, second, third and so on) to help us to measure how far into the month today is.

Ask questions about the calendar such as: *What was yesterday's date? What is tomorrow's date? What date is your birthday? What is your favourite date in the year?* Emphasise the link between the number on the calendar and the way we say the date.

 **Main activity**

Show students the calendar page on page 27 of the Student Book. Check that students understand that the top row shows the days of the week in abbreviated form and that the day of the week at the top of the column means that all the dates in that column are the same day of the week.

Ask students to work in pairs to read and answer each question. They may find it useful to refer to the ordinal number table on page 26 of the Student Book to help them read and answer the questions.

As pairs work, walk around the classroom and ask students questions such as: *How did you know which day of the week that date was? How would you read this date?*

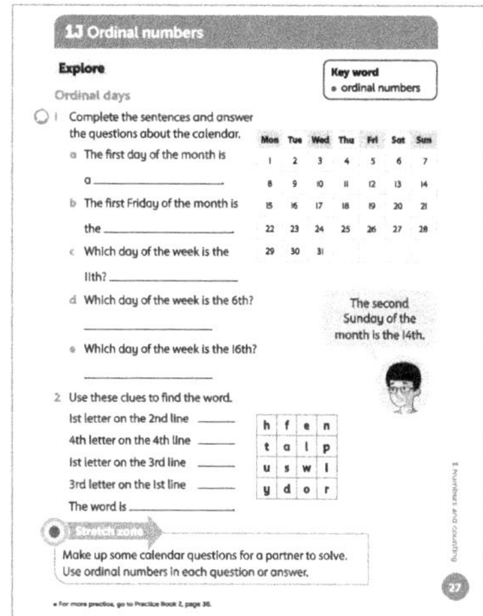

Differentiation

Supporting: Help students to think about how they can use what they already know (e.g. 'I know how 4 relates to fourth so 24 must be twenty-fourth.')

Consolidating: Ask students to find key dates for you and ask them questions about that date.

Extending: Encourage students to pose each other problems based around key dates in the year.

Stretch zone: *Make up some calendar questions for a partner to solve. Use ordinal numbers in each question or answer.*

Students should be encouraged to use the patterns in the calendar, for example, all Tuesdays are seven days apart.

 **Reflection time**

Display the large version of the current month calendar and ask pairs to make up questions for each other to answer. For example: *What date is a week after the 5th? What date is two days after/before the 17th?*

Use a year calendar to identify key dates in the year. Count the number of days until a key date.

Practice Book: Students can complete page 36 of the Practice Book. They can do this directly after the main activity, as homework, or as the focus of a separate mathematics session to help consolidate their learning and build fluency. Ask students to look at question 1 and find the fifth mouse. Explain that they should start from the left as they do when reading in English. Then ask them to show you which mouse is the fifth by holding up their books and pointing. *Has everyone chosen the same mouse? Why not?* Discuss how the order you count the objects makes a difference. Before they begin counting the objects, they need to choose whether they will count in rows or columns. They should use the same approach for all questions and record their choice at the top of their page, writing either 'row' or 'column'.

Differentiated outcomes	
All students	should use ordinal numbers for dates of 20 or less.
Most students	will understand how a calendar is read.
Some students	may know key dates in the annual calendar and be able say and record them as an ordinal numbers correctly.

Student Book page 27

1 a Monday

   b 5th

   c Thursday

   d Saturday

   e Tuesday

2 t, r, u, e

Practice Book page 36

Check that students have coloured the correct item in each group, depending on whether they counted in rows or columns.

Stretch zone: Check that students have described the positions of the objects in their drawing accurately.

1 Numbers and counting

Big idea

- I can count in twos, fives and tens to help me make sensible estimates. I can use < and > signs to compare numbers.

Global skills

- **Interpersonal skills:** communication
- **Self-development skills:** reflecting on learning

Key vocabulary

- tens, ones, place value

Resources

- base-10 equipment, small bag big enough to hold seven pieces
- photos of animals in the wild in numbers that are uncountable

Language support

Revisit all the key vocabulary for this unit to check understanding and pronunciation.

 ## Introductory activity

Show students a small bag that is big enough to hold seven pieces of base-10 equipment. Explain that you cannot remember how many ones-cubes and tens-rods you put in the bag. Ask, *What numbers could be shown using the cubes and rods in the bag?* Take feedback to work out all the different numbers that are possible. Ask students how they know if they have found all the possible numbers. Model how to work systematically, writing the numbers in order on the board. Start with the possibility of the bag holding 7 cubes, making 7. The next possible number is 1 rod and 6 cubes, making 16, and so on.

 ## Main activity

Ask students to imagine that they are in a wildlife park. *What might you see?* Take their suggestions, introducing any new vocabulary, recording it on the board for reference. Choose any wildlife they may have mentioned, for example parrots, and say, *How many parrots might you expect to see? Would they be easy to count? Why or why not?* Agree that, in the wild, animals will move around a lot and look similar. *What might be a good way to describe how many?* Agree that estimating would be useful.

Show students photos of wildlife taken from the internet or books and ask them to think about how many of the different animals they can see. Take estimates and invite students to explain their strategy for estimating. For example, they could look for possible groups of 10 and then think about how many groups of 10 they can see.

Students then complete the activity on page 28 of the Student Book individually. Once they have completed question 1, they should explain their strategy to a partner.

Differentiation

Supporting: Suggest strategies for estimating and support students to make a rough count of the animals.

Consolidating: Ask students to describe how they made their estimate.

Extending: Ask students to choose several groups of objects around the classroom that it would make sense to estimate the number of rather than count exactly. Can they explain why?

Stretch zone: *Estimate how many people there are in your classroom.*

How did you estimate? Did everyone use the same strategy?

Discuss the estimates and ask some students to share their strategy as part of reflection time.

 ### Reflection time

Ask pairs to share their estimates for how many of the different animals there are in the photo on page 28 of the Student Book and record them on the board. Assign pairs a different animal type to count and then record these amounts on the board. Find the total number of animals together, building the total, animal type by animal type, with tens-rods and ones-cubes. How did the actual total compare to their estimates? Would they describe their estimates as reasonable? Why or why not?

Differentiated outcomes	
All students	should make reasonable estimates.
Most students	will make reasonable estimates and be able to explain how they reached their estimates.
Some students	may begin to refine their estimation strategies based on previous estimation.

1 Numbers and counting

Global skills

- **Creative skills:** exploring
- **Interpersonal skills:** communication
- **Self-development skills:** reflecting on learning

Student Book

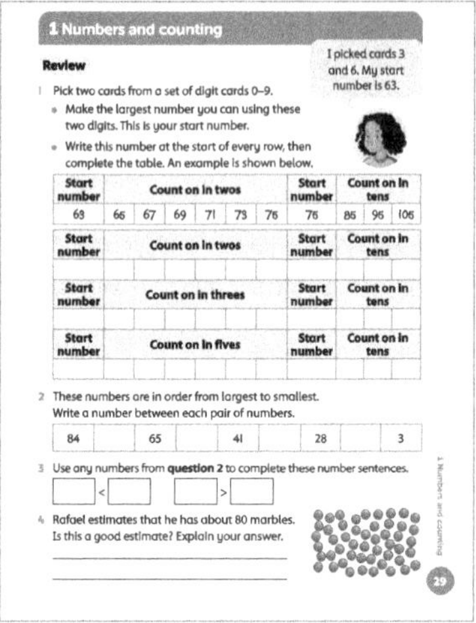

With young children, assessment activities are most effective when carried out as an everyday classroom activity. You could have number cards available for students with both the digits and the numbers in words so that they can refer to these to support them.

Watch as students make the numbers using the digit cards. Listen as they count to check that they say the number words in order and do not omit any numbers.

Encourage students to use a number line and a 100-square for answering the questions. Extend by doing a range of similar tasks, for example, *If I start at 35 and count on in threes, will I get to 48?* Some students may go beyond 100 in this activity so you should support them to do so as needed.

Question 4 can be assessed orally, with students explaining their estimates.

Answers

Student Book page 29

1   Check that the students have counted on correctly from their starting numbers.

2   Check that the numbers inserted make a sequence from higher to lower numbers.

3   Check that the number sentences using < and > are correct.

4   Check that the explanation for Rafael's estimate is reasonable.

Practice Book

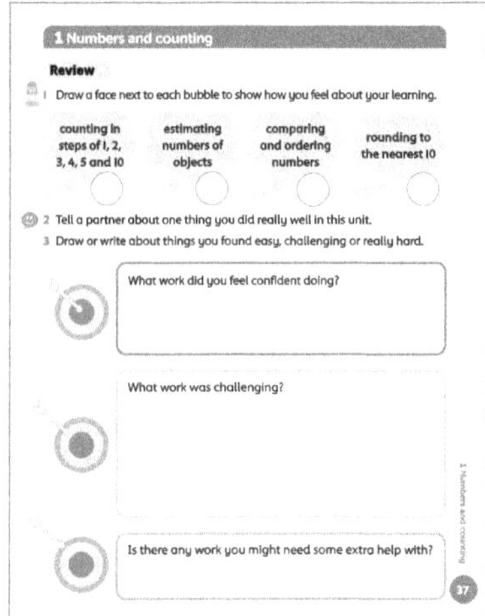

It is appropriate to complete this Practice Book review as a whole-class discussion. You may choose to keep a record of the class discussion or a copy of the review page for your own records. The review provides an opportunity for students to reflect on their learning from the unit, to discuss any areas of mathematics that they feel went particularly well, and any areas that they feel less confident about. Ensure all students have a copy of the Student Book as a reminder of the areas of mathematics that they have worked on in this unit.

Allow students plenty of time for discussion before asking them to complete the Practice Book page individually, and then, if appropriate, to share their responses with the rest of the class. If students complete this self-assessment at home, encourage them to discuss this with adults. Make a note of areas that students still feel unsure about.

As counting is at the heart of many of the other units, you can revisit these areas regularly. You can also build counting into everyday practice. For example, you can count how many students are in the class each day; count objects as you give them out, and so on.

Additional material

There are additional end-of-unit assessments available on the *Oxford Owl* website.

"

2 Exploring numbers

Overview

Big Idea

The Big Idea for this unit is that there are many patterns within numbers. Once we recognise these patterns, we can use them to help us develop a sense of number, and therefore to calculate. Patterns such as even and odd continue throughout the full range of whole numbers.

We use place value to help us compare and order numbers: for example, by understanding the value of each digit we can arrange numbers in order of size. We know that 75 is bigger than 29 because the 7 tens in 75 are more than the 2 tens in 29. If the number of tens is the same in two numbers, for example, 52 and 58, then we compare the ones to see that 8 is bigger than 2 so 58 is bigger than 52.

Look out for

- **Students who partition, say, 14 as 'one' and 'four'.** We need to take care to say numbers correctly to help avoid this confusion. 14 is not a one and a four, because $1 + 4 = 5$. It is a ten and a four, which can be demonstrated by using place-value cards, base-10 equipment or ten-frames and counters.

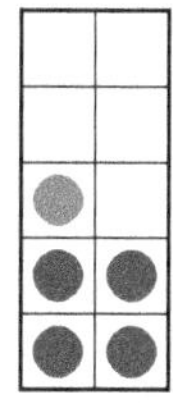 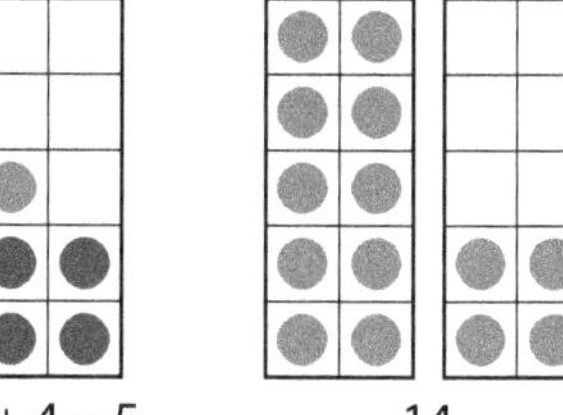

$1 + 4 = 5$ 14

Possible misconceptions

- **Students may order 2-digit numbers by looking at the ones first rather than the tens.** We need to show students how to read numbers from left to right, ordering by the highest-value digit first. In order to do this, they need to develop a good understanding of place value. They need to understand that a 2-digit number such as 26 is not made from a two and a six but a twenty and a six. Early and repeated use of place-value cards and base-10 equipment will help students to develop an understanding of place value.
- **Students may understand adding 1 more as adding the digit '1' to a number.** For example, they understand '1 more' than 4 as 41. Model making 1 more using practical resources such as base-10 equipment and then show finding 1 more of the same number using a 100-square. Contrast this with their given answer.

Key vocabulary

- place value, tens, ones, partition
- row, column, exchange
- even, odd, double, next to, between, order, less than, greater than

Coverage in lessons

Learning focus	Learning outcomes (the ENC objectives)
Digit values	Recognise the place value of each digit in a 2-digit number (tens, ones).
	Identify, represent and estimate numbers using different representations, including the number line.
1 more, 1 less, 10 more, 10 less	Recognise the place value of each digit in a 2-digit number (tens, ones).
	Identify, represent and estimate numbers using different representations, including the number line.

2 Exploring numbers

Engage
Student Book page 30

Big question

- What do we know about numbers?

Global skills

- **Creative skills:** exploring
- **Real-world skills:** presenting information
- **Interpersonal skills:** communication/teamwork

Key vocabulary

- odd, even, greater than, less than

Resources

- selection of pictures of numbers – door numbers, packaging, bus numbers and so on

Language support

Listen for correct pronunciation of -teen and –ty numbers. Support the correct pronunciation of –ty when saying 20, 30, 40, 50, 60, 70, 80 and 90 on a displayed 100-square or number line.

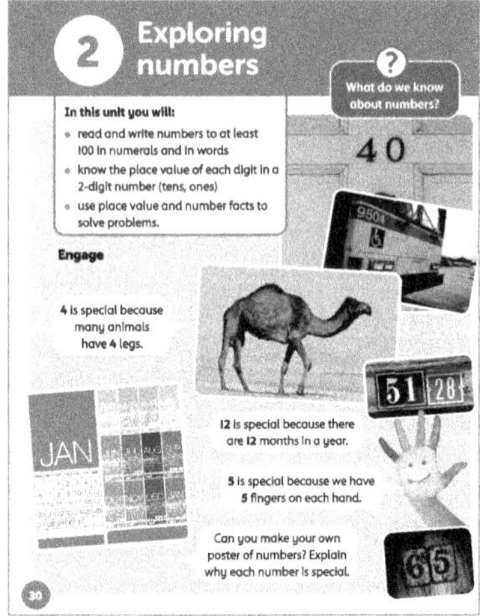

Introductory activity

Ask students to talk to a partner about where they see numbers and which numbers are special to them. You could give them some sentence starters (e.g. A special number to me is ______________. It is special because ______________.)

Draw up a list of ideas on the board. Show your selection of pictures to keep the discussion going if necessary. Ask students to talk to their partner again, this time about what numbers are used for. Draw up a further list of ideas on the board. You could group these ideas under headings such as 'Labels', 'How many', 'How much' and so on.

Main activity

Organise students into small groups of no more than six. Ask students to look at the pictures and statements on Student Book page 30 (display on the IWB if possible) and ask students to look at the list of reasons given in the introductory activity for why particular numbers are special. Ask them to think about these in relation to the numbers shown on page 30. Ask, for example, *Can you describe these numbers as being odd or even? 1-digit or 2-digit? Are they found in nature, such as a number of legs or wings?*

They should then go on to make a poster of four numbers of their own, for example 7, 10, 5 and 11, explaining why they are special. They should include pictures that show different ways that the numbers can be made, for example, showing a foot with five toes for 5.

Differentiation

Supporting: Use posters that have lists of odd and even numbers to help less-confident students describe a number that is special to them.

Consolidating: Prompt students by asking, *Is it odd? Is it more or less than 50? What is 1 more? 1 less?* and so on for other possible properties of numbers.

Extending: Encourage students to develop posters showing a detailed list of the properties of a particular number including odd or even, how many tens or ones, and multiples. These posters can then be referred to throughout the unit.

Reflection time

Ask each group to share a reason for each of their four numbers. Ask them to describe each of their numbers in different ways, for example, whether it is odd or even, whether its digits are odd or even, and whether it has a special significance to them in some way.

2A Digit values

Discover Student Book page 31 • Practice Book page 38

Specific learning foci
- Count, read and write numbers to at least 100 and back again.
- Know what each digit represents in 2-digit numbers; partition into tens and ones.

Global skills
- **Interpersonal skills:** communication/teamwork

Key vocabulary
- place value, tens, ones, partition

Resources
- sets of place-value cards to 100, from Resource sheet 1 (one for each pair)
- large 100-square for front of class
- base-10 equipment

Language support
Display a list of teen numbers near the 100-square, headed '-teen numbers'. Then listen carefully to how each individual says the numbers out loud. Correct any confusions between '-teen' and '-ty'.

 **Introductory activity**

You will need 19 students. Give each student a different place-value card up to 100. Call out a number such as 35 and ask students with the matching place-value cards to come to the front of the class and use their cards to make that number. For example, for 35 the students holding the 30 card and the 5 card come to the front of the class. As each number is made, ask students to say how it is made, for example, 35 is made from 30 and 5.

 **Main activity**

Give pairs of students a set of place-value cards to 90 and ask them to create nine different 2-digit numbers. Display the large 100-square, choose a row and ask students to say which number they made on that row. Ask, *Are there other possible answers?* (They can only make one number on each row, as they only have one of each of the multiples-of-10 place-value cards.) Repeat with another row.

Students then complete questions 1–3 on page 31 in their pairs.

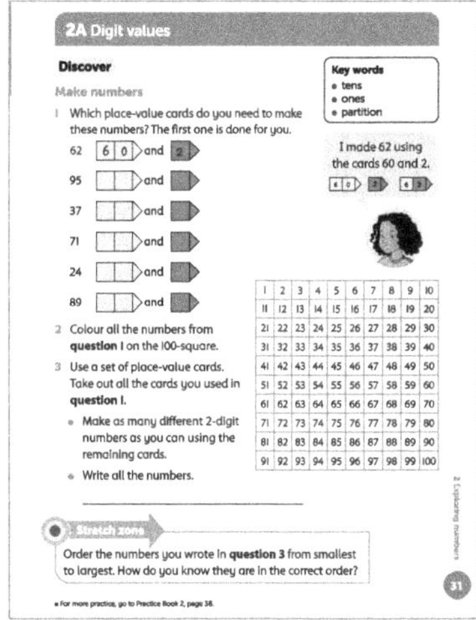

Differentiation
Supporting: Help students to make the numbers using base-10 equipment.

Consolidating: Ask students to explain why the same digit in a different place has a different value.

Extending: Encourage students to use place-value cards to make 3-digit numbers.

Stretch zone: *Order the numbers you wrote in question 3 from smallest to largest. How do you know they are in the correct order?*

This activity is designed to challenge students to think carefully about the place value of each digit so that they can order the set of 2-digit numbers they made earlier. Encourage students to think in a structured way to help them to find the correct order.

 **Reflection time**
Ask students, *What do you notice about the numbers you coloured on the 100-square?* (There is never more than one number in each row and each column.)

Why is there only one number in each row and column? (The students only had one of each **tens** card and **ones** card and did not use them all.)

Do you know the names of any 3-digit numbers? Ask students who made 3-digit numbers to share them, saying the names out loud and making them with place-value cards.

Practice Book: Students complete page 38 of the Practice Book. They can do this directly after the main activity, as homework, or as the focus of a separate mathematics session to help students consolidate their learning and build fluency. Look at the example together and say 53 out loud. *What do you notice about how we say 53 and how we **partition** it into tens and ones?* (We say the tens first and then the ones.) *Is it the same for a number like nineteen in question 3?* (No) *What other numbers are like this?* (11–18)

Differentiated outcomes	
All students	should be able to make 2-digit numbers and identify them on the 100-square. They should name the numbers with support.
Most students	will know the names of all 2-digit numbers and partition them into tens and ones.
Some students	may know the names of some 3-digit numbers and partition these into hundreds, tens and ones.

Answers

Student Book page 31

1 60 and 2   70 and 1

   90 and 5   20 and 4

   30 and 7   80 and 9

2 Check that students have coloured the correct numbers.

3 Check that students have made numbers appropriately from the place-value cards they have remaining.

Practice Book page 38

1 70 and 8

2 60 and 1

3 10 and 9

4 30 and 7

5 40 and 6

6 90 and 9

7 20 and 4

8 80 and 2

9 50 and 5

Stretch zone: 19, 24, 37, 46, 53, 55, 61, 78, 82, 99

2A Digit values

Explore 1 Student Book page 32 • Practice Book page 39

Specific learning focus

- Relate numbers of ones-cubes and tens-rods to written numbers.

Global skills

- **Creative skills:** investigating
- **Interpersonal skills:** communication/teamwork

Key vocabulary

- tens, ones, partition

Resources

- base-10 equipment (tens-rods and ones-cubes)
- feely bags (one per pair)

Language support

Make some example 'rod and cube' cards so that there is a visual image of numbers for students. You can also support them with numbers written in words, for example.

 **Introductory activity**

Show students the base-10 equipment (tens-rods and ones-cubes). *What is the value of one cube?* (1) *What is the value of one rod?* (1 ten) Show the class three rods and eight cubes. Ask: *How many rods do I have?* (3) *How much do they represent?* (30) *How many cubes do I have?* (8) *How much do they represent?* (8)

How much do these rods and cubes represent together? (38)

Repeat using different numbers of ten-rods and ones-cubes, finding how much the ten-rods represent and then how much the ones-cubes represent, before adding them to make the total.

 **Main activity**

Give each pair a set of 9 tens-rods and 9 ones-cubes in a feely bag. Ask them to pick some items from the bag, then use them to show a number and record the number. For example, they pick 4 tens-rods and 6 ones-cubes, then record that they have 46 in total. They should then replace the ten-rods and ones-cubes in the bag and repeat. Some numbers will be 2-digit numbers, some will be a multiple of ten and others will be single-digit numbers.

What would you pick from the feely bag to make a number that ends in zero? (rods only) *What if you only picked cubes?* (They would make a 1-digit number.)

Ensure that students understand that, if they only pick ones-cubes, the more ones-cubes they pick the bigger the number but it will still not be as big as just one tens-rod.

Ask students to complete page 32 of the Student Book individually. Once they are finished, ask them to consider the question in the second speech bubble and explain their thinking to a partner. Do they agree? Can they give examples to support their thinking?

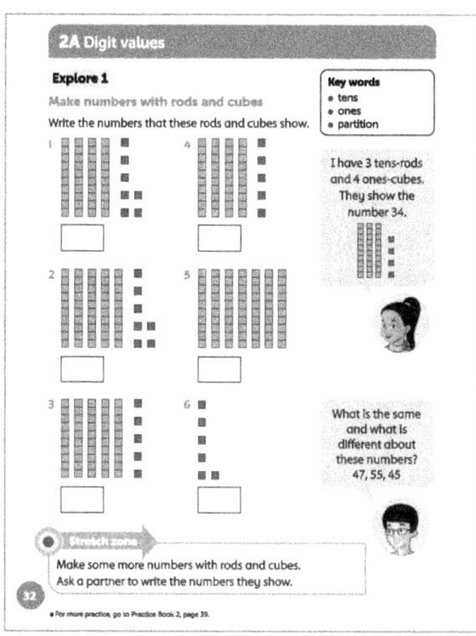

Differentiation

Supporting: Ask students how many cubes have the same value as a rod. Model how to write the numbers.

Consolidating: Ask students to tell you how many tens and ones for each of their numbers before they write them down.

Extending: Ask students to say what is the same and different about the numbers represented in the Student Book.

Stretch zone: *Make some more numbers with rods and cubes. Ask a partner to write the numbers they show.*

Check that students are able to correctly write down the number made by their partner's tens-rods and ones-cubes. Challenge them to represent a number in a specific range, for example greater than 25 but less than 30.

 **Reflection time**

Write the numbers 72 and 27 on the board. Ask students, *Which number needs more rods?* (72) *Which number needs more cubes?* (27)

Help students complete the sentence: 'I show 72 with __ rods and __ cubes.'

Repeat with another pair of numbers selected by a student – the numbers should have the digits reversed (e.g. 38 and 83). You can also ask students to use pairs such as 3 and 30.

Practice Book: Students can complete page 39 of the Practice Book. They can do this directly after the main activity, as homework, or as the focus of a separate mathematics session to help students consolidate their learning and build fluency. Encourage students to count the tens-rods in tens and cubes in ones. They should say each part aloud and then the whole number (e.g. 50 and 3, 53).

Differentiated outcomes	
All students	should be able to name any 2-digit number and model it with tens-rods and ones-cubes.
Most students	will use the fact that 1 ten has the same value as 10 ones.
Some students	may partition 2-digit numbers confidently, compare numbers and describe the difference.

Answers

Student Book page 32

1 47

2 57

3 55

4 45

5 70

6 6

Practice Book page 39

1 70 and 3 = 73

2 70 and 4 = 74

3 20 and 4 = 24

4 30 and 4 = 34

Stretch zone: Students' answers will vary but will likely mention that the cubes help you to see how many tens, and the cubes help you to see how many ones.

2A Digit values

Explore 2
Student Book page 33 • Practice Book page 40

Specific learning foci

- Count, read and write numbers to at least 100 and back again.
- Know what each digit represents in 2-digit numbers; partition into tens and ones.

Global skills

- **Creative skills:** exploring
- **Interpersonal skills:** communication/teamwork

Key vocabulary

- tens, ones, place value, partition, exchange

Resources

- place-value counters, one set per pair – each set contains ten ones-counters and nine tens-counters
- place-value cards (from Resource sheet 1)
- base-10 equipment
- 100-squares (one for each pair)

Language support

As students make numbers, check that the place-value cards and counters match. Listen to students as they count the counters. Challenge them to count the ones first, then count the tens. Check that they are saying the number names correctly.

Introductory activity

Show students the base-10 equipment. *What is the value of one cube? What is the value of one rod?* (Each cube represents 1 and each rod represents 10.) Give each student several place-value counters and ask them what they notice about the counters (The yellow counters are labelled 10 and the red counters are labelled 1.) Explain that a ones-counter represents 1 and has the same value as a ones-cube. A tens-counter represents 10 and has the same value as a tens-rod. *How many ones-counters have the same value as one tens-counter?* Check that students understand that 10 ones and 1 ten have the same value.

Show students an arrangement of place-value counters (e.g. 5 tens and 4 ones). Ask *How many tens can you see?* (5) *Let's count up in tens. 10, 20, 30, 40, 50.* Choose a '50' place-value card. *This represents the five tens. How many ones can you see?* (4) *Let's count. 1, 2, 3, 4.* Choose a '4' place-value card. Combine the cards and ask *What number is this?* (54) Record 54 on the board. *5 tens-counters and 4 ones-counters make 54.*

Ask students to hold up a tens-counter and 2 ones-counters, and model this with them. Say *1 ten and 2 ones is 12.* Hold up 2 tens and 3 ones. Say *2 tens and 3 ones is 23.*

Continue with further examples.

 Main activity

Ask one student in each pair to think of a 2-digit number and make it using the place-value counters. For example, they could use 1 tens-counter and 3 ones-counters to make 13. Their partner should say what the number is and make it using base-10 equipment. They can then reverse roles.

Ask students to then use their place-value counters to help them identify the numbers on page 33 of the Student Book page. In their pairs, students should check one another's diagrams for the final two questions for accuracy.

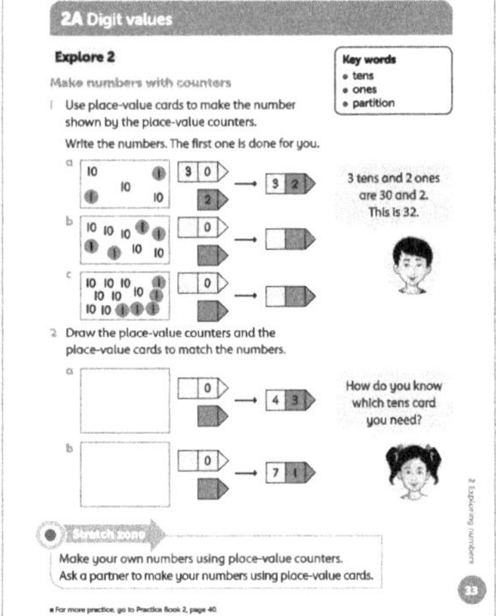

Differentiation

Supporting: Show students how to model 2-digit numbers using place-value counters, counting the tens 'in tens' and then counting the ones to find the number.

Consolidating: Model the use of the language: 57 is made up of 5 tens and 7 ones, then ask students to say how many tens and how many ones for their numbers.

Extending: Ask students *Make the number that is 8 less. What happens with the place-value counters?*

Stretch zone: *Make your own numbers using place-value counters. Ask a partner to make your numbers using place-value cards.*

Students can take turns at doing this, and then reverse the process, making a number in place-value cards first, and then with place-value counters.

 Reflection time

Give each pair of students a 100-square. Ask students to come to the front of the class and say one of the numbers that they have drawn for question 2 in the Student Book. Pairs should make this number using place-value counters, taking it in turns to make the number and to check the answer.

How many tens? How many ones? Can you find that number on a 100-square?

Practice Book: Students then complete page 40 of the Practice Book. They can do this directly after the main activity, as homework, or as the focus of a separate mathematics session to help students consolidate their learning and build fluency. Students may benefit from working with ones-cubes and tens-rods alongside the images of place-value counters.

Differentiated outcomes	
All students	should name correctly any 2-digit number and model it with place-value counters with teacher support.
Most students	will name correctly any 2-digit number and model it with place-value counters.
Some students	may explain confidently that 10 ones have the same value as 1 ten.

2B 1 more, 1 less, 10 more, 10 less

Discover Student Book pages 34–35 • Practice Book pages 41–42

Specific learning foci

- Find 1 or 10 more/less than any 2-digit number.
- Relate counting on/back in tens to finding 10 more/less than any 2-digit number.

Global skills

- **Creative skills:** exploring
- **Real-world skills:** presenting information
- **Interpersonal skills:** communication/teamwork

Key vocabulary

- 1 more, 1 less, 10 more, 10 less, row, column

Resources

- large 100-square for front of class
- base-10 equipment
- digit cards 0–9

Language support

Look out for opportunities to model using the words 'more' and 'less' correctly. When a student gives a number in answer to a question, ask a different student to tell you the number that is 1 more or 1 less.

Answers

Student Book page 33

1 b 50 and 4 = 54

  c 80 and 5 = 85

2 a 4 tens and 3 ones

  b 7 tens and 1 one

Practice Book page 40

1 40 and 7 = 47

2 30 and 3 = 33

3 10 and 6 = 16

4 80 and 3 = 83

Stretch zone: Students' answers will vary but will probably say that the tens-counters show how many tens, and the ones-counters show how many ones.

 **Introductory activity**

Use base-10 equipment to build a number, say 52. Ask students how much the tens-rods represent (50), and then the ones (2). *What would have to be changed to make the number **1 less** than 52?* (Remove a cube.) *What would have to be changed to make the number **10 more**?* (Add a rod.) Select students to come to the front of the class and make 2-digit numbers, then describe what they would change to make the number that is **1 more**, 1 less, 10 more, **10 less**.

Write a 2-digit multiple of 10 on the board and ask students to make this with base-10 equipment, for example, 40 (4 rods). *What do you need to do to find the number that is 1 less?* (Take 1 away.) Ask students to work in pairs to explore this, then select a pair to share with the class how they changed a rod for 10 cubes so that they could take away one cube to get 39.

Students can complete page 34 of the Student Book at this point or later in the lesson.

 **Main activity**

Highlight a number (e.g. 45) on a large 100-square and ask students to tell you the number that is 1 more, then highlight their answer (46). Repeat several times and then ask students to describe the pattern that they see. They should notice that 1 more is the next number to the right. Choose a number from the far right-hand **column** of the 100-square and say, *There is no number to the right? Where is the number that is 1 more?* Some students should suggest that the number is at the start of the next **row**, because of the way the numbers are arranged. If they were in one long line, as on a number line, 1 more would always be to the right because it is the next counting number. If necessary, you could show this with a paper 1–30 number track, cutting it up to reorganise it in three rows of ten, for additional visual support.

Repeat the activity on the 100-square for 1 less, then for 10 more and 10 less.

Check that students recognise that the number that is 10 more is below the chosen number, on the next row but in the same column, and the number that is 10 less is on the previous row but in the same column. You may need to count on or back 10 in ones from the first number together for all students to recognise this.

Ask students to complete the activity on pages 34 and 35 of the Student Book. Draw students' attention to how to record each number in the table by working through the example (47) before they begin.

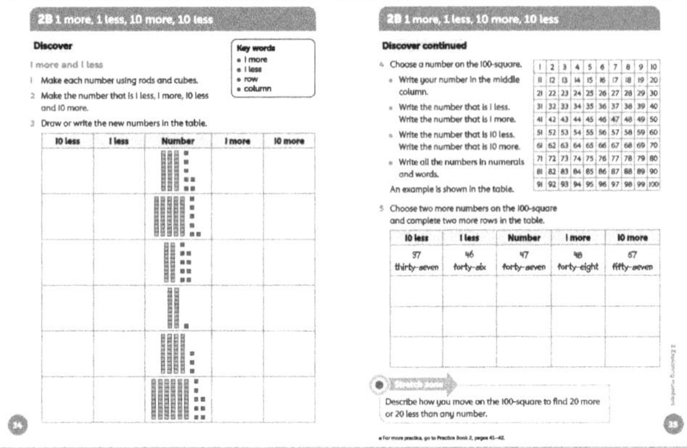

Differentiation

Supporting: Encourage students to continue to use base-10 equipment.

Consolidating: Ask students to explain how the digits change when they find 1 or 10 more or less.

Extending: Ask students to choose 2-digit numbers and then carry out a two-step calculation, for example find 10 more and then 1 less.

Stretch zone: *Describe how you move on the 100-square to find 20 more or 20 less than any number.*

Students should be able to reason how they can find 20 more by moving down 2 squares on the 100-square, and similarly for 20 less by moving up 2 squares.

 Reflection time

Ask students to explain how they found 1 more or 1 less than the given numbers, using either the base-10 equipment or the 100-square. Ask, for example, *What happened to the ones digit when you found 1 more or 1 less? What happened to the tens digits? Is it always the same? Think about the numbers 20 and 18. What happened to the digits when you found 10 more or 10 less?*

Practice Book: Students complete pages 41 and 42 of the Practice Book. They can do this directly after the main activity, as homework, or as the focus of a separate mathematics session to help students consolidate their learning and build fluency. This activity may work best if split over more than one session or worked on over an extended period of time. Students will need digit cards to complete the activity.

Differentiated outcomes	
All students	should use base-10 equipment to find 1 more or less and 10 more or less.
Most students	will find 1 more or less and 10 more or less using mental methods.
Some students	may find 10 more and 1 less and see the link with finding 9 more.

Answers

Student Book pages 34–35

1–3

10 less	1 less	Number	1 more	10 more
27	36	37	38	47
46	55	56	57	66
19	28	29	30	39
11	20	21	22	31
33	42	43	44	53
58	67	68	69	78

4, 5 Check that students have correctly found 1 less, 1 more, 10 less and 10 more for their chosen numbers.

Practice Book pages 41–42

Check that students have correctly found 1 less and 1 more for the first set of numbers, and 10 less and 10 more for the second set.

Stretch zone: The 100-square helps to find 1 more/less and 10 more/less because of the way the numbers are in rows and columns.

2B 1 more, 1 less, 10 more, 10 less

Explore Student Book pages 36–37 • Practice Book pages 43–44

Specific learning foci

- Find 1 or 10 more/less than any 2-digit number.
- Know what each digit represents in 2-digit numbers, partition into tens and ones.

Global skills

- **Creative skills:** exploring
- **Real-world skills:** presenting information
- **Interpersonal skills:** communication/teamwork

Key vocabulary

- more, less, sequence, 1 more, 1 less, 10 more, 10 less

Resources

- large 100-square for front of class and individual 100-squares
- four cards or pieces of paper labelled '1 more', '1 less', '10 more', '10 less'
- set of digit cards 0–9
- base-10 equipment

Language support

Model the correct use of more and less. Where you would like students to explain, give a model sentence with missing words to help them.

 **Introductory activity**

Invite four students to come to the front of the class. Give them each one of the cards, '1 more', '1 less', '10 more' and '10 less'. Ask a student to call out a number for the four students to answer in turn with '1 more', '1 less', '10 more' and '10 less'. For example, the student says 19 and the four students with the cards answer in turn: 20, 18, 29 and 9.

After students have called out three numbers, invite four different students to take over in holding the cards.

Alternatively, the four students holding the cards could each choose a secret number, then follow the instruction on their card and say the result. The class use the information to work out each student's original number.

 **Main activity**

Explain to students that a **sequence** is a set of numbers where each number is made from the one before it by making the same change, for example, finding 1 more.

Write a sequence on the board with one 'mystery' number (e.g. 44, 45, 46, ___, 48, 49). Ask, *What do you notice about these numbers? What is the same? What is different? Look at the tens. What about the ones?*

Ask students whether they can see how the numbers are changing (1 more each time). *Can you see what the mystery number is?* Add the missing number (47) to the sequence.

Repeat with a second sequence, but with two numbers missing.

Ask students to complete question 1 of page 36 of the Student Book in pairs. Students can use a 100-square to help them identify the sequences. Once most pairs have completed the sequence, invite different pairs to say what the missing numbers are and explain how they know by describing the pattern in the sequence.

Direct students to look at question 2 on page 36 of the Student Book. Ask them to look at the first number in each square (93, 94, 95). *How is 93 different to 94?* (It is 1 less.) *How is 95 different to 94?* (It is 1 more.) Help students by talking through finding the next number in the top row of each box. Starting with 76, can they find 1 less and 1 more? They should continue to find 1 more and 1 less for each number in the centre box to complete the boxes on either side.

Ask students to complete questions 2–6 on pages 36–37 of the Student Book.

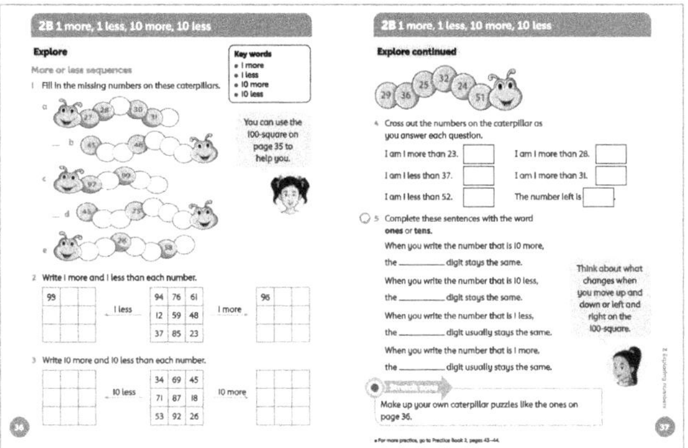

Differentiation

Supporting: Model question 4 step by step using a 100-square to narrow down possible numbers.

Consolidating: Focus on moving over tens boundaries and developing mental methods. Remind students that the ones digit does not change for a number that is 10 more or 10 less.

Extending: Ask students to take one of the statements from question 5 and give examples of when it is true or not true.

Stretch zone: *Make up your own caterpillar puzzles like the ones on page 36.*

Encourage to students first to make a number sequence with numbers going up or down in tens or ones.

Work through question 5 on page 37 of the Student Book with students. Display a large 100-square and ask them to explain each sentence using the square to support their thinking. They can also use tens-rods and ones-cubes to demonstrate finding 1 more or less and 10 more or less.

Practice Book: Students can complete pages 43–45 of the Practice Book. They can do this directly after the main activity, as homework, or as the focus of a separate mathematics session to help students consolidate their learning and build fluency. This activity may work best if split over more than one session or worked on over an extended period of time. Students will need a 100-square for reference to complete the activity.

Differentiated outcomes	
All students	should use a 100-square to find 1 more/1 less and 10 more/10 less than a given number.
Most students	will carry out calculations mentally to find 1 more/1 less and 10 more/10 less.
Some students	may make general statements about how numbers change when you find 1 more/1 less and 10 more/10 less, and give examples.

Answers

Student Book pages 36–37

1 a 27, 28, 29, 30, 31, 32

b 45, 46, 47, 48, 49, 50, 51

c 97, 98, 99, 100, 101

d 45, 55, 65, 75, 85, 95, 105

e 8, 18, 28, 38, 48, 58, 68

2

93	75	60		94	76	61		95	77	62
11	58	47	← 1 less	12	59	48	1 more →	13	60	49
36	84	22		37	85	23		38	86	24

3

24	59	35		34	69	45		44	79	55
61	77	8	← 10 less	71	87	18	10 more →	81	97	28
43	82	16		53	92	26		63	102	36

4 24   29   36   32   51   25

5 When you write the number that is 10 more, the **ones** digit stays the same.

When you write the number that is 10 less, the **ones** digit stays the same.

When you write the number that is 1 less, the **tens** digit usually stays the same.

When you write the number that is 1 more, the **tens** digit usually stays the same.

Practice Book pages 43–44

1

1 less		Pattern		1 more	
71	72	72	73	73	74
81	82	82	83	83	84

2

1 less		Pattern		1 more	
17	18	18	19	19	20
27	28	28	29	29	30

3

1 less		Pattern		1 more	
65	66	66	67	67	68
75	76	76	77	77	78

4

1 less		Pattern		1 more	
20	21	21	22	22	23
30	31	31	32	32	33

5

1 less		Pattern		1 more	
77	78	78	79	79	80
87	88	88	89	89	90

1

10 less		Pattern		10 more	
62	63	72	73	82	83
72	73	82	83	92	93

2

10 less		Pattern		10 more	
8	9	18	19	28	29
18	19	28	29	38	39

3

10 less		Pattern		10 more	
56	57	66	67	76	77
66	67	76	77	86	87

4

10 less		Pattern		10 more	
11	12	21	22	31	32
21	22	31	32	41	42

5

10 less		Pattern		10 more	
68	69	78	79	88	89
78	79	88	89	98	99

Stretch zone: Students might say they notice that on the 1 more, 1 less patterns the ones digits change, but on the 10 more, 10 less patterns the tens digits change.

2 Exploring numbers

Connect Student Book page 38

Big idea

Numbers have properties. The numbers in this unit are made up of some tens and some ones. Numbers can be big or small.

Global skills

Creative skills: exploring
Real-world skills: presenting information
Interpersonal skills: communication / teamwork
Self-development skills: reflecting on learning

Key vocabulary

- double, more than, less than, <, >, tens, ones, place value, partition, 1 more, 1 less, 10 more, 10 less

Resources

- base-10 equipment, Numicon shapes, place-value cards

Language support

Display large spider diagrams of particular numbers, such as 8 and 27, so that students can see and remind themselves of the vocabulary of numbers.

 **Introductory activity**

Remind students that they now know a lot about numbers and their properties. Ask questions such as: *What do we call any number that ends in 0, 2, 4, 6 or 8? What word do we use to describe any number that is twice the size of another? What sign do we write to describe a number that is less than another number?*

Ask students to look through Unit 2 of the Student Book as well as relevant pages in Unit 1 to remind themselves of the patterns and properties they have explored to this point. Remind them of any points covered in the prevous year as well. Draw up a list on the board, with an example number next to each property. This could include, for example:

Between   12 is between 13 and 14
Odd   15
Even   10
Double   8 is double 4
1 less   9 is 1 less than 10
10 more   29 is 10 more than 19
Greater than 40 > 16

 **Main activity**

Write the number 8 on the board and ask students to tell you things they know about the number. Construct a spider diagram using students' ideas. Ask students to think about what number 8 means to them and give examples, for example, 8 slices of their favourite pizza, 8 legs on a spider and so on. Can they represent 8 using base-10 equipment, Numicon shapes, place-value cards, or show it on a number line, for example?

Ask students to complete spider diagrams for 48 and 60 on Student Book page 38 to show their understanding of properties of numbers and number patterns. If students prefer, they can work in pairs or groups to create their diagrams on large sheets of paper.

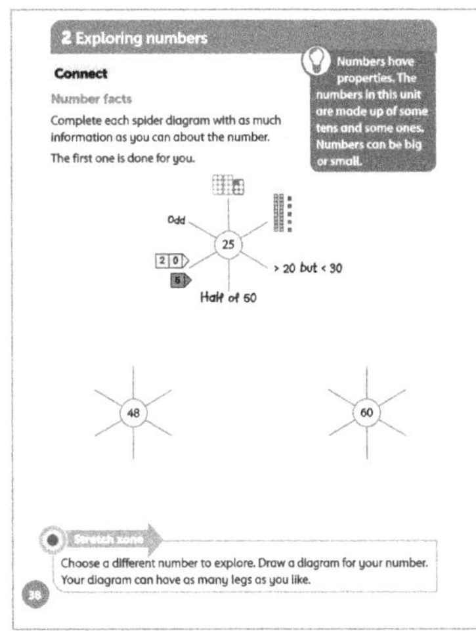

Differentiation

Supporting: Use the Student Book to help students focus on key points.

Consolidating: Ask students to describe their choices of information.

Extending: Challenge students to find more than six ways to describe 48 and 60.

Stretch zone: *Choose a different number to explore. Draw a diagram for your number. Your diagram can have as many legs as you like.*

Students should be encouraged to choose a 2-digit number to explore.

Differentiated outcomes	
All students	should use a range of properties they have learned in Units 1 and 2 to make a spider diagram with some direction.
Most students	will use a wide range of properties they have learned in Units 1 and 2.
Some students	may record many properties for numbers drawing on information beyond this unit and the last.

 Reflection time

Challenge the class to come up with 24 facts about 24. Go round the class, collecting information and constructing a spider diagram. If the class is running out of ideas and no one has suggested a negative fact such as 24 is not double 10, introduce this idea yourself to ensure that the students do achieve 24 facts.

---

2 Exploring numbers

Review Student Book page 39 • Practice Book page 45

Global skills

- **Creative skills:** problem solving
- **Interpersonal skills:** communication
- **Self-development skills:** reflecting on learning

Student Book

With young children, assessment activities are most effective when carried out as an everyday classroom activity. Students should be able to build and describe numbers using tens and ones, describe the place value of the digits and find 1 more, 1 less, 10 more and 10 less than a number.

It may help to provide students with 100-squares, place-value counters and base-10 equipment. Some students may need help reading and understanding question 6. Read the problem aloud, explaining the problem and any unfamiliar vocabulary, using drawings for example, to support their understanding.

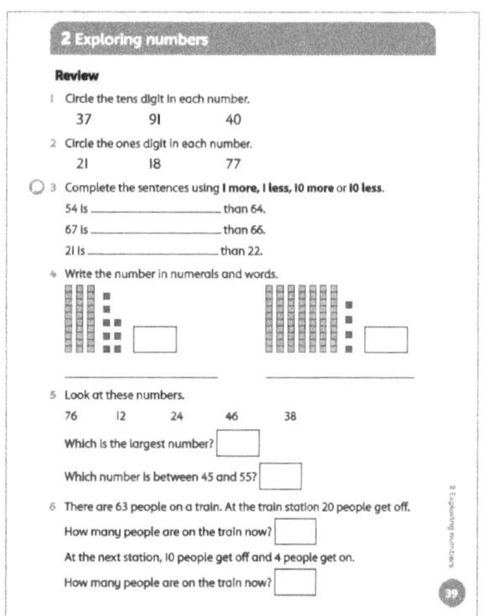

Answers

Student Book page 39

1 3   9   4

2 1   8   7

3 54 is 10 less than 64.

67 is 1 more than 66.

21 is 1 less than 22.

4 38 thirty-eight   74 seventy-four

5 The largest number is 76.
46 is between 45 and 55.

6 43 people are on the train.

37 people are on the train.

Practice Book

It is appropriate to complete this Practice Book review as a whole-class discussion. You may choose to keep a record of the class discussion or a copy of the review page for your own records. The review provides an opportunity for students to reflect on their learning from the unit, to discuss any areas of mathematics that they feel went particularly well, and any areas that they feel less confident about. Ensure all students have a copy of the Student Book as a reminder of the areas of mathematics that they have worked on in this unit.

Allow students plenty of time for discussion before asking them to complete the Practice Book page individually, and then, if appropriate, to share their responses with the rest of the class. If students complete this self-assessment at home, encourage them to discuss this with adults. Make a note of areas that students still feel unsure about.

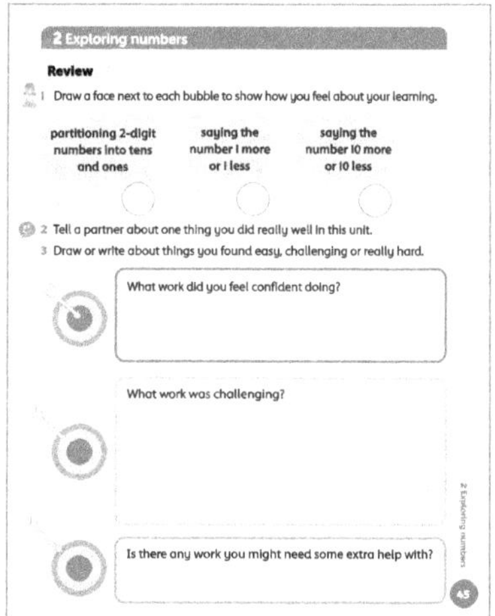

Additional material

There are additional end-of-unit assessments available on the *Oxford Owl* website.

Overview

Big Idea

The Big idea for this unit is that we can add numbers together and that it does not matter which order we add them in. This means that addition is 'commutative', because we can commute (change) the order of the numbers being added; we will always get the same total.

Each number has a specific set of number bonds or pairs that can be added to make that number. Reordering the pair of numbers does not make it a different pair. So the number bonds (or pairs) for 10 are: 10 and 0, 9 and 1, 8 and 2, 7 and 3, 6 and 4, 5 and 5; for example, $8 + 2 = 10$, and $2 + 8 = 10$.

It is important not to give students 'rules' that do not always apply. We may want to say that 'adding always makes numbers bigger' but students will find out that this is not true when they meet negative numbers. We should always take care not to teach anything that needs to be corrected later.

Look out for

- **Students who do not see how bonds to 10 help us to know bonds with a total of 20.** For example, seeing that $6 + 4 = 10$ enables us to write $16 + 4 = 20$. Look for opportunities to show these connections using concrete resources.

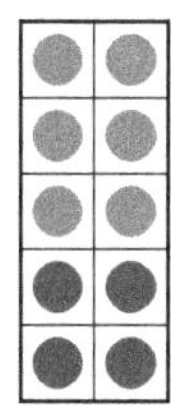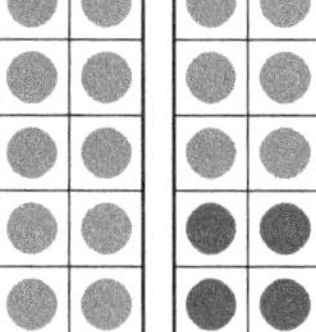

$6 + 4$ $16 + 4$

Possible misconceptions

- **Students may understand that addition is commutative, but may not realise that subtraction is not.** For example, $6 + 4$ and $4 + 6$ have the same answer, but $10 - 4$ and $4 - 10$ do not.

- **Students may believe that only one number belongs after the equals sign.** For example, students may be confused by number sentences such as $10 = 6 + 4$ or $6 + 4 = 7 + 3$. Give students examples of calculations presented in a variety of ways and emphasise that the equals sign means whatever is on one side is equal to whatever is on the other side. The equals sign does not always mean 'the answer'.

Key vocabulary

- number bond, number fact
- multiple, mulitples of 10
- commutative, predict
- total, addition, subtraction
- fact family
- number sentence
- grid, row, column

Coverage in lessons

Learning focus	Learning outcomes (the ENC objectives)
Bonds for 10	Recall and use addition and subtraction facts to 10 fluently.
Bonds for 20 and 100	Recall and use addition and subtraction facts to 20 fluently, and derive and use related facts up to 100.
Fact families for 20	Recall and use addition and subtraction facts to 20 fluently.
Fact families for 100	Recall and use addition and subtraction facts to 20 fluently, and derive and use related facts up to 100.

3 Number bonds and fact families

Big Question

- What are number bonds? What are fact families?

Global skills

- **Creative skills:** exploring/investigating
- **Interpersonal skills:** communication/teamwork

Key vocabulary

- number bond

Resources

- number rods and interlocking cubes

Language support

Encourage students to say the full number sentence, not just, for example, '6 and 4', as this links the three numbers together. Make some very simple card jigsaws of number bonds for 10 to help students practise. As they match the two pieces together, they can say the number bond, for example, $6 + 4 = 10$.

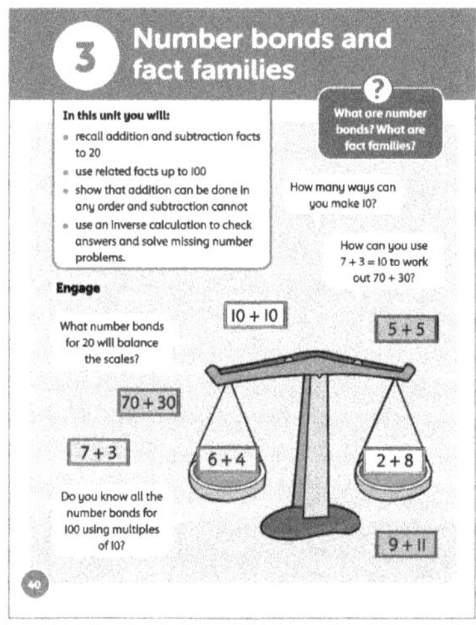

Introductory activity

Ask students to think of as many different examples of things that come in pairs as they can. If necessary, prompt with questions: *Do we ever buy one sock or one shoe? Does a pair of trousers count as one thing or two?* They should discuss ideas with a partner.

Share pairs' thoughts and then use Student Book page 40 to continue the discussion. If possible, you could display it on the IWB. Explain that when we have pairs of numbers, and we add them together, we can make a **number bond.** The pairs of numbers that add to make a number like 10 are the number bonds for 10. *How many number bonds can you see on page 40? Which are number bonds for 10? Can you see a number bond for 100? How do you know?*

Main activity

Ask students to think about the number bonds for 2 and to discuss them in pairs and write them down. ($0 + 2 = 2$ and $1 + 1 = 2$). *Are these the only ways you can make 2? How do you know?* (Because only numbers that are 2 or less can add to make 2, and the only numbers that are 2 or less are 0, 1 and 2.) Ask, *If there are two number bonds for 2, are there three number bonds for 3, four for 4 and so on?* Pairs of students should explore number bonds for the numbers from 3 to 9 and record them on a large sheet of paper. They could use concrete resources such as number rods or interlocking cubes to model number bonds.

Differentiation

Supporting: Give students concrete resources such as cubes to help them find the pairs of numbers, for example, by building a number and taking none away, then 1 away, then 2, and so on, recording the number bond each time.

Consolidating: Support students in developing a systematic approach to finding most of the number bonds for each number.

Extending: Encourage students to work systematically to find all the number bonds and explore the patterns made by the bonds.

Reflection time

Ask students to share their findings from making number bonds for every number from 3 to 9. Discuss how they found the bonds. *Did you find all the possibilities? How do you know?* Ask students to share examples of how they represented their bonds, concretely or as a written list. Did they use a strategy to help them?

3A Bonds for 10

Specific learning focus

- Find and start to recall numbers bonds to 10.

Global skills

- **Creative skills:** exploring
- **Interpersonal skills:** communication/teamwork

Key vocabulary

- number bond, total, number sentence, subtraction

Resources

- sets of number cards 0–10: enough for one per student
- cubes, ten-frames and counters

Language support

Make sure that students say the full number bond, for example $12 + 8 = 20$ as this helps to link the three numbers together. Ensure that students can 'read' the symbols of addition (+) and equals (=).

 **Introductory activity**

Give each student a number card from 0 to 10. Shout out the number '6'. Students should find a partner so that the **total** for their two cards is 6, if possible.

Repeat for 8, 11, 14. After each round, notice which students do not have a partner. Discuss the reasons for this.

Finally, ask students to find a partner to make a total of 10. (Try to organise the cards so that everyone has a partner for this.)

 **Main activity**

Give each pair of students a pile of 10 cubes between them and ask them to look at page 41 of the Student Book. Ask one student to come out to demonstrate with you how to do this activity. Ask them to count out 10 cubes and then hide some of the cubes in their hands. You can model to the class how to count the remaining cubes and then work out how many were hidden and record it as an addition sentence. Review how to record totals as an addition sentence, linking each part of the sentence to the whole number of cubes (10) and the parts (e.g. 6 and 4), and what the symbols '+' and '=' represent. Students repeat the activity in pairs,

explaining to each other how they find each number bond and recording their results in the table on page 41 of the Student Book as a **number sentence** and as a drawing of cubes.

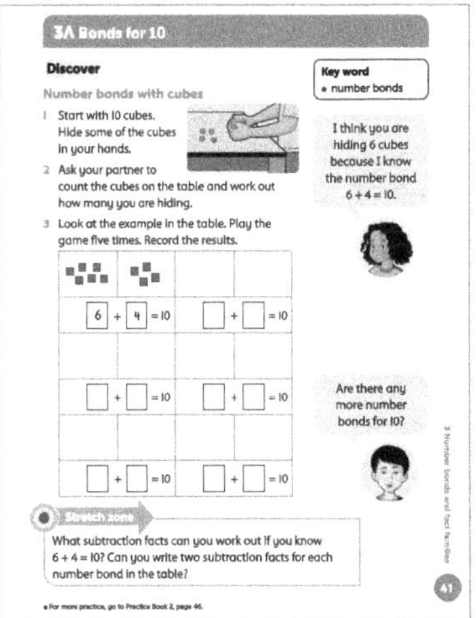

Differentiation

Supporting: Give students cubes to help with finding the bonds.

Consolidating: Support students in developing a systematic approach to finding all the bonds.

Extending: Encourage students to explore the patterns made by the number bonds.

Stretch zone: *What subtraction facts can you work out if you know $6 + 4 = 10$? Can you write two subtraction facts for each number bond in the table?*

Answer will vary depending on which bonds students recorded. Check whether students have worked systematically in recording the **subtraction** facts. For $6 + 4 = 10$, students should write $10 - 6 = 4$ and $10 - 4 = 6$. The subtraction facts for the other number bonds are: $10 - 0 = 10$, $10 - 10 = 0$; $10 - 9 = 1$, $10 - 1 = 9$; $10 - 8 = 2$, $10 - 2 = 8$; $10 - 7 = 3$, $10 - 3 = 7$; $10 - 5 = 5$.

 **Reflection time**

Ask students to name the number bonds for 10, beginning with the example given in the Student Book ($6 + 4 = 10$). Represent the bond visually, for example, using a ten-frame and counters or a part whole model, or both.

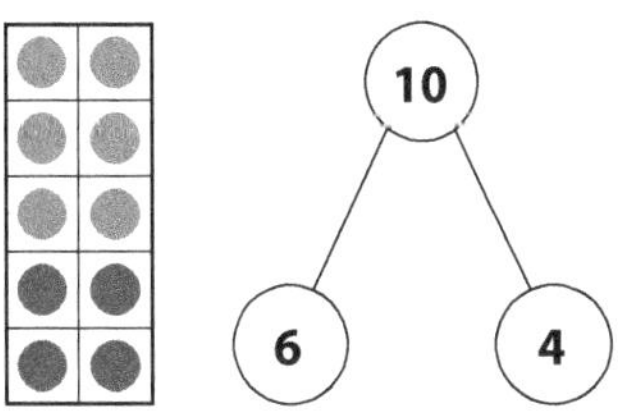

Ask questions such as: *How many bonds for 10 are there? Did you find all the possible answers? How do you know?* Build up the pattern with ten-frames and counters to check.

Discuss matching bonds (1 + 9 and 9 + 1). *What happens when we add numbers in a different order?* Agree, and model as necessary, that it does not matter what order we add numbers in, we always get the same answer.

Practice Book: Students complete page 46 of the Practice Book. They can do this directly after the main activity, as homework, or as the focus of a separate mathematics session to help students consolidate their learning and build fluency. Look at the example together. *How could we know that we have the correct number of dots in the second frame to make ten?* Agree that if you put the seven dots in the first frame it would be full because three squares are already filled.

Differentiated outcomes	
All students	should find three different number bonds for 10.
Most students	will find all of the possible number bonds for 10.
Some students	may use the number bonds to find the associated subtraction facts.

3A Bonds for 10

Explore Student Book page 42 • Practice Book page 47

Specific learning focus
- Write the number bonds to 10.

Global skills
- **Creative skills:** exploring/investigating
- **Real-world skills:** presenting information
- **Interpersonal skills:** communication/teamwork

Key vocabulary
- row, number bond, commutative, predict, total

Resources
- number cards 0–10, ten-frames and counters, interlocking cubes

Language support
Support students with the pronunciation of each number bond and help them to hear the patterns in the pairs of number bonds for each fact, for example, *6 plus 4 equals 10* and *4 plus 6 equals 10.*

Student Book page 41
Observe students as they work through the activity and check that they correctly record the numbers of cubes that are covered and remaining.

Practice Book page 46
1 $4 + 6 = 10$ (check for 6 counters filled)

2 $7 + 3 = 10$ (check for 3 counters filled)

3 $6 + 4 = 10$ (check for 4 counters filled)

4 $5 + 5 = 10$ (check for 5 counters filled)

Stretch zone: Check that students have drawn a correct pattern to show $6 + 7 = 13$.

 Introductory activity

Revise the number bonds to 10 from the previous lesson. Arrange students so that they are standing in a circle. Pass an object such as a ball to a student and say a number (e.g. 3) and the student passes it back to you saying the complement (number bond) to 10 (i.e. 7). Continue round the circle and choose students randomly, asking them to sit down once they have given a correct answer.

 Main activity

Give each pair of students a set of number cards 0–10. Explain that they are going to use them to make number bonds for different numbers, 10 or less. One student chooses a card and then together they use the other cards to make as many number bonds as they can for that number. They can use equipment such as cubes or ten-frames and counters to check. For example, if the number 6 is chosen, they can make 1 + 5, 2 + 4. Explain that they cannot use 3, because they need another 3 to make 6 and they only have one 3 card.

They can take turns at choosing the target number and record the different number bonds they make in each case.

Students should then complete the activity on page 42 of the Student Book, making number bonds to 10 and recording them in the table. Look at the example for question 4. *What does this example show us?* Agree that it shows that you can add the parts in any order and the whole will be the same, 10. Say to students that this shows that addition is **commutative**.

While students are working, move around the pairs and support them where necessary. Encourage students to cross out the numbers they have used and work towards the middle of the numbered **row** so that they generate an ordered list. Ask them to check the total of each pair of numbers.

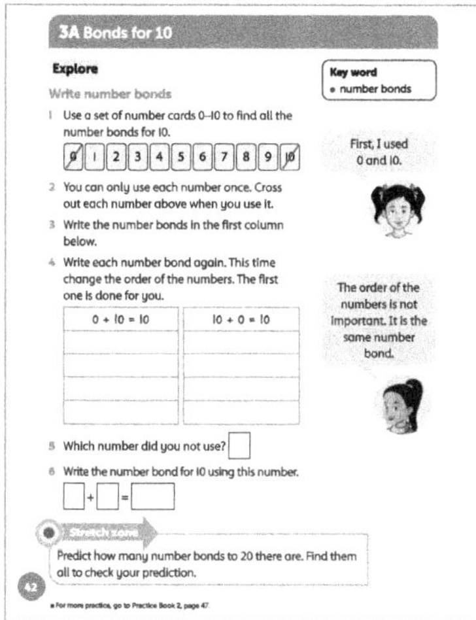

Differentiation

Supporting: Work with pairs using cubes or ten-frames and counters to help them find the number bonds.

Consolidating: Encourage students to identify the reverse pairings by looking at the groupings of cubes they can make.

Extending: Ask students to list the number bonds systematically: $0 + 10 = 10$, $1 + 9 = 10$, $2 + 8 = 10$ and so on.

Stretch zone: *Predict how many number bonds to 20 there are. Find them all to check your prediction*

Explain that **predict** means to have a reasonable guess, based on what they know about number bonds they have found for other numbers. Look to see whether students are working systematically and suggest ways for them to do this. For example, they might like to start with 20 and see what needs to be added ($20 + 0 = 20$) then go to 19, then 18 and so on.

 **Reflection time**

Discuss with students how they checked that they had found all the number bonds. Ask them to explain their method. Ask students to look at the number bonds they made with number cards for even numbers and the number bonds for odd numbers. What do they notice? They should see (if not, prompt them) that, for even numbers, there is always a number not used, which is half of the total.

Practice Book: Students can complete page 47 of the Practice Book. They can do this directly after the main activity, as homework, or as the focus of a separate mathematics session to help students consolidate their learning and build fluency. Challenge students to try to recall the number bond to 10 using the only the row of cubes on the left before they colour the second row of cubes to make 10. For example, for question 1, they ask themselves, 'What number should I add to 1 to make 10?'

Differentiated outcomes	
All students	should find the number bonds for 10.
Most students	will find the reverse number bonds to 10.
Some students	may order their list of number bonds to 10.

Answers

Student Book page 42

Check that the number bonds to 10 have been used correctly, with each number bond and its reverse being listed, for example, $7 + 3 = 10$ and $3 + 7 = 10$.

Practice Book page 47

Check that the correct number of cubes have been coloured to make 10.

1 9

2 1

3 2

4 8

Stretch zone: Students can describe how they recall the number facts that add to 10.

Discover Student Book page 43 · Practice Book page 48

Specific learning focus

- Find all pairs of multiples of 10 with a total of 100 and record the related addition.

Global skills

- **Creative skills:** exploring/investigating
- **Real-world skills:** presenting information
- **Interpersonal skills:** communication /teamwork

Key vocabulary

- number bonds, total, multiples of 10, predict

Resources

- counting stick, sticky notes with multiples of 10 (0–100) marked on them
- base-10 equipment

Language support

Students could make large colourful versions of the number bonds for display. Listen carefully as students say the number bond they have made. Check for correct pronunciation, especially of '–ty' numbers. Remind students to say the whole number bond so that the three numbers become linked together.

Introductory activity

Ask students to count with you from 0 to 100 in tens. You could use a counting stick and mark the multiples of ten with sticky notes. Count forward and back. Cover up some of the numbers and repeat the count.

Remind students that these numbers are called **multiples of 10**, because they are made up of chunks of 10. Return to the counting stick and count each multiple of ten in order (i.e. 10–100) saying *1 ten, 2 tens, 3 tens* and so on and then back.

Main activity

Remind students how they used the numbers 0 to 10 to find all the number bonds for 10. Explain that this time they have a set of multiples of 10, and they are going to find number bonds for 100, but that what they did before will help them.

Ask students to work, in pairs, on the activity in the Student Book page 43 so that they can explain to each other how they find each number bond. Look at the example for question 4 on both pages 42 and 43.

What is the same, what is different? Students should notice that both the bonds are made out of the first and last cards from the row of cards at the top of the pages, that the numbers are made of the digits 1 and 0, that both use the same three numbers to make two addition sentences with the parts in different orders. The example on page 42 has 10 as a part and as the whole but the example on page 43 has 100 as a part and the whole.

While they are working on these activities move around the pairs and ask, *If I know that 6 and 4 make 10 how can this help me to predict what 60 and 40 will be?* Draw out the link by saying, *6 and 4 make 10. 6 tens and 4 tens make 10 tens. 6 tens is 60, 4 tens is 40 and 10 tens is 100. 60 and 40 makes 100.*

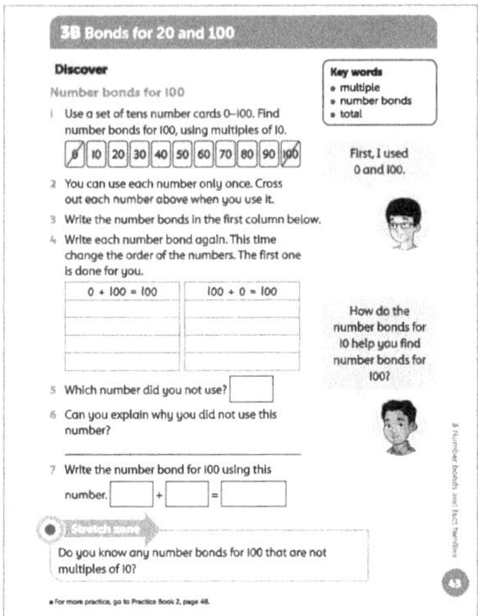

Differentiation

Supporting: Give students base-10 equipment to help them to find the bonds, swapping ones-cubes for tens-rods to show the link between number bonds to 10 and number bonds to 100.

Consolidating: Encourage students to choose a number bond to 100 and explain how they know it is correct.

Extending: Ask students to explore the patterns made by the number bonds and extend these to investigate pairs of multiples of 100.

Stretch zone: *Do you know any number bonds for 100 that are not multiples of 10?*

Students may be prompted with 99 + 1 first, so if they know that 99 + 1 = 100, can they see what needs to be added to 98 to make 100? Then try 97 and so on.

 ### Reflection time

Ask students to name the multiple of 10 number bonds for 100, beginning with the example given in the Student Book. Check that everyone has correctly identified the final number bond. Finish the session by asking: *How was this the same and how was it different to finding number bonds to 10?* Focus on how the patterns in the number system repeat themselves again and again. Model with ones-cubes and tens-rods to represent it visually.

Practice Book: Students complete page 48 of the Practice Book. They can do this directly after the main activity, as homework, or as the focus of a separate mathematics session to help students consolidate their learning and build fluency. Show students how to count on along the line in jumps of 10 to find how many to make 100.

Differentiated outcomes	
All students	should find some number bonds for 100.
Most students	will find all the number bonds for 100.
Some students	may work systematically to find all possible number bonds and see the connection between number pairs to 100 and number pairs to 10.

Student Book page 43

Observe students as they work through the activity and check that they correctly use the bonds for 100.

3, 4 $0 + 100 = 100$, $100 + 0 = 100$

$10 + 90 = 100$, $90 + 10 = 100$

$20 + 80 = 100$, $80 + 20 = 100$

$30 + 70 = 100$, $70 + 30 = 100$

$40 + 60 = 100$, $60 + 40 = 100$

5 50 is not used.

6 There is only one 50 card in the set but two 50 cards are needed.

7 $50 + 50 = 100$

Practice Book page 48

1 70	**3** 90	**5** 50	**7** 20	**9** 40
2 40	**4** 20	**6** 50	**8** 90	**10** 70

Stretch zone: Students may notice that the number bonds for 100 have a similar pattern to number bonds for 10.

3B Bonds for 20 and 100

Explore Student Book page 44 • Practice Book page 49

Specific learning focus

- Find and begin to recall all number bonds for 20.

Global skills

- **Creative skills:** exploring/investigating
- **Real-world skills:** presenting information
- **Interpersonal skills:** communication/ teamwork

Key vocabulary

- number bonds, multiples, total

Resources

- concrete resources (e.g. Numicon shapes, rods and cubes, ten-frames and counters, interlocking cubes)

Language support

Students could make large colourful patterns of the number bonds for display. Refer to the display whenever students use number bonds to 20. Listen as students say the number bond they have made. Check for correct pronunciation, especially of '-teen' numbers. Remind students to say the whole number bond so that the three numbers are linked together.

 **Introductory activity**

Start with a pile of 20 cubes. Ask a student to come to the front of the class and move some of the cubes (e.g. 7) into a separate pile. Count the cubes in each pile with the student and write the addition fact for the piles: $7 + 13 = 20$. Push the piles back together and invite another student to remove some cubes and repeat.

For each of the addition facts recorded, students should notice that the total is always 20, so the numbers in the piles represent number bonds for 20. Reinforce by drawing part-whole models to represent each bond.

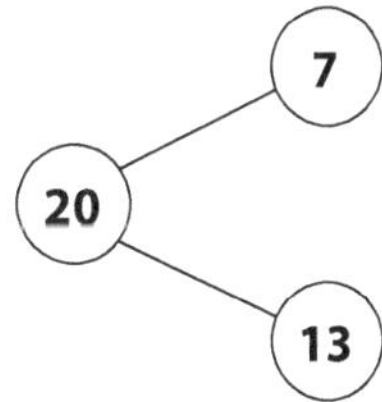

 **Main activity**

Tell students that now that they have worked on their number bonds for 10 and 100, they will work on their number bonds to 20. Display page 44 of the Student Book and explain that, as they did in the last lesson, they are going to work in pairs to find all number bonds to 20 using the number cards at the top of the page and record them on page 44 of the Student Book. Students can use concrete resources (e.g. Numicon shapes, interlocking cubes or ten-frames and counters) to help them to work out each bond or to check their number bonds when they have finished.

While they are working on these activities, move around the classroom and ask pairs: *How is this the same and how is it different to number bonds for 10 and 100?* Encourage students to cross out the numbers they have used and work towards the middle of the numbered row so that they create an ordered list. Remind them to check the total of each number bond.

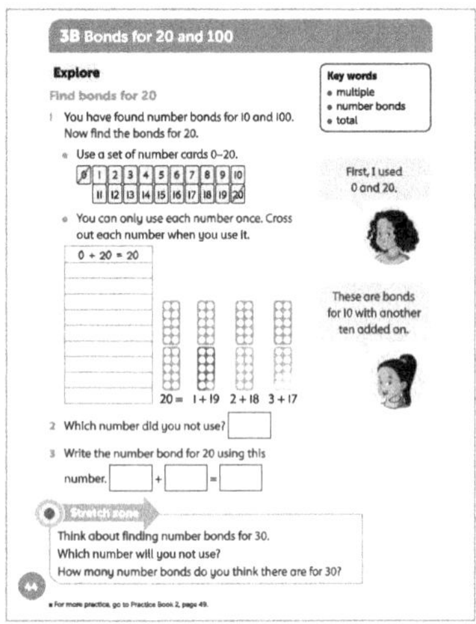

Differentiation

Supporting: Give students concrete resources to help them to find the bonds.

Consolidating: Say one part of a number bond for 20 (e.g. 15), and ask students to tell you the other part. Can they describe and show you with tens-rods, ones-cubes and so on how they know they are correct?

Extending: Ask students to think about how knowing our number bonds to 10 helps us to know our number bonds to 20. Ask them to explain it to someone else and use concrete resources to support their explanation.

Stretch zone: *Think about finding number bonds for 30. Which number will you not use?*
How many number bonds do you think there are for 30?

15 would not be used in number bonds for 30 if each number is used once only.

Ask students to first estimate and then check. Students should be able to extend the process from their work with number bonds to 20. Point them in the direction of this strategy, if necessary.

 Reflection time

Ask students to think about what is the same and different about the number bonds for 20 compared with the number bonds for 10. Can they explain the patterns and use concrete resources to represent their thinking? For example,

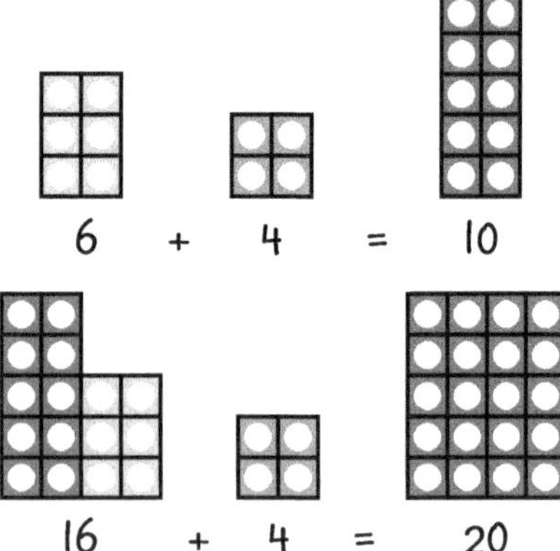

Practice Book: Students complete page 49 of the Practice Book. They can do this directly after the main activity, as homework, or as the focus of a separate mathematics session to help students consolidate their learning and build fluency. Provide students with a 0–20 number line or concrete resources to find numbers bonds for 20 or to check their answers.

Differentiated outcomes	
All students	should find some number bonds for 20.
Most students	will find all of the number bonds for 20.
Some students	may work systematically to find all possible number bonds for 20 and see the link between number bonds for 20 and number bonds for 10.

Answers

Student Book page 44

1 $0 + 20 = 20$ $5 + 15 = 20$

    $1 + 19 = 20$ $6 + 14 = 20$

    $2 + 18 = 20$ $7 + 13 = 20$

    $3 + 17 = 20$ $8 + 12 = 20$

    $4 + 16 = 20$ $9 + 11 = 20$

2   10

3 $10 + 10 = 20$

Practice Book page 49

1 $7 + 13 = 20$ **7** $19 + 1 = 20$

2 $11 + 9 = 20$ **8** $2 + 18 = 20$

3 $4 + 16 = 20$ **9** $6 + 14 = 20$

4 $14 + 6 = 20$ **10** $16 + 4 = 20$

5 $18 + 2 = 20$ **11** $9 + 11 = 20$

6 $1 + 19 = 20$ **12** $13 + 7 = 20$

Stretch zone: Students' answers will vary.

3C Fact families for 20

Discover
Student Book page 45 • Practice Book page 50

Specific learning foci

- Find and begin to learn by heart all numbers bonds for 10 and for 20.
- Partition all numbers to 20 into pairs and record the related addition and subtraction facts.

Global skills

- **Creative skills:** exploring
- **Real-world skills:** presenting information
- **Interpersonal skills:** communication/teamwork

Key vocabulary

- number bonds for 10, number bonds for 20, addition, subtraction, number facts, fact family

Resources

- cubes or counters
- number cards 0–20 (one set per pair)

Language support

Ask different students to explain how they found the rest of the fact family. Talking this through helps to reinforce the relationship between the addition and subtraction facts. They could also produce posters of fact families to add to the classroom display, decorating them in whatever way they choose.

 **Introductory activity**

Remind students that they found all the number bonds for 20 in the previous lesson. Revise these by holding up a number card less than 20 and asking students to tell you the number that goes with it to make 20. Invite different students to give you one of the number bonds for 20. After recording several number sentences (e.g. $4 + 16 = 20$), ask students whether you have collected all of the number bonds for 20 and how they know. Continue until all the number bonds have been identified and students are confident that they have found all of them. Continue to support with visual representations such as those on page 44 of the Student Book.

 **Main activity**

Draw an empty part-whole diagram on the board and ask students to give you a number between 10 and 20 (e.g. 14). *This is our whole. Where should I record it?* Invite a student to add it to the diagram. Ask students to tell you two numbers that add to make 14. Take suggestions

and record one suggestion in your part-whole diagram (e.g. 8 and 6). You may also want to model with concrete resouces, for example, counting out 14 cubes and splitting them into one part of 8 and a second part of 6.

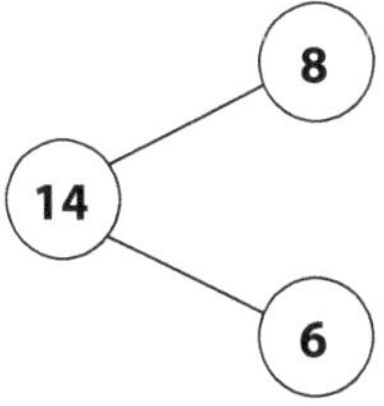

Say, *Which is the whole? Which are the parts? Can you use these numbers to make an **addition**? Can you use the same numbers to make another addition fact? What about a **subtraction**?* Remind students of how to record a subtraction sentence, linking it to the part-whole diagram, stressing that we start with the whole and subtract one part to find the other part. Give students time to work in pairs to come up with four number sentences using cubes for support if they choose.

Take their suggestions and record them on the board, for example:

$8 + 6 = 14$ $14 - 8 = 6$
$6 + 8 = 14$ $14 - 6 = 8$

Explain to students that the addition and subtraction facts that are made from the three numbers in the number bond are called a **fact family**.

Now ask pairs to choose their own three numbers that form a number bond to 20, record them in a part-whole diagram and write the fact family for this number bond.

Next, talk through how to play the game on page 45 of the Student Book. Explain that each pair has a set of number cards 0–20, one student chooses a card and the number on this card will be the answer. Both need to write down as many **number facts** as they can for that number in one minute. Explain that these can be a fact family but also any other addition or subtraction sentence.

As they play, move around the class and ask, for example, *How do you know your number facts are all correct? Is there a pattern in your number facts? Did you use a fact family to make some of your sentences?*

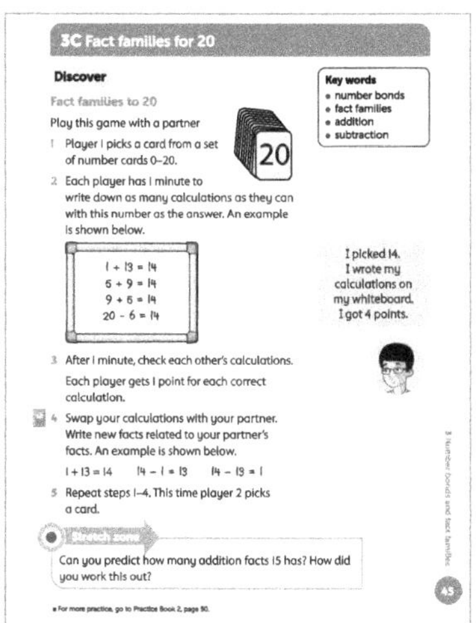

Differentiation

Supporting: Provide concrete resources to help students create fact families and number facts.

Consolidating: For each addition sentence that students record, ask them to tell you a related subtraction sentence.

Extending: Challenge students to make a fact family for every number sentence they think of.

Stretch zone: *Can you predict how many addition facts 15 has? How did you work this out?*

This activity will extend some students to forming number bonds to 15, which is not a multiple of 10. See whether they are using a similar strategy for finding bonds to multiples of 10. There should 15 number bonds in total.

 Reflection time

Ask a pair to share one of their numbers and the number sentences they came up with for that number as the answer, as well as the related facts. Record these on the board and say, *Is this all of the possible number sentences? How can we check?* Students may suggest using cubes and thinking about fact families. Use a similar model to the one on page 50 in the Practice Book to help identify each number bond systematically and then corresponding facts, for example:

<table>
<tr><td colspan="2" align="center">20</td></tr>
</table>

$1 + 19 = 20$

$19 + 1 = 20$

$20 - 19 = 1$

$20 - 1 = 19$

The rest of the class can build on this, working in pairs, until you have all the number bonds for that number.

Practice Book: Students can complete page 50 of the Practice Book. They can do this directly after the main activity, as homework, or as the focus of a separate mathematics session to help students consolidate their learning and build fluency. Students can draw on their work during the reflection time activity to complete the questions.

Differentiated outcomes	
All students	should find some number sentences for their chosen number.
Most students	will find several number sentences for their chosen number and apply their knowledge of commutativity to find them.
Some students	may draw upon their knowledge of fact families to find number facts for their chosen number.

Answers

Student Book page 45

Observe students as they work through the activity and mark their answers to the questions in the Student book, checking that they have written the correct numbers as they play the game.

Practice Book page 50

Check that students have coloured the correct number of squares for each number bond to 20.

Stretch zone: Drawings will vary but should represent a number bond for 15.

3C Fact families for 20

Specific learning foci

- Find number bonds for 20 and numbers up to 20.
- Record the related addition and subtraction facts.

Global skills

- **Creative skills:** investigating
- **Real-world skills:** presenting information
- **Interpersonal skills:** communication/teamwork

Key vocabulary

- number bonds, addition, subtraction, number facts, fact family

Resources

- cubes or counters

Language support

Provide a simple speaking frame such as 'I know ______ because ______' to support students to explain how they knew that they had completed all the possible fact family houses. You could model an answer for one number or number bond before asking students to use the same structure for a different number or number bond. Ask short, closed questions to prompt the next part of a longer response.

Introductory activity

Remind students that they found the fact family for every number bond for 20 in the previous lesson. Ask each pair to list quickly all the number bonds for 17. Challenge different pairs of students to tell the class the whole fact family for one of the number bonds.

Main activity

Draw a copy of a fact family house on the board from page 46 of the Student Book. Explain the words 'roof' and 'floor' if necessary. Alongside the house, draw an empty part-whole diagram. Write the numbers 1, 19 and 20 in the three boxes on the roof, putting 20 at the apex. Also draw a part-whole diagram for 1, 19 and 20.

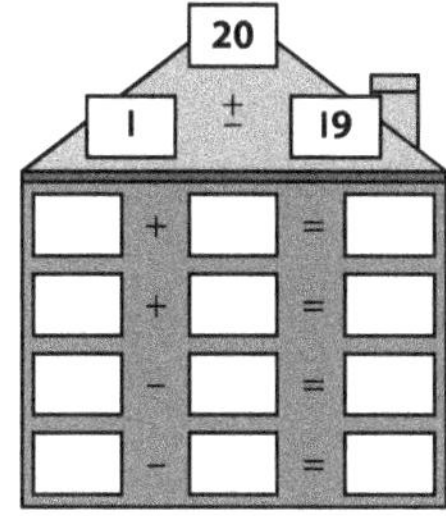
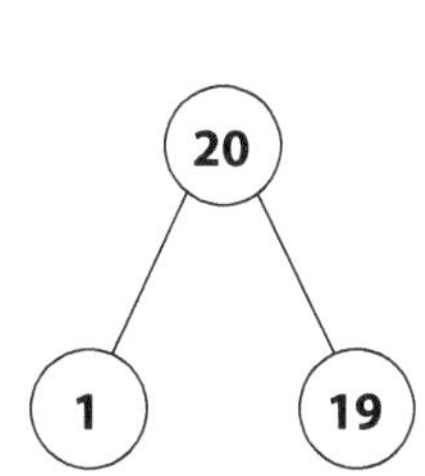

Say, *We have written our three numbers. What do you notice about these numbers? Which is the whole? Which are parts? Can you use these numbers to make an addition sentence? Can you use the same numbers to make another addition sentence? What about a subtraction sentence?*

Take suggestions from students and record each fact (1 + 19 = 20, 19 + 1 = 20, 20 − 1 = 19, 20 − 19 = 1) in the fact family house.

What do we call the addition and subtraction facts that are made from the three numbers in a number bond? (a fact family) *We have just completed a fact family house.*

Ask students to look at page 46 of the Student Book and fill in the blue house using the information from the board. Ask them to look at the number bonds now for the first two houses. *What do you notice? What number bond do you think we should put in the roof of our next house?* Agree 2, 18 and 20. Ask students to continue to work systematically to fill each of the fact family houses on pages 46 and 47 of the Student Book.

Students then work in pairs, each choosing a number between 14 and 19, finding all the number bonds for that number and then writing the fact family for each number bond. Once they have recorded the fact family for each of their number bonds, they should swap books with their partner and check that their partner has found the fact families for all their number bonds. Students then complete the sentences at the bottom of Student Book page 47.

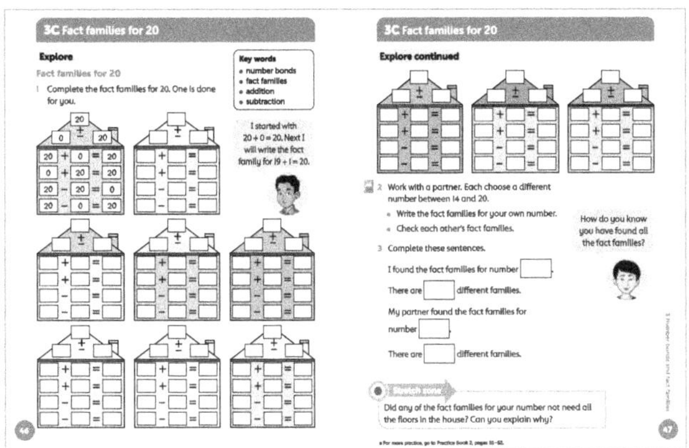

Differentiation

Supporting: Help write the answers for students and give students concrete resources to create fact families.

Consolidating: Encourage students to work systematically to find how many fact families there are.

Extending: Students can extend their understanding by finding fact families for all the number bonds for numbers between 21 and 30.

Stretch zone: *Did any of the fact families for your number not need all the floors in the house? Can you explain why?*

Students should show that they understand that the number bond 10 + 10 = 20 does not have a reverse addition (like 11 + 9 = 9 + 11), and only has one subtraction fact 20 − 10 = 10. Ask whether they can think of another number for which this is the case.

Draw up a list of how many fact families there are for each number from 10 to 19. Ask students whether they can see a pattern. Challenge them to think about odd and even numbers to help explain the pattern. For an odd number, each number bond makes a fact family of two additions and two subtractions (four facts). For an even number, one number bond makes a fact family of only one addition and one subtraction (two facts).

Practice Book: Students can complete page 51 and 52 of the Practice Book. They can do this directly after the main activity, as homework, or as the focus of a separate mathematics session to help students consolidate their learning and build fluency. Encourage students to choose different numbers from the ones they worked on in the Student Book for the second part of the activity. After they have chosen their number, they need to find six number bonds for it, writing them in the roofs of the houses, then write the fact families for each number bond.

Differentiated outcomes	
All students	should find some members of the fact families for number bonds for 20.
Most students	will complete all the fact families for number bonds for 20.
Some students	will predict and find how many fact families there are for numbers between 21 and 30.

Student Book pages 46–47

1 $20 + 0 = 20, 0 + 20 = 20, 20 - 20 = 0, 20 - 0 = 20$

$19 + 1 = 20, 1 + 19 = 20, 20 - 19 = 1, 20 - 1 = 19$

$18 + 2 = 20, 2 + 18 = 20, 20 - 18 = 2, 20 - 2 = 18$

$17 + 3 = 20, 3 + 17 = 20, 20 - 17 = 3, 20 - 3 = 17$

$16 + 4 = 20, 4 + 16 = 20, 20 - 16 = 4, 20 - 4 = 16$

$15 + 5 = 20, 5 + 15 = 20, 20 - 15 = 5, 20 - 5 = 15$

$14 + 6 = 20, 6 + 14 = 20, 20 - 14 = 6, 20 - 6 = 14$

$13 + 7 = 20, 7 + 13 = 20, 20 - 13 = 7, 20 - 7 = 13$

$12 + 8 = 20, 8 + 12 = 20, 20 - 12 = 8, 20 - 8 = 12$

$11 + 9 = 20, 9 + 11 = 20, 20 - 11 = 9, 20 - 9 = 11$

$10 + 10 = 20, 20 - 10 = 10$

2 Answers will vary as students choose a number between 14 and 19. Check that they have written all the number bonds for that number and have written the fact family for each number bond.

3 Answers will depend on students' choice of number.

Practice Book pages 51–52

$19 + 1 = 20$	$20 - 1 = 19$	$20 - 19 = 1$
$18 + 2 = 20$	$20 - 2 = 18$	$20 - 18 = 2$
$17 + 3 = 20$	$20 - 3 = 17$	$20 - 17 = 3$
$16 + 4 = 20$	$20 - 4 = 16$	$20 - 16 = 4$
$15 + 5 = 20$	$20 - 5 = 15$	$20 - 15 = 5$
$14 + 6 = 20$	$20 - 6 = 14$	$20 - 14 = 6$
$13 + 7 = 20$	$20 - 7 = 13$	$20 - 13 = 7$
$12 + 8 = 20$	$20 - 8 = 12$	$20 - 12 = 8$
$11 + 9 = 20$	$20 - 9 = 11$	$20 - 11 = 9$
$10 + 10 = 20$	$20 - 10 = 10$	$20 - 10 = 10$

Check that students have completed fact families correctly for the number bonds for their chosen number between 11 and 19.

Stretch zone: Students' answers will vary depending on the fact families for the number they chose.

3D Fact families for 100

Specific learning focus

- Find all number bonds to 100 using multiples of 10 and record the fact families.

Global skills

- **Creative skills:** exploring
- **Interpersonal skills:** communication/teamwork

Key vocabulary

- number bonds for 10, number bonds for 100, addition, subtraction, number facts, multiples of 10, fact family, column

Resources

- large 100-square for front of class
- counting stick, sticky notes labelled 0–10 and labelled with multiples of 10 from 0 to 100
- base-10 equipment (ones-cubes and tens-rods)
- number cards with multiples of 10 from 0 to 100 (one set per pair)

Language support

Display some of the sets of four numbers with a sentence saying 'I know that this number does not belong here because the fact family is

____ + ____ = ____!'

 **Introductory activity**

Using a large 100-square for support, count forward and back together in tens from any single-digit number. Ask students to talk about the patterns they notice.

End the counting by focusing on counting down and back in the right-hand **column** 10, 20, 30, …. Ask students to explain how that count was different from the others.

Repeat this count using a counting stick.

Point to the left-hand end of the stick (viewed from the class) and say, *This is zero.*

Point to the right-hand end of the stick (viewed from the class) and say, *This is 10.*

Count along from zero to the fourth mark on the stick and ask what number this shows (4). Label this with a sticky note. Ask how many more marks there are to get to 10 (6) and ask what number bond this shows (4 + 6 = 10).

Point to the left-hand end of the stick (viewed from the class) and say, *This is zero.*

Point to the right-hand end of the stick (viewed from the class) and say, *This is 100.*

Point to the fourth division and ask what number bond this shows (40 + 60 = 100). Label 40 and 60 with the sticky notes. Repeat with a few more examples.

 **Main activity**

Ask each pair to write down a number bond for 10. Ask pairs whether they can use this bond to work out a number bond for 100. For example:

6 + 4 = 10

60 + 40 = 100

Talk through how the 6 ones have changed to 6 tens, and the 4 ones to 4 tens. In each case, model this with base-10 equipment, swapping each ones-cube for a tens-rod. Each number is ten times bigger, so it is a multiple-of-10 number bond for 100. Ask pairs to find all the bonds of multiples of 10 for 100, then to move on to finding the fact family for one bond of their choice (e.g. 70 + 30 = 100, 30 + 70 = 100, 100 − 70 = 30, 100 − 30 = 70).

They should then play the game on page 48 of the Student Book in pairs. Explain that they will have a set of tens cards from 0 to 100. One of the pair picks a card, for example, 40. Their partner finds the card that makes the total 100 (60), then they take turns to choose a card and find the card for the number bond. Encourage them to describe the card they have chosen to make the bond by using it in a corresponding sentence (e.g. The card was 20 so I chose 80 because 20 add 80 equals 100).

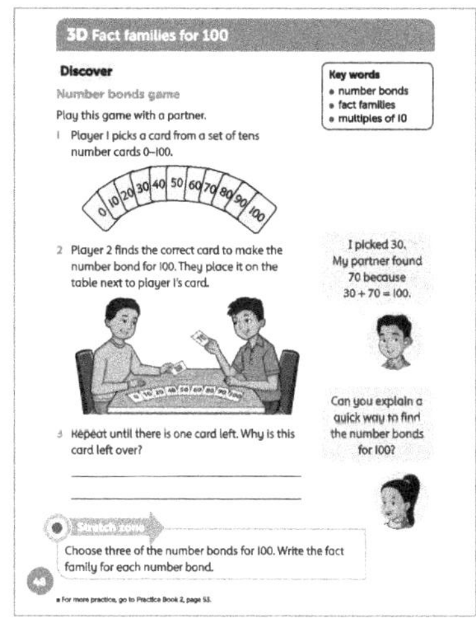

Differentiation

Supporting: Give students base-10 equipment for support in counting in tens to 100 and making number bonds for 100.

Consolidating: Ask students to explain their choice of second card to make a bond for 100 using a subtraction sentence as well as an addition sentence.

Extending: Ask students to generalise and explore the fact families to 100.

Stretch zone: *Choose three of the number bonds for 100. Write the fact family for each number bond.*

This activity will help students reinforce their learning of how fact families for any number bond are related. Check that they correctly write down the fact families. Can they explain how they know they are correct?

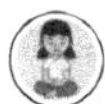 **Reflection time**

Ask different students to explain their thinking as they matched the numbers to form number bonds for 100. Use the questions in the Student Book margin, for example asking a student to explain how they knew to find 70 if their partner picked 30, or explain whether they have found any quick ways to find the number bonds.

Practice Book: Students can complete page 53 of the Practice Book. They can do this directly after the main activity, as homework, or as the focus of a separate mathematics session to help students consolidate their learning and build fluency. This activity is suitable to follow the first part of the main activity.

<table>
<tr><td colspan="2">Differentiated outcomes</td></tr>
<tr><td>All students</td><td>should make some number bonds for 100.</td></tr>
<tr><td>Most students</td><td>will find all number bonds to 100 in tens using their knowledge of bonds to 10.</td></tr>
<tr><td>Some students</td><td>will generalise number bonds to 100.</td></tr>
</table>

Answers

Student Book page 48

3   50 is left over, because two 50s are needed to make 100.

Practice Book page 53

1 $2 + 8 = 10$ $20 + 80 = 100$

2 $4 + 6 = 10$ $40 + 60 = 100$

3 $5 + 5 = 10$ $50 + 50 = 100$

4 $1 + 9 = 10$ $10 + 90 = 100$

5 $9 + 1 = 10$ $90 + 10 = 100$

Stretch zone: Answers will vary depending on the number bond for 30 chosen.

3D Fact families for 100

Explore 1   Student Book page 49 • Practice Book page 54

Specific learning focus

- Find all number bonds to 100 using multiples of 10 and record the fact families.

Global skills

- **Creative skills:** exploring/investigating
- **Real-world skills:** presenting information
- **Interpersonal skills:** communication/teamwork

Key vocabulary

- number bonds for 100, addition, subtraction, number facts, multiple of 10, fact family

Resources

- large 100-square for front of class
- counting stick
- concrete resources: Numicon shapes, base-10 equipment, ten-frames, cubes and counters

Language support

Ask different students to explain how they found the complete fact family. Provide some key words for them to use such as 'number bond', 'reversed', 'swapped', 'addition' and 'subtraction'. All these words should be familiar to students by now, but writing the list will remind students to use them. You could also provide a talking frame such as 'I started with ____ then I ______. Next I ______.'

 **Introductory activity**

Remind students that they found the number bonds for 100 using multiples of 10 in the previous lesson. Ask each pair to quickly list the number bonds and then ask pairs to explain how they knew which numbers went together. Ask different pairs of students to give you a number bond until you have revised them all.

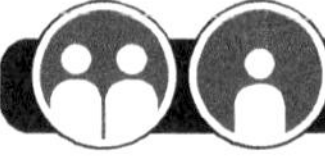 **Main activity**

Show students page 49 of the Student Book and ask them to think about the statement in the first speech bubble. *Can you find the number bonds for 100 by multiplying each number in the number bonds for 10 by 10? Is that correct? How do you know? Can you give examples?*

Give students time to discuss this in pairs. Give them access to concrete resources such as Numicon shapes, base-10 equipment, ten-frames and counters to support their thinking.

Students may see that, with a fact such as 4 + 6 = 10, making each number ten times bigger gives 40 + 60 = 100, which is the same as multiplying each number by 10.

Ask students to recall the fact family for 4 + 6 = 10 (6 + 4 = 10, 10 − 4 = 6, 10 − 6 = 4). *Can you make the fact family for 40 + 60 = 100 using this to help you?*

Students complete page 49 of the Student Book individually. Students can continue to use concrete resources to make bonds for 10 and 100 but encourage them to recall the bonds in the first instance if they can and use the equipment to check, particularly for number bonds for 10.

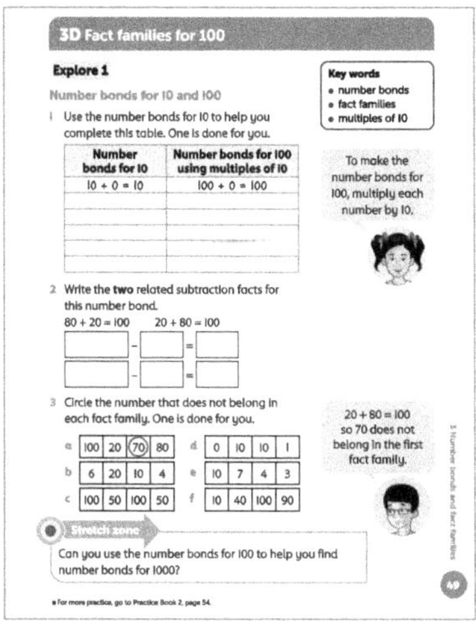

Differentiation

Supporting: Use a counting stick and base-10 equipment to model bonds for students.

Consolidating: Ask students to explain why a number does not belong in one of the number sets in question 3. Can they give you examples of facts to support their answer?

Extending: Ask students to generalise and explore fact families to 1000.

Stretch zone: *Can you use the number bonds for 100 to help you find number bonds for 1000?*

This activity allows students to extend their thinking to form number bonds for larger numbers. Ask students to think about the relationship between tens and hundreds first, in other words that 100 is 10 times bigger than 10. Then consider how many times 1000 is bigger than 100 (10 times).

 Reflection time

Choose one of the number bonds for 10 and its related number bond for 100, e.g. 3 + 7 = 10 and 30 + 70 = 100.

Ask students to write all the related subtraction facts. Can they explain why they know they are correct? Encourage them to think about the part-whole diagrams from earlier lessons to give their answer. (For example, '30 and 70 are

the parts and together they make the whole, 100. When I subtract, I take away a part from the whole to make the other part, so I know they are correct.')

Repeat with other bonds.

Practice Book: Students can complete page 54 of the Practice Book. They can do this directly after the main activity, as homework, or as the focus of a separate mathematics session to help students consolidate their learning and build fluency. Encourage students to refer back to the example each time to ensure that they are structuring their number sentences correctly. To complete the Stretch zone, encourage students to think about their number bonds for 20 but explain that they will be thinking about this more in the next lesson.

Differentiated outcomes	
All students	should count up in tens to 100 and find number bonds for 100 and as well as 10.
Most students	will find all number bonds for multiples of 10 to 100 using their knowledge of bonds for 10.
Some students	may use their knowledge of the relationship between 100 and 1000 to find number bonds for 1000.

Answers

Student Book page 49

1 10 + 0 = 10   100 + 0 = 100

9 + 1 = 10   90 + 10 = 100

8 + 2 = 10   80 + 20 = 100

7 + 3 = 10   70 + 30 = 100

6 + 4 = 10   60 + 40 = 100

5 + 5 = 10   50 + 50 = 100

2 100 − 80 = 20   100 − 20 = 80

3 **a** 70   **b** 20   **c** 100   **d** 1   **e** 4   **f** 40

Practice Book page 54

1 80 + 20 = 100   100 − 20 = 80   100 − 80 = 20

2 70 + 30 = 100   100 − 30 = 70   100 − 70 = 30

3 60 + 40 = 100   100 − 40 = 60   100 − 60 = 40

4 50 + 50 = 100   100 − 50 = 50   100 − 50 = 50

5 40 + 60 = 100   100 − 60 = 40   100 − 40 = 60

6 30 + 70 = 100   100 − 70 = 30   100 − 30 = 70

7 20 + 80 = 100   100 − 80 = 20   100 − 20 = 80

8 10 + 90 = 100   100 − 90 = 10   100 − 10 = 90

Stretch zone: Students' answers will vary.

3D Fact families for 100

Explore 2 Student Book page 50 • Practice Book page 55

Specific learning focus

- Find fact families for number bonds for 100 in multiples of 10.

Global skills

- **Creative skills:** exploring
- **Real-world skills:** presenting information
- **Interpersonal skills:** communication/teamwork

Key vocabulary

- number bonds, fact families, multiples of 10

Resources

- large 100-square for front of class, small 100-squares
- number cards 10–100 in multiples of 10
- digit cards 0–9, base-10 equipment

Language support

Provide the speaking frame 'I know _______ because _______' to support students to try to explain how they knew that they had completed all the possible fact family houses for number bonds of multiples of 10 to 100. You could model an answer for one number bond before asking students to use the same structure for a different number bond. Ask short, closed questions to prompt the next part of a longer response.

 Introductory activity

Ask each pair to list quickly all the number bonds for 100 made from multiples of 10, for example, 80 + 20 = 100. Challenge different pairs of students to record a set of three numbers in a part-whole diagram on the board and then use this to tell the class the fact family for those numbers. Students can then complete the fact family houses on page 50 of the Student Book.

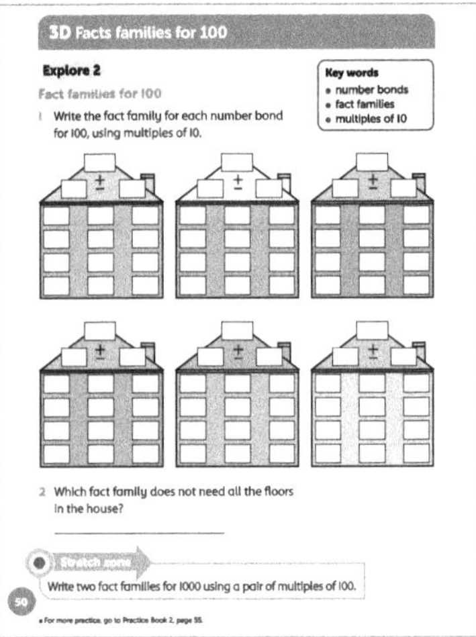

 Main activity

Direct students to look at the Practice Book page 55. Explain that you will be using fact families for 20 to help them find fact families for 200. Start by writing 17 on the board and asking students to give you the number that will add to 17 to make 20 (3). Complete the number bond as 17 + 3 = 20 and ask students for the rest of the fact family for this bond (3 + 17 = 20, 20 − 17 = 3, 20 − 3 = 17).

Now write 170 and see whether students can tell you which number they can add to that to make 200. They might see a pattern in the numbers from 17 + 3 = 20, or they may be able to count on in tens from 170 to get to 200 (30). Together complete the fact family for this bond on the board, then ask students to write the fact family for this in the first house on the Practice Book page.

Choose another number from 0 to 20 (e.g. 14) and repeat the process, arriving at the fact family for 140 + 60 = 200. Continue to complete the fact family houses on page 55 of the Practice Book as a class.

Differentiation

Supporting: Encourage students to work systematically and always record facts in the same order.

Consolidating: Ask students whether they can count in tens from 100 to 200 to help them.

Extending: Students can create their own 2-digit number, write a number bond for it and the corresponding fact family.

Stretch zone: *Write two fact families for 1000 using a pair of multiples of 100.*

You may want to support students to find a number bond for 1000 building on facts for 10 and 100: for example, 7 + 3 = 10, 70 + 30 = 100, 700 + 300 = 1000. Use base-ten equipment to represent the link and then ask students to write the other three related facts for each number bond they have chosen.

 Reflection time

Draw up a list of how many fact families there are for each multiple of 10 from 100 to 200. Ask students whether they can see a pattern. Challenge them to explain the pattern. It may be useful to think about odd and even numbers of bonds to help explain the pattern: for example, there are 11 number bonds for 100 (in tens) and 11 fact families, but one of the fact families has only two facts instead of four (as 50 + 50 = 100 can't be reversed to make a different fact). There are 12 number bonds for 110 and 12 fact families, each with four facts.

Practice Book: Students complete page 55 of the Practice Book as the focus of the main activity, working on it together as a class.

Differentiated outcomes	
All students	should find some facts in each family for multiples-of-10 number bonds for 100.
Most students	will complete all the fact families for multiples-of-10 number bonds for 100.
Some students	may find a link between a number bond and how many number facts the fact family has.

Answers

Student Book page 50

1 $100 + 0 = 100$, $0 + 100 = 100$, $100 - 100 = 0$, $100 - 0 = 100$

$90 + 10 = 100$, $10 + 90 = 100$, $100 - 90 = 90$, $100 - 10 = 90$

$80 + 20 = 100$, $20 + 80 = 100$, $100 - 80 = 20$, $100 - 20 = 80$

$70 + 30 = 100$, $30 + 70 = 100$, $100 - 70 = 30$, $100 - 30 = 70$

$60 + 40 = 100$, $40 + 60 = 100$, $100 - 60 = 40$, $100 - 40 = 60$

$50 + 50 = 100$, $100 - 50 = 50$

2 The number bond $50 + 50 = 100$ only has two number facts so does not use all the floors in the house. It does not have a reverse addition (such as $40 + 60 = 60 + 40$), and only has one subtraction fact $100 - 50 = 50$.

Practice Book page 55

Check that students have correctly completed six different fact families for number bonds for 200 as part of the whole-class work, for example, $60 + 140 = 200$.

Stretch zone: At this stage, students may describe the relationship between number bonds for 20 and 200 as 'adding a zero to each part of a number bond to 20 gives the answer to number bonds for 200'.

3 Number bonds and fact families

Connect 1   Student Book page 51

Big Idea

Number bonds are any two numbers that add together to make another number. Fact families contain all possible addition and subtraction facts for a set of three numbers.

Global skills

- **Creative skills:** problem solving/investigating
- **Real-world skills:** presenting information
- **Interpersonal skills:** communication / teamwork

Key vocabulary

- number bonds to 20, number sentence, grid

Resources

- set of digit cards 1–9
- collection of boxes with between 11 and 20 spaces to hold items (e.g. egg boxes or similar); squared paper
- cubes and counters

Language support

Listen to how students describe their investigation. Support them to be systematic by suggesting that they change only one thing at a time. This way, the vocabulary used will become repetitive and students will begin to understand the key words.

 **Introductory activity**

Draw a large 2×6 **grid** on the board.

Ask one student to pick a digit card (e.g. 4) and to colour in that number of squares on the grid. Ask students to discuss in pairs and come up with a number story and write a matching number sentence. For example, 'There are 4 coloured squares and 8 squares that are not coloured. That makes 12 squares altogether': $4 + 8 = 12$. Ask pairs to share their stories and number sentences with the class.

Repeat the activity four times, encouraging students to come up with subtraction as well as addition sentences: for example: 'There are 12 squares. I coloured 5 of them, so there are 7 left uncoloured.': $12 - 5 = 7$.

 **Main activity**

Ask students to look at the activity on page 51 of the Student Book.

Organise students into groups of four. Allocate or allow the group to choose a box size to work on.

As students explore the activity in the Student Book, ask questions such as: *How can I tell which type of cake this number means? How do you know you have all the possibilities?*

When a group has found all the possibilities with one box, ask them to explore a different-sized box.

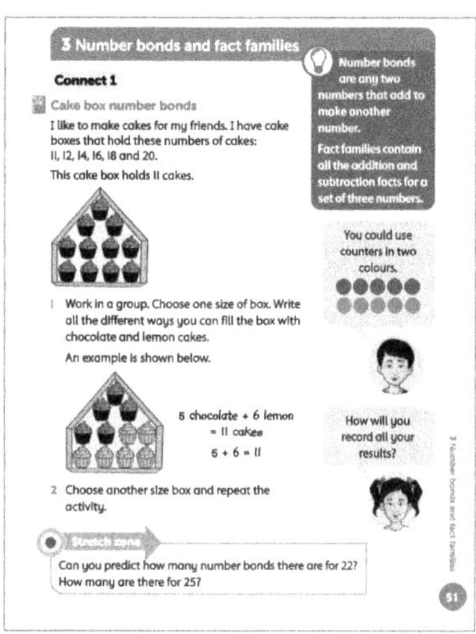

Ask students to imagine there are three different kinds of cakes – chocolate, lemon and orange. *How many different ways could I fill a box of 11 cakes now?*

Differentiated outcomes	
All students	should suggest some possibilities for number pairs.
Most students	will begin to work systematically to record all possible answers.
Some students	will explain how they know they have all the possibilities.

Answers

Student Book page 51

Check that students have correctly organised the cake boxes and put the correct number of cakes in each box. For example, for 11 and 12 cakes:

1 1 chocolate + 10 lemon   6 chocolate + 5 lemon

2 chocolate + 9 lemon   7 chocolate + 4 lemon

3 chocolate + 8 lemon   8 chocolate + 3 lemon

4 chocolate + 7 lemon   9 chocolate + 2 lemon

5 chocolate + 6 lemon   10 chocolate + 1 lemon

2 1 chocolate + 11 lemon   7 chocolate + 5 lemon

2 chocolate + 10 lemon   8 chocolate + 4 lemon

3 chocolate + 9 lemon   9 chocolate + 3 lemon

4 chocolate + 8 lemon   10 chocolate + 2 lemon

5 chocolate + 7 lemon   11 chocolate + 1 lemon

6 chocolate + 6 lemon

Differentiation

Supporting: Encourage students to use cubes or counters to help them to find possible combinations, and to use a systematic approach to find all bonds.

Consolidating: Ask students to explain how they plan to record their results. Then ask, *Why? How will it help you to find all possible combinations?*

Extending: Ask students to explore the patterns made by the number bonds and investigate other bonds. Can they use what they have learned so far to predict how many bonds there would be for 13 cakes? 15 cakes?

Stretch zone: *Can you predict how many number bonds there are for 22? How many are there for 25?* Check that students can extend their learning to find how many number bonds there are for various numbers.

 Reflection time

At the end of the lesson ask groups to report their findings to each other. Draw up a table of how many possibilities there are for filling each box size. Ask students to discuss the question: *Is there a pattern?*

3 Number bonds and fact families

Connect 2   Student Book page 52

Big Idea

- Number bonds are any two numbers that add together to make another number. Fact families contain all the addition and subtraction facts for a set of three numbers.

Global skills

- **Creative skills:** exploring/investigating
- **Real-world skills:** presenting information
- **Interpersonal skills:** communication/teamwork
- **Self-development skills:** reflecting on learning

Key vocabulary

- addition, subtraction, fact family, multiple

Resources

- none required

Language support

You may need to explore each of the fact family statements with students to be sure that they are clear about what the statement is saying. Go through the model answer on page 52 of the Student Book with them and talk about how it supports the answer that the statement is false.

Introductory activity

Ask students in pairs to define a 'fact family'. Take feedback from the pairs and agree as a class on the most appropriate definition. This might come from combining several definitions.

Main activity

Ask students to look at page 52 of the Student Book and to discuss the statements about fact families. Look at the first question together and agree that the statement is false. *Can you give any more examples of why it is false?*

In groups, they should decide whether each of the two remaining statements is true or false, and then write at least three examples to support their answer. For question 1, students should write any fact family that has more than two facts, for example: $3 + 7 = 10$, $7 + 3 = 10$, $10 - 7 = 3$, $10 - 3 = 7$. In question 2, encourage students to represent a fact family using a part-whole diagram and then link each part with a fact from a fact family.

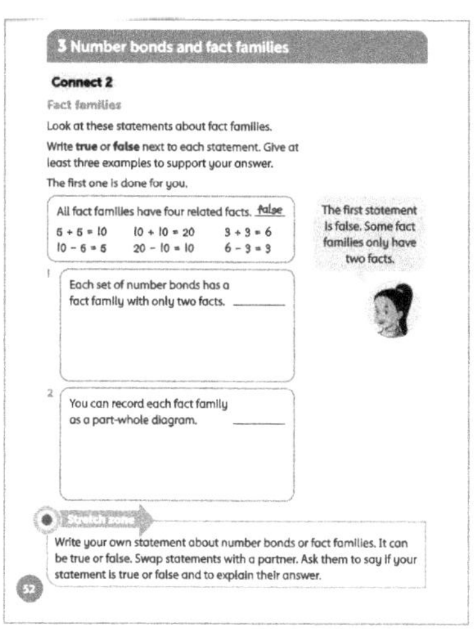

Differentiation

All students should work in mixed-attainment groups for this activity. You may have to help write the answers for those less confident in writing but all students should join in the discussion. For consolidation, ask students to explain to you the conclusions made by the group.

Stretch zone: *Write your own statement about number bonds or fact families. It can be true or false. Swap statements with a partner. Ask them to say if your statement is true or false and to explain their answer.*

Encourage students to think about what they know about fact families and number bonds. Ask them to try changing true statements to make them false. Check that students have written suitable statements and listen to their explanations.

Reflection time

Invite different groups to explain how they decided if a statement was true or false. Check whether the other groups reached the same conclusion and share examples used.

'Each set of number bonds has a fact family with only two facts' is false. For example, the set of number bonds for 5 have four facts in every fact family:

$5 + 0 = 5, 0 + 5 = 5, 5 - 0 = 5, 5 - 5 = 0$

$4 + 1 = 5, 1 + 4 = 5, 5 - 1 = 4, 5 - 4 = 1$

$3 + 2 = 5, 2 + 3 = 5, 5 - 2 = 3, 5 - 3 = 2$

It is only when there is an odd number of number bonds that one of the number bonds has a fact family of only two facts.

'You can record each fact family as a part-whole diagram' is true.

Different ways of recording fact families include spider diagrams and tables.

Differentiated outcomes	
All students	should join in the discussion and write the answers with help.
Most students	will give correct 'true/false' answers to both questions.
Some students	may give correct answers and provide examples to justify the answer.

Answers

Student Book page 52

1 False: Even numbers (e.g. 6) have one number bond ($3 + 3 = 6$) that has only two facts in its fact family. Odd numbers (e.g. 5) have four facts in the fact family for every number bond (e.g. there are four facts for each of the bonds: $5 + 0 = 5$, $4 + 1 = 5$, $3 + 2 = 5$).

2 True: For example, for the number bond $4 + 6 = 10$, all four facts ($4 + 6 = 10$, $6 + 4 = 10$, $10 - 6 = 4$, $10 - 4 = 6$) can be represented by the following part-whole diagram:

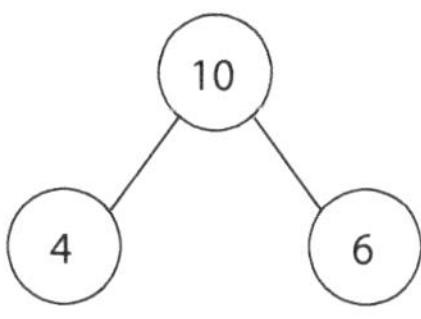

3 Number bonds and fact families

Review Student Book page 53 • Practice Book page 56

Global skills

- **Creative skills:** problem solving/exploring
- **Interpersonal skills:** communication
- **Self-development skills:** reflecting on learning

Student Book

With young children, assessment activities are most effective when carried out as an everyday classroom activity. Students should be able to work on developing their knowledge and understanding of number bonds and then represent them as diagrams or in sentences. They should be able to write fact families for given number bonds, sometimes using concrete resources to help them.

Watch as students increase their knowledge and use it to form number bonds and fact families for larger numbers. When assessing students, ask questions such as: *How did you complete the number bond? How did you check that you are correct? Have you found all the facts for this family? How do you know?*

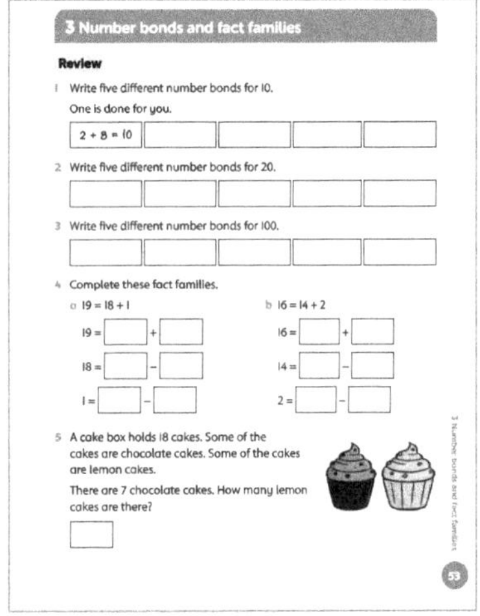

Answers

Student Book page 53

1 $2 + 8 = 10$ $3 + 7 = 10$ $4 + 6 = 10$

$5 + 5 = 10$ $1 + 9 = 10$

2 $1 + 19 = 20$ $2 + 18 = 20$ $3 + 17 = 20$

$4 + 16 = 20$ and so on

3 $10 + 90 = 100$ $20 + 80 = 100$ $30 + 70 = 100$

$40 + 60 = 100$ $50 + 50 = 100$ and so on

4 $19 = 18 + 1$ $16 = 14 + 2$

$19 = 1 + 18$ $16 = 2 + 14$

$18 = 19 - 1$ $14 = 16 - 2$

$1 = 19 - 18$ $2 = 16 - 14$

5 11 lemon cakes

Practice Book

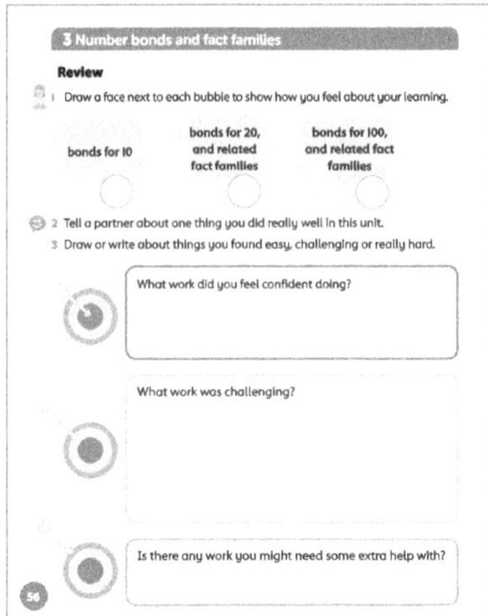

It is appropriate to complete this Practice Book review as a whole-class discussion. You may choose to keep a record of the class discussion or a copy of the review page for your own records. The review provides an opportunity for students to reflect on their learning from the unit, to discuss any areas of mathematics that they feel went particularly well, and any areas that they feel less confident about. Ensure all students have a copy of the Student Book as a reminder of the areas of mathematics that they have worked on in this unit.

Allow students plenty of time for discussion before asking them to complete the Practice Book page individually, and then, if appropriate, to share their responses with the rest of the class. If students complete this self-assessment at home, encourage them to discuss this with adults. Make a note of areas that students still feel unsure about.

Additional material

There are additional end-of-unit assessments available on the *Oxford Owl* website.

4 Addition

Big idea

By this stage, students will have begun to develop a sense of number. They will be aware that they can add numbers in any order but cannot subtract numbers in any order. Addition and subtraction can be modelled on number lines and the recognition of number patterns can be developed using the 100-square.

Students need to recognise place value and begin to memorise key facts such as number bonds for 10, 20 and 100 in order to carry out mental and written calculations efficiently. When choosing a calculation method, students should understand that there can be many ways to do a calculation and that the numbers involved may help determine which is the most efficient. Students need to be able to express their mathematical thinking clearly to discuss methods, because not all methods are useful in all situations.

Try to avoid making statements that are not always true or taking generalisations too far. When talking through a method for subtraction such as 15 – 7, do not say '5 – 7 doesn't go'. Instead, you could count back on a number line or subtract 5 and then a further 2 using number bonds for 10. Other common expressions such as 'always take the smallest from the largest' and 'always put the biggest number in your head and add on the smallest' are not always appropriate approaches so be wary of stating them as generalisations.

Look out for

- **Students who do not associate reciting counting on or back in tens (4, 14, 24, 34… or 85, 75, 65, 55…) with adding or subtracting tens.**
- **Students who find the move into calculating difficult and want to continue to count everything.** It is important that they learn number bonds and when to use them; recognising number bonds for 10 will help with addition and subtraction as well as finding a missing number.
- **Students who see numbers as in fixed positions in a calculation and find it hard to move them around to make calculations easier.** When students are calculating, allow time for them to consider and reflect on the most efficient way to calculate an answer. Encourage discussion in pairs to expose students to different approaches to solving calculations and to consolidate their own thinking.

Blindly following a rule leads to misunderstandings. It is useful for students to carry out the same calculation in different ways so that they see that various methods are appropriate. Build in time for students to talk about what they and other students are doing to help them become more flexible in their thinking.

Possible misconceptions

- **Students may think that the equals sign is a symbol used to show what the answer is to a calculation.** They need to develop some flexibility of thinking so that they can recognise that '=' means that both sides of the sign have the same value rather than 'here comes the answer'.
- **Students may assume that there is only one way to solve a problem, either using subtraction or addition.** Provide students with a variety of word problems to solve that can be solved in more than one way. Allow time for students to talk about what they and other students are doing to help them become more flexible in their thinking.
- **Students may use the starting number when counting on along a number line.** Provide additional practice counting along a number line, emphasising counting each jump rather than the numbers.

Key vocabulary

- add, adding, addition, tens, ones, total, partition, strategy, start, end, calculation, same, different
- number bonds, +, –
- subtract, subtraction, count on, count back, number line, 2-digit number, more, less, single-digit number, number sentence
- re-grouping, altogether, difference, equal, take away, missing number, balance

Learning focus	Learning outcomes (the ENC objectives)
Adding small numbers	Recall and use addition and subtraction facts to 20 fluently. Show that addition of two numbers can be done in any order (commutative). Solve problems with addition: using concrete objects and pictorial representations, including those involving numbers, quantities and measures; applying their increasing knowledge of mental and written methods.
Add and subtract 1-digit numbers	Add and subtract 1-digit numbers using concrete objects, pictorial representations and mentally.
Add two 2-digit numbers	Add and subtract numbers using concrete objects, pictorial representations, and mentally, including two 2-digit numbers.
Finding the difference	Recall and use subtraction facts to 20 fluently, and derive and use related facts up to 100.
Missing numbers	Recognise and use the inverse relationship between addition and subtraction and use this to check calculations and solve missing number problems.

4 Addition

Big question

What is the best way to add or subtract numbers?

Global skills

- **Creative skills:** exploring
- **Interpersonal skills:** communication/teamwork

Key vocabulary

- add, addition, total, partition, tens, ones, strategy, calculation, same, different

Resources

- large number cards 15, 27, 30 and 41
- mini whiteboards and markers
- base-10 equipment or Numicon shapes
- number lines

Language support

Listen carefully to the way the students describe their strategies. They may say, 'I just added them.' Question them carefully so that they get used to describing their thinking. *Which number did you start with? Did you imagine a number line? Could you have used partitioning to add the numbers together? How did you know you had the right answer?*

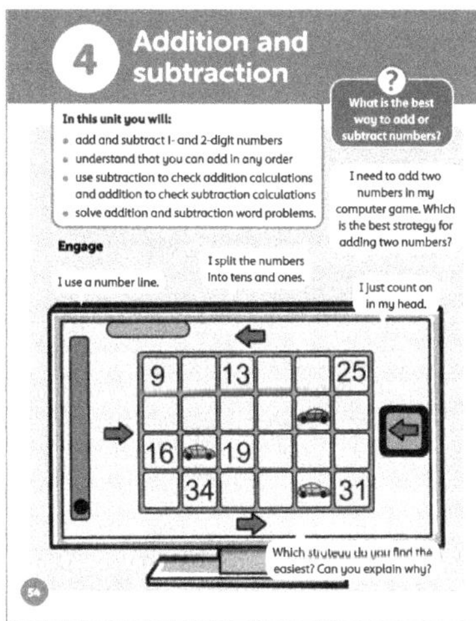

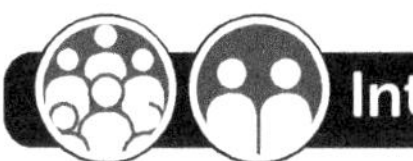 **Introductory activity**

Use large numbers cards 15, 27, 30 and 41 for this activity. Hold up the '15' card. Ask pairs to write as many **addition calculations** as they can in one minute with the **total** 15. Take feedback from the pairs until you have

five different calculations. Make sure that you discuss the strategy for adding each time. Ask each student: *How did you work that out?* Ask whether anyone else had the same calculation and to explain how they worked it out. This is a good way to start thinking about different strategies.

Repeat for the other numbers on the cards.

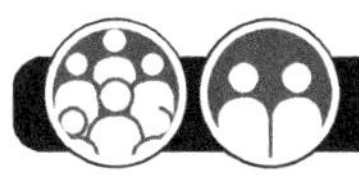 **Main activity**

Ask students to look at page 54 of the Student Book (display on an IWB, if possible). Ask pairs to look at the speech bubbles. Explain that each of the **different** methods, or ways, used for **adding** is called a **strategy**. Ask students to discuss in their pairs which of the strategies mentioned they are familiar with, which they prefer to use and why.

One student in each pair selects two of the numbers and adds them together. They then describe how they carried out the calculation to their partner. They should repeat this, taking turns, until they have described five calculations each. As students work on this, make a note of the different strategies students are using so that you can model a range of strategies in reflection time.

Differentiation

Supporting: Model calculating totals of pairs of the numbers in the game using tens-rods and ones-cubes (to represent **tens** and **ones**) and a number line. Encourage students to carry out addition calculations with 9 and 31, 16 and 13.

Consolidating: Support students in describing their strategies for a wider range of numbers, using the questions outlined in the Language support section above.

Extending: Ask students to try different strategies to add the same pair of numbers. They should decide which is the most effective for this particular pair of numbers and explain why. Encourage them to calculate with all the numbers.

 **Reflection time**

Ask pairs of students to come to the front of the class and explain the strategy they used for one of their calculations. Make sure that you include examples of counting on, partitioning and using a number line. You can have concrete resources such as base-10 equipment or Numicon shapes available and have number lines drawn for students to use. If students haven't used these methods, introduce them as alternative strategies and explain that later lessons will support the students in developing new strategies for addition.

4A Adding small numbers

Discover Student Book page 55 • Practice Book page 57

Specific learning foci

- Add three small numbers together.
- Understand that addition can be done in any order.

Global skills

- **Creative skills:** exploring/investigating
- **Real-world skills:** presenting information
- **Interpersonal skills:** communication/teamwork

Key vocabulary

- addition, add, number bonds

Resources

- sets of large number cards 0–10: set of small digit cards 1–9 per pair of students
- base-10 equipment, cubes or counters
- three dice for each pair of students

Language support

Display a chart of the number bonds for 10. Include a speech bubble saying, 'We can use our number bonds to 10 to help us add bigger numbers.' Remind students that they have done this previously.

Introductory activity

Ask 11 students to come to the front of the class. Give each student a large number card 0–10. Ask the rest of the students to suggest which of the students can pair with each other so that their number cards add together to make 10. Repeat until the student holding card 5 is left over. Write a list of the pairs of numbers on the board: 1 and 9, 2 and 8, 3 and 7, 4 and 6.

Now say, *Oh dear, 5 is left over and on its own. Let's ask number 5 to join one of the pairs.*

Ask the class to suggest a pair for 5 to join. The student holding 5 should join the selcted pair, say 6 and 4. Then write '6 + 4 + 5 =' on the board and ask the 11 students to return to their places. *In pairs, can you work out the total?* Provide students with concrete resources (e.g. tens-rods and ones-cubes), paper and pencils and observe them working. Ask students to share their strategies but don't comment on them at this stage. Move the student holding number 5 to join another pair and write the **addition** on the board, for example 8 + 2 + 5. Ask pairs to work out that total, then repeat a third time. *What do you notice? Why?* Students should notice that the total is always 15 because the other pairs make ten and 10 + 5 = 15.

Main activity

Write the numbers 8, 7 and 2 on the board. Ask students to work in pairs to calculate the total. Ask them to write down as many different addition calculations as they can using 8, 7 and 2. There are six possibilities:

8 + 7 + 2	7 + 2 + 8
8 + 2 + 7	2 + 7 + 8
7 + 8 + 2	2 + 8 + 7

Ask students what they notice about the answers to these calculations. They should recognise that the totals are all the same (17). *Which are easier to calculate? Can you explain why?* Students are likely to suggest 8 + 2 + 7 and 2 + 8 + 7 because 8 and 2 make 10. *If we know that all these addition sentences make 17, what does that tell us about the order we can add the numbers?* Agree that they can be added in any order. *This means that we can add numbers in the order that we find easiest, not just the order they are written.*

Ask students to look at page 55 in the Student Book together. Talk through how to play the games: the first uses a set of dice to provide three digits to add mentally and write the answer; the second involves picking three digits from a set of cards and adding them. In both cases, students work in pairs, taking turns to choose the digits and scoring points for whoever records the calculation faster. As students are playing, ask questions such as: *Will you/did you add the numbers in order? Are there any pairs that are easier to add first than others? Can you see any pairs of numbers that make 10?*

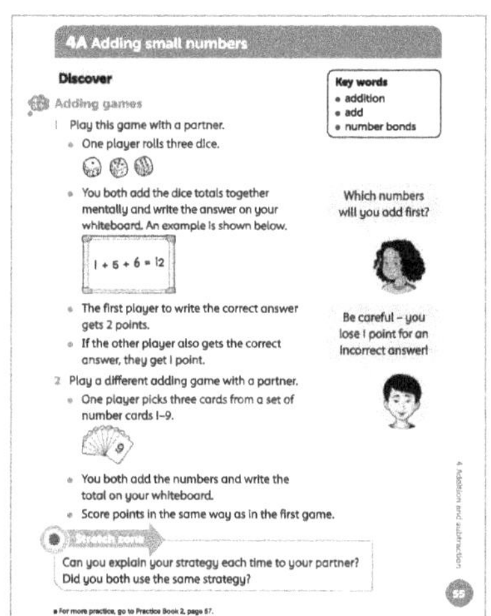

Differentiation

Supporting: Use concrete resources to support students with the additions, explaining how to total the numbers and modelling the mathematical talking.

Consolidating: Ask students to identify any bonds to 10 to help them to calculate the totals.

Extending: Encourage students to calculate in different ways, for example, find bonds to 10, or add the larger digits first.

Stretch zone: *Can you explain your strategy each time to your partner? Did you both use the same strategy?*

Help students to see that there are different strategies they can use when adding, and to look for easy pairs of numbers first, for example, finding any bonds to 10 and then counting on to the total.

 **Reflection time**

Ask different students to tell you which **number bond** for 10 they noticed in any particular question. Check the different ways that students completed one of the totals in the second part of the activity. This will help students to see that there are often different ways of doing the same thing.

Students could take turns to shuffle a set of digit cards and turn one card over at a time until they spot a number bond for 10. They then find the total of the cards turned over.

Practice Book: Students complete page 57 of the Practice Book. They can do this directly after the main activity, as homework, or as the focus of a separate mathematics session to help students consolidate their learning and build fluency. Students may struggle to add the numbers mentally. Encourage them to try, talking through their strategy out loud, and then check with concrete resources such as cubes or counters.

Differentiated outcomes	
All students	should calculate up to 20 using concrete resources.
Most students	will notice bonds to 10 if they appear in order such as $8 + 2 + 3$.
Some students	may notice bonds to 10 in number sentences such as $2 + 3 + 8$.

Answers

Student Book page 55

Check that students have added the numbers from the dice and cards correctly and recorded their examples.

Practice Book page 57

Check that students' additions are correct. The totals may be greater than 10.

4A Adding small numbers

Explore 1   Student Book page 56 • Practice Book page 58

Specific learning foci

- Add four or five small numbers together.
- Understand that addition can be done in any order but that subtraction cannot.

Global skills

- **Creative skills:** exploring
- **Interpersonal skills:** communication/teamwork

Key vocabulary

- addition, add, number bonds

Resources

- set of large digit cards 0–9
- concrete resources (e.g. Numicon shapes, interlocking cubes, ten-frames, number rods, base-10 equipment – tens-rods and ones-cubes)
- 0–20 number lines

Language support

Include an example of adding three numbers together using number bonds for 10 in a classroom display. Show at least two different ways of finding the total. Draw a speech bubble saying 'I can make 10 with 3 numbers' alongside a visual record, linking to a concrete representation such as interlocking cubes, number rods or Numicon shapes.

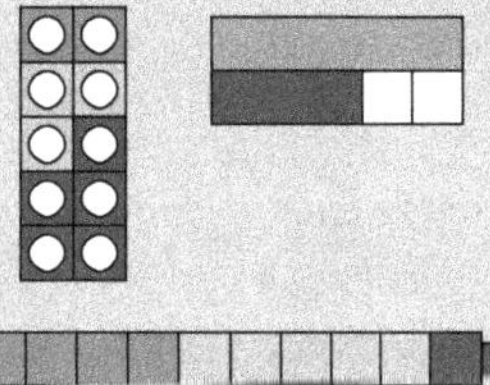

Introductory activity

Invite five students to the front and give each student a digit card, such as 1, 2, 7, 8 and 9. Ask the other students to spot number bonds for 10, pairing up students holding these digits. Find the total of the five numbers. Move on to ask whether students can also see three numbers that make 10 (1, 2, 7). Explain that sometimes looking for three numbers to make 10 may help them. Repeat with five different students after shuffling the cards, recording the outcomes in a list on the board.

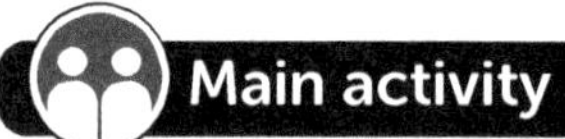

Main activity

Remind students that when looking for bonds for 10 they can put the numbers in any order. Ask students to look at page 56 of the Student Book (display it on the IWB if possible) and look at the example in question 2. *Which numbers are circled? Why?* Write the calculation again, re-ordering the numbers, making clear that students can do this in their heads. Model this with concrete resources (e.g. Numicon shapes, ten-frames or number rods) showing how the total is the same regardless of the order. Ask students to work in pairs on the Student Book activities. Encourage them to look together to spot bonds for 10 to simplify their calculations. Provide concrete resources and/or 0–20 number lines for students to check their calculations.

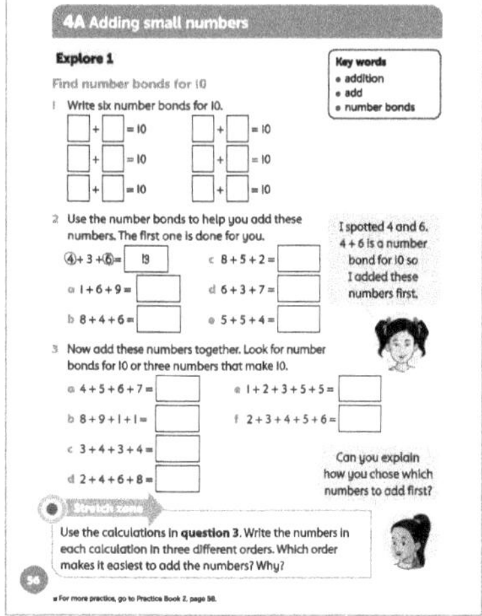

Differentiation

Supporting: Give students concrete resources to support addition and consolidate bonds for 10.

Consolidating: Ask students to find bonds to 10 when adding three numbers.

Extending: Encourage students to try to add four or more small numbers together, looking for bonds to 10 to help them.

Stretch zone: *Use the calculations in question 3. Write the numbers in each calculation in three different orders. Which order makes it easiest to add the numbers? Why?*

This activity reinforces the idea of adding a set of numbers in different ways. Check that students have understood that the order of addition makes no difference to the total.

 ## Reflection time

Remind students that the total is the same, whichever order they add the numbers in. Ask different pairs to share how they found the totals. Ask other students whether they added in a different way and to share their strategy.

Challenge some students to add five digit cards together, looking for number bonds for 10 and any three numbers that make 10.

Practice Book: Students can complete page 58 of the Practice Book. They can do this directly after the main activity, as homework, or as the focus of a separate mathematics session to help students consolidate their learning and build fluency. They should use concrete resources to help them to add five numbers together. An adult can record the answers for the Stretch zone, if necessary.

Differentiated outcomes	
All students	should add three numbers using concrete resources.
Most students	will reorder calculations to make them easier by finding bonds for 10.
Some students	may calculate mentally using number bonds for 10 to help.

Answers

Student Book page 56

1 $5 + 5 = 10$ $8 + 2 = 10$

   $6 + 4 = 10$ $9 + 1 = 10$

   $7 + 3 = 10$ $10 = 0 = 10$

2 **a** 16   **c** 15   **e** 14

   b 18   **d** 16

3 **a** 22   **c** 14   **e** 16

   b 19   **d** 20   **f** 20

Practice Book page 58

Answers will vary. Check that students have used the correct numbers and that the additions they have made are correct.

Stretch zone: Students should be able to explain that adding to ten first is easier as you only need to add 1 to the tens column and then count up the remaining numbers.

4A Adding small numbers

Explore 2
Student Book page 57 • Practice Book page 59

Specific learning focus

- Understand that changing the order of addition does not change the total.

Global skills

- **Creative skills:** exploring
- **Interpersonal skills:** communication/teamwork

Key vocabulary

- addition, add, total, +, =, number bonds

Resources

- number cards 0–10
- tabletop 0–20 number lines
- concrete resources such as Numicon shapes, base-ten apparatus, ten-frames and counters or interlocking cubes
- five different-coloured large hoops, five small bean bags

Language support

Display the words and symbols associated with addition, for example, 'add', 'total', '+', '=' and refer to these as students are working.

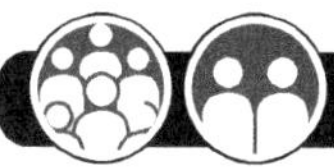 **Introductory activity**

Choose four number cards from a set of 1–9, for example, 1, 4, 5, 7. Ask students to work in pairs or small groups and to use these four numbers to make two pairings, for example 1 and 5, 4 and 7, and then find these two totals: $1 + 5 = 6$ and $4 + 7 = 11$. They then need to add the two numbers to find the overall total.

Now ask them to make new pairings within the same group of numbers, for example, $1 + 4$, $5 + 7$, giving 5 and 12. Finally, make the other pairings: $1 + 7$, $4 + 5$ giving 8 and 9.

Did you notice anything about the totals of the numbers each time? Students should notice that the total is the same each time: $6 + 11 = 17$, $5 + 12 = 17$, $8 + 9 = 17$. Ask why they think this is. Students may suggest that, comparing the final additions, as one number increases, the other goes down by the same amount. Prompt students if necessary to think about the fact that all the additions result from adding the same four numbers, but in different orders.

Main activity

Play a bean bag game like the one in question 3 on page 57 of Student Book. Start with three bean bags and three hoops and invite a student to throw the bags into the hoops. The class should mentally add the numbers scored and you can check that they have the correct total.

Now add an extra hoop, perhaps with a score of 6 points, and ask a student to throw four bean bags. Ask students to add the scores to get the total. Encourage them to look for number bonds to 10. Take feedback from students on their totals and how they worked them out.

Ask pairs of students to complete the activities on page 57 of the Student Book. Encourage them to mentally add the numbers if possible, looking for number bonds to 10 to help them find the totals. Make concrete resources available to support students with their calculations. They work on questions in pairs, discussing their thinking as they work.

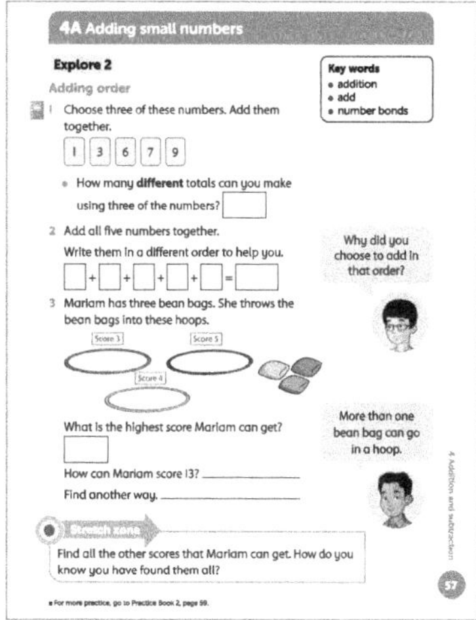

Differentiation

Supporting: Continue to have concrete resources available to support addition. Ten-frames and counters and number rods can help consolidate bonds to 10, and help with small-digit additions.

Consolidating: Ask students to explain how they did the calculation.

Extending: Encourage students to try to make different totals using six digit cards and choosing three at a time.

Stretch zone: *Find all the other scores that Mariam can get. How do you know you have found them all?*

Support students to find the possible totals by being systematic and recording their solutions in an organised way. Model this by writing the totals on the board.

 **Reflection time**

Ask pairs of students to share with the class what five-number additions they made. Ask them to explain how they chose the different pairs from their digits. Were they systematic? Did they notice anything about the pairs of answers they came up with?

Practice Book: Students can complete page 59 of the Practice Book. They can do this directly after the main activity, as homework, or as the focus of a separate mathematics session to help students consolidate their learning and build fluency. Encourage them to use concrete resources and work systematically. For example, they might start with 7 and then use that to help them find all possible numbers for 8.

Differentiated outcomes	
All students	should understand that they can re-order calculations to simplify adding and use some number bonds to 10.
Most students	will re-order calculations and use number bonds to 10.
Some students	may find pairs of number bonds to 10 when adding 5 or more numbers.

4B Adding and subtracting 1-digit numbers

Discover
Student Book page 58 • Practice Book page 60

Specific learning focus

- Add and subtract a single digit to and from a 2-digit number using a number line.

Global skills

- **Creative skills:** exploring
- **Real-world skills:** presenting information
- **Interpersonal skills:** communication/teamwork

Key vocabulary

- addition, subtract, subtraction, number line, 2-digit number, tens, ones, count on, count back, more, less, column

Resources

- 0–100 and 0–30 number lines (one for each student)
- tens-rods and ones-cubes, interlocking cubes

Language support

Display a large 0–100 number line at children's eye level throughout the year. Include an example of an addition, showing the jumps and circling the start and end numbers. Add the number sentence and a speech bubble, saying 'We can use a number line to help us to add.'

Introductory activity

Give each student a 0–100 number line. Call out a number (e.g. 14) for students to put a finger on, then ask them to imagine that they are jumping along the **number line**. Ask them to **count on** or **count back** 1, 2, 3 or 4 from a particular number and tell you where they landed. Repeat several times, varying the language you use. Include **more**, **less**, count on, count back and so on. Ask students to explain how they used the number line to carry out the calculations.

Main activity

Draw or display a 0–30 number line on the board. Write $23 + 6 = [\ \ ]$ beneath this. *Talk to a partner. What do you think 23 and 6 make?*

Give each pair a 0–30 number line and ask them to add 6 to 23 by jumping on 6 from 23 on the number line to find the total. Agree the total (29) and complete the number sentence as 23 + 6 = 29 on the board. Give students tens-rods and ones-cubes to check each calculation. Invite students to describe any other strategies they used to find the total (e.g. I added the ones in my head counting on 3 from 6 to make 9 and then adding 20 to make 29.) Repeat with a subtraction example.

Students should then work on the calculations on page 58 of the Student Book in pairs and then discuss their answers, sharing strategies.

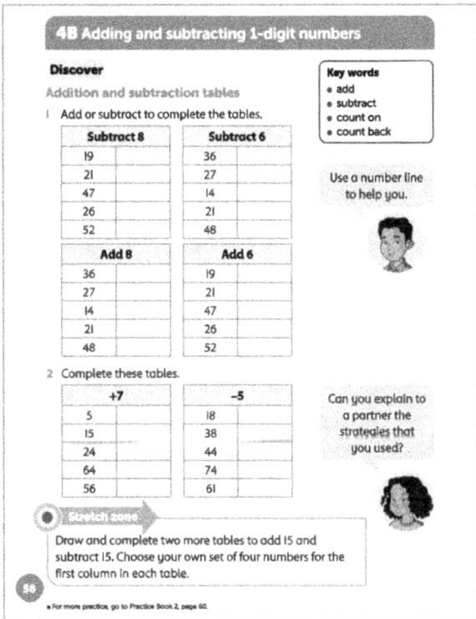

Differentiation

Supporting: Model the use of the number line, taking care with the counting on and back.

Consolidating: Ask students to add the numbers on the number line in a different order. *What does this show?*

Extending: Explore the strategy with adding or subtracting small **2-digit numbers** near a multiple of 10, for example, adding 21 or subtracting 28.

Stretch zone: *Draw and complete two more tables to add 15 and subtract 15. Choose your own set of four numbers for the first **column** in each table.*

Can students tell you how they chose their numbers? Did they use any strategies in particular to find their answers?

 **Reflection time**

Talk through each of the calculations, checking the totals. Ask students to explain to you how they did the calculations in each table. For example, to add 8, did they add 10 then subtract 2?

Practice Book: Students can complete page 60 of the Practice Book. They can do this directly after the main activity, as homework, or as the focus of a separate mathematics session to help students consolidate their learning and build fluency. Encourage students to consider what strategy they could use to solve each calculation. Can they show this strategy on the number line? For example, 'I know double 8 is 16. 7 is 1 less than 8 so 8 add 7 will be 1 less than 16. I start at 8 on my number line, do one jump to show + 8, landing on 16, and then jump back 1 to 15.'

Differentiated outcomes	
All students	should add single-digit numbers to 2-digit numbers by counting on with support of a number line and/or concrete resources.
Most students	will sometimes use a strategy other than counting on to add single-digit numbers to 2-digit numbers with support of a number line and/or concrete resources.
Some students	may use a mental calculation strategy other than counting on to add single-digit numbers to 2-digit numbers and use a number line or a concrete resource to check.

Answers

Student Book page 58

1   Subtract 8: 11, 13, 39, 18, 44

   Subtract 6: 30, 21, 8, 15, 42

   Add 8: 44, 35, 22, 29, 56

   Add 6: 25, 27, 53, 32, 58

2   +7: 12, 22, 31, 71, 63

   − 5: 13, 33, 39, 69, 56

Practice Book page 60

Check that students have marked the number lines with the jumps correctly for each calculation.

13 + 6 = 19

8 + 7 = 15

18 + 7 = 25

13 + 5 = 18

23 + 5 = 28

Stretch zone: Check that students have represented 18 + 7 correctly with a diagram of tens-rods and ones-cubes.

4B Adding and subtracting 1-digit numbers

Explore 1 — Student Book page 59 • Practice Book page 61

Specific learning foci

- Add and subtract a single-digit number to and from a 2-digit number by counting on and back using a number line.
- Understand that addition can be done in any order, but that subtraction cannot.

Global skills

- **Creative skills:** exploring
- **Interpersonal skills:** communication/teamwork

Key vocabulary

- addition, subtraction, number line, 2-digit number, tens, ones, start, end, number sentence

Resources

- large 0–30 number line for front of class

Language support

Add an example of a subtraction to the displayed number line, showing the jumps and circling the start and end numbers. Include the number sentence. Add a speech bubble saying, 'We can use a number line to help us to subtract.'

 **Introductory activity**

Draw the following addition table on the board.

+	4	7	9
17			
26			
32			

Ask students to discuss in pairs ways to find the totals. Take feedback on the different strategies that they used. Look for patterns such as pairs to a multiple of ten (17 + 3 and 24 + 6) or other patterns. (For example, the total when adding 7 will be 2 more than the total when adding 5.)

Repeat for a subtraction square. (The same numbers are used deliberately.)

−	4	7	9
17			
26			
32			

Again, look for patterns in the answers.

 **Main activity**

Show a 0–30 number line on the board. Mark a jump on the number line starting from 12 (circle the 12) and finishing on 21 (circle 21). Tell students that we are going to write a **number sentence** for the calculation shown on the number line. Start by asking what number it **starts** with (12). Then ask what number it **ends** on (21), and say that this is the answer to the calculation. Write on the board 12 + __ = 21. Now ask students to count the size of the jump (9). You can now complete the sentence as 12 + 9 = 21.

Ask students to look at the worked example on page 59 in the Student Book. *Can you explain what you see? What strategy would you use to add 12 and 7?*

Ask students to work individually on the question on page 59 and then check their answers with a partner. If they get different answers, they should check, agree the correct answer and be able to tell you the error that was made.

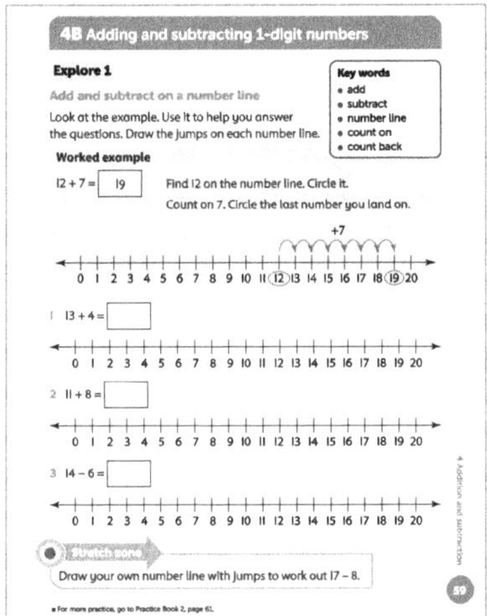

Differentiation

Supporting: Model the calculations on a number line. Focus on the start number and the jump.

Consolidating: Ask students to explain their strategies to you as they use the number line.

Extending: Students could make up their own addition or subtraction problems for partners to solve using the number line.

Stretch zone: *Draw your own number line with jumps to work out 17 − 8.*

Check that students have drawn and used an appropriate number line correctly to show how they calculated 17 − 8. For example, a 0–20 line will be suitable. Some students may be able to see that counting back 7 easily to get to 10 and then counting back 1 more gives the answer as 9, which is more efficient than counting back in ones from 17 to get to 9.

 ## Reflection time

Ask students to share their solutions by modelling the calculations on the class number line. Ask students to think about how they might have done the calculation without a number line.

Show students a marked number line with only the jump shown and remind them how to find the number sentence for the calculation it shows.

Practice Book: Students can complete page 61 of the Practice Book. They can do this directly after the main activity, as homework, or as the focus of a separate mathematics session to help students consolidate their learning and build fluency. Encourage students to highlight or circle the start and end numbers on their number lines to make it easier to see the size of the jump.

Differentiated outcomes	
All students	should calculate accurately using the number line.
Most students	will calculate confidently and accurately using the number line and be able to explain their strategy.
Some students	may be able to develop their own addition and subtraction problems for partners to solve using a number line.

Answers

Student Book page 59

1 $13 + 4 = 17$　　**2** $11 + 8 = 19$　　**3** $14 - 6 = 8$

Practice Book page 61

1 $15 + 7 = 22$　　**3** $19 + 7 = 26$　　**5** $20 + 9 = 29$

2 $17 + 7 = 24$　　**4** $16 + 9 = 25$

Stretch zone: Ask students to explain what makes their calculations easy or hard.

4B Adding and subtracting 1-digit numbers

Explore 2　Student Book page 60 · Practice Book page 62

Specific learning foci

- Add and subtract a single digit to and from a 2-digit number.
- Understand that addition can be done in any order, but that subtraction cannot.

Global skills

- **Creative skills:** exploring
- **Real-world skills:** presenting information
- **Interpersonal skills:** communication

Key vocabulary

- addition, subtraction, number line, 2-digit number, tens, ones, single-digit number

Resources

- large 0–30 number line for front of class; small 0–30 number line for each pair
- sets of tens-rods and ones-cubes (one set for each pair), interlocking cubes

Language support

Revise and reinforce the language or 'more' or 'less' by asking questions such as *Can you show me 6 more than 29? Can you show me 10 less than 30?*

 ## Introductory activity

Ask a student to call out a number less than 10 (e.g. 7). Say, *What is 7 less than 10? How would you record this as a number sentence?* Give students a minute to discuss with a partner and then take the answer (3). Then record it as a subtraction sentence, both $3 = 10 - 7$ and $10 - 7 = 3$ so that students become more used to seeing different arrangements of number sentences.

Repeat several times, speeding up the practice as you go along, including all bonds for 10 more than once. Include 'more than' questions as well.

 ## Main activity

Write the calculation $6 + 11$ on the board and represent it with, for example, tens-rods and ones-cubes or interlocking cubes. Ask *What do you notice about 11?* (It is 1 more than 10.) Model using the materials to show that adding 11 can be done by adding 1 and then adding 10, so $6 + 11$ becomes $6 + 1 = 7$, then $7 + 10 = 17$.

Now ask students to work out $8 + 11$ in the same way, by adding 8 and 1 to make 9, then $9 + 10$ for a total of 19.

Show students the calculation $20 - 9$ on the board. *What can you tell me about 9?* They should say that it is 1 less than 10. Use this information to explain how to **subtract** 9 by subtracting 10 and then adding 1 back on. Model this with base-10 equipment or interlocking cubes to show $20 - 9$ as $20 - 10 = 10$ and then $10 + 1 = 11$. Model on a number line, explaining each step of the calculation.

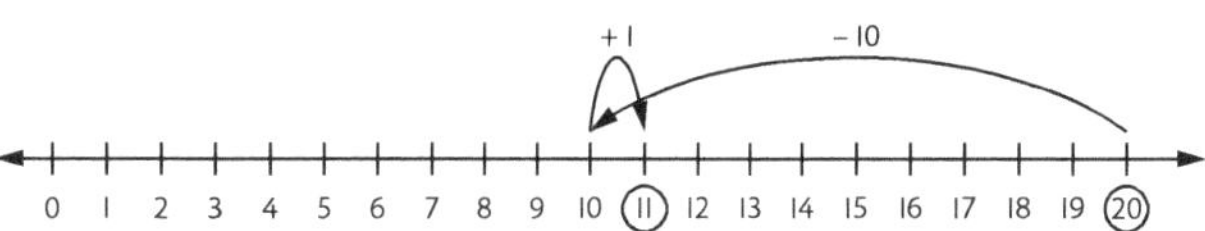

Now ask students to work out 37 − 9 in the same way, by subtracting 10 from 37 to get 27, then adding 1 back to 27 for a total of 28.

Now write 18 + 7 = __ and ask students to add 7 to 18 by jumping on 7 from 18 on their number lines and then to check using tens-rods and ones-cubes. Write down the number sentence as 18 + 7 = 25. Invite a pair to model how they swapped 10 ones-cubes for a tens-rod. *What size were your jumps? Did you do one big jump of 7? 7 jumps of one? Any other size of jumps?* If no students suggest it, model doing one jump from 18 to 20 and then another jump of 5 to 25.

Repeat with at least two more examples, this time subtracting a **single-digit number** from a 2-digit number. Include an example that crosses the tens boundary (e.g. 34 − 7).

Ask students to complete page 60 in the Student Book. Provide tens-rods and ones-cubes or similar concrete resources for students to represent the calculations. They should work individually and then check their answers with a partner. If they get different answers, they should check, agree the correct answer and be able to tell you the error that was made.

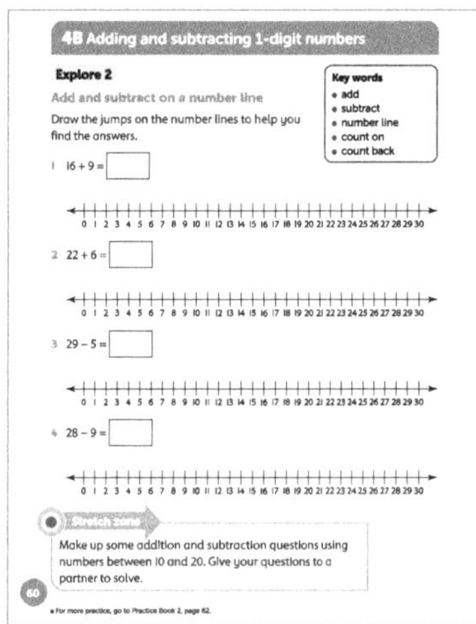

Differentiation

Supporting: Ask questions such as *What is your start number? Circle it on the number line. What is your jump number?* and so on.

Consolidating: Ask students to explain their strategies to you.

Extending: Students could make up their own addition or subtraction problems for partners to solve using the number line, perhaps up to 50. They might present the problems as a mystery number problem, for example, 'I make a jump of 6 and end at 12. Where did I start?'

Stretch zone: *Make up some addition and subtraction questions using numbers between 10 and 20. Give your questions to a partner to solve.*

Encourage students to use a number line to show how they solved the questions and explain their strategy to their partner.

Reflection time

Ask students to share their solutions, and to explain any errors that they made. Have a number line displayed so they can show their jumps. Ask whether anyone did different jumps. *Did anyone calculate the answer mentally first? Can you explain your strategy?*

Practice Book: Students can complete Practice Book page 62. They can do this directly after the main activity, as homework, or as the focus of a separate mathematics session to help students consolidate their learning and build fluency.

Differentiated outcomes	
All students	should calculate accurately using concrete resources alongside a number line.
Most students	will calculate confidently and accurately using a number line.
Some students	may calculate using a mental method and use the number line to check.

Answers

Student Book page 60

1 25

2 28

3 24

4 19

Practice Book page 62

1 16

2 14

3 15

4 20

5 21

Stretch zone: Students may represent this by showing 2 tens-rods and 9 ones-cubes, then circling 8 ones-cubes that are to be subtracted. The uncircled tens-rods and ones-cubes will show the answer of 21.

4C Adding two 2-digit numbers

Specific learning foci

- Know what each digit represents in 2-digit numbers.
- Add pairs of 2-digit numbers by partitioning numbers into tens and ones.

Global skills

- **Creative skills:** exploring
- **Real-world skills:** presenting information
- **Interpersonal skills:** communication/teamwork

Key vocabulary

- add, addition, total, number line, 2-digit number, partitioning, tens, ones, re-grouping, altogether

Resources

- sets of tens-rods and ones-cubes (one set for each pair)
- digit cards 1–9 and number cards 10–90

Language support

Make a display of several 2-digit numbers showing how they are made up of tens-rods and ones-cubes representing tens and ones. Use the sentence '___ is made up of ___ rods and ___ cubes, which represents ___ tens and ___ ones'. For example, '46 is made up of 4 rods and 6 cubes, which represents 4 tens and 6 ones'.

 Introductory activity

Write on the board 23 + 14. Ask students, *How can we make 23 and 14 using rods and cubes?* Students should be able to represent 23 as 2 rods and 3 cubes, and 14 as 1 rod and 4 cubes, and then describe them both. Display these in a diagram on the board, if possible. Now ask, *How many rods do you have in total? What do you need to do?* Students should see that they need to look at the tens-rods and ones-cubes separately, then combine their rods from the two numbers and say that there are 3 rods in total. Now ask, *How many cubes in total?* and students should be able to combine the cubes and say that there are 7 in total.

Now ask, *What do the tens-rods represent?* (3 tens or 30) *What do the ones-cubes represent?* (7 ones or 7) *What is that **altogether**?* (37) Add '= 37' to your calculation on the board. Explain that when you 'split' a number into two or more parts, you are **partitioning**. *We partitioned each number into tens and ones.*

Repeat with another calculation, not crossing the tens boundary.

Main activity

Write 35 + 58 on the board and ask students to make the two numbers using tens-rods and ones-cubes.

Tell students to use partitioning to find the total. In the example, there are 8 tens-rods and 13 ones-cubes altogether. See whether the students can work out how to combine these to make the final answer by grouping 10 cubes and swapping them for a tens-rod, giving 9 rods and 3 cubes, which represent 35 + 58 = 93. Discuss how this example was the same and how it was different from the examples in the introductory activity. (They both involved partitioning two 2-digit numbers but in the second calculation students had to swap 10 cubes for a rod because there were more than 10 ones.)

Ask students to look at the worked example on page 61 of the Student Book, displaying on the IWB if possible. *Do you need to swap any cubes for rods in this example?* (No) *Why not?* Students should be able to explain that you do not need to because the ones do not make more than 10. Students work in pairs to complete the questions on page 61 of the Student Book. Encourage students with a good understanding of place value to choose two 2-digit numbers that will require them to cross the tens boundary. Depending on the 2-digit number they choose, they may cross the hundreds boundary. Support as necessary. Have 100-flats on hand to model the calculation.

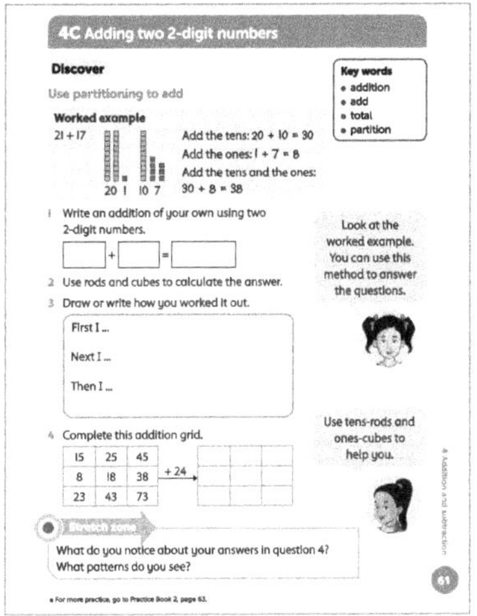

Differentiation

Supporting: Help students count the tens-rods and ones-cubes and write down the partitions and support them with the re-grouping.

Consolidating: Ask students to add two 2-digit numbers by forming the partitions and writing down the totals.

Extending: Ask students whether they could do any of the calculations mentally and explain how they did them.

Stretch zone: *What do you notice about your answers in question 4? What patterns do you see?*

Check whether students can notice how the ones digits in each row remain the same as each other when the numbers in the row are all added to 24.

 ## Reflection time

Ask some students to describe how they completed an addition. Get them to model their calculation with tens-rods and ones-cubes and then verbalise the process of forming the numbers, partitioning the tens and ones, adding them and **re-grouping** (recombining) them for the total.

Practice Book: Students can complete page 63 of the Practice Book. They can do this directly after the main activity, as homework, or as the focus of a separate mathematics session to help students consolidate their learning and build fluency. Go through the worked example together, emphasising how students can record their partitioning with four parts, first recording the tens and then the ones (i.e. $24 + 32 = 20 + 30 + 4 + 2$). Students then continue to work with tens-rods and ones-cubes to represent each calculation.

Differentiated outcomes	
All students	should complete calculations that do not involve re-grouping using tens-rods and ones-cubes.
Most students	will partition 2-digit numbers, successfully re-grouping using ones-cubes and tens-rods.
Some students	may add pairs of 2-digit numbers without using tens-rods and ones-cubes to calculate.

Student Book page 61

1   Check that students have written a 2-digit + 2-digit addition correctly.

2   Check that numbers have been made correctly using tens-rods and ones-cubes and that any re-grouping has been done correctly.

3   Check that the method is recorded appropriately.

4   39, 49, 69

    32, 42, 62

    47, 67, 97

Practice Book page 63

1   $16 + 23 = 10 + 20 + 6 + 3 = 39$

2   $24 + 35 = 20 + 30 + 4 + 5 = 59$

3   $35 + 27 = 30 + 20 + 5 + 7 = 62$

4   $18 + 19 = 10 + 10 + 8 + 9 = 37$

5   $34 + 36 = 30 + 30 + 4 + 6 = 70$

Stretch zone: Students may answer, for example, that they added the tens and then the ones and then added the two to find the total.

4C Adding two 2-digit numbers

Explore 1   Student Book page 62 • Practice Book page 64

Specific learning foci

- Know what each digit represents in 2-digit numbers; partition into tens and ones.
- Add pairs of 2-digit numbers using partitioning into tens and ones.

Global skills

- **Creative skills:** exploring/investigating
- **Real-world skills:** presenting information
- **Interpersonal skills:** communication/teamwork

Key vocabulary

- add, addition, total, number line, 2-digit number, partitioning, tens, ones, re-grouping

Resources

- tabletop 0–100 number lines
- base-10 equipment: tens-rods and ones-cubes (one set for each pair)
- digit cards 0–9 (one set for each pair)

Language support

Display a worked example using partitioning to add two 2-digit numbers. Include a speech bubble saying 'We can add numbers together by partitioning them into tens and ones.'

 ## Introductory activity

Give out base-10 equipment to each pair. Call out a number (e.g. 36). One student in each pair uses the tens-rods and ones-cubes to make the number you call out.

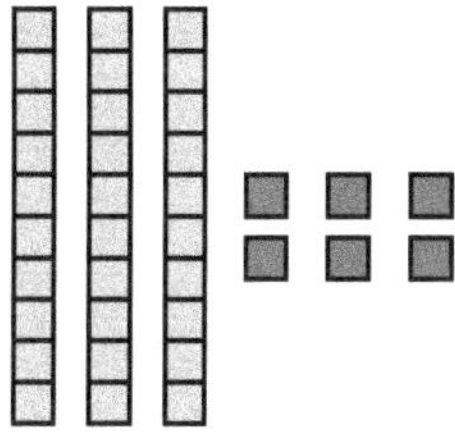

Their partner should check they are correct and say why or why not. (36 has 3 tens and 6 ones. They have used 3 tens-rods and 6 cubes so this shows 36.)

Repeat for: 14, 84, 55, 91, 29.

Draw a part-whole diagram on the board, recording '35' as the whole and placing question marks in each part.

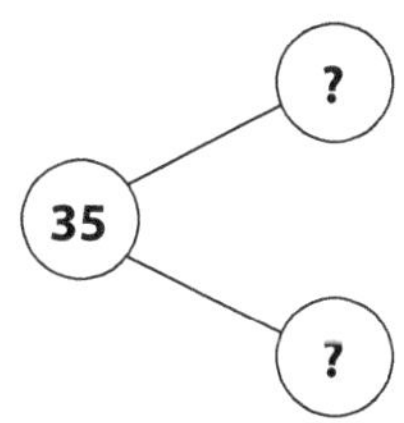

Say, *35 is the whole. What could the parts be? Think about different ways you could split, or partition, 35 into two parts*. Give students time to discuss in pairs before they share their ideas with the class. They can use concrete resources such as ones-cubes and tens-rods (or number lines) to help them partition in different ways.

Record their suggestions on the board and ask pairs to show or explain how the the parts they chose make 35. For example, they might think about what number is 1 less than 35 and split 35 into 34 and 1. They may randomly point to a number less than 35 on a number line and then find the size of the jump to 35. They might make 35 with ones-cubes and tens-rods and record the parts as 30 + 5 = 35. Draw part-whole diagrams for each suggestion they propose.

Students complete the activity on page 62 of the Student Book in pairs, choosing the number together and coming up with a calculation independently. They then share their calculations, describing how they made them and checking to see if both are correct.

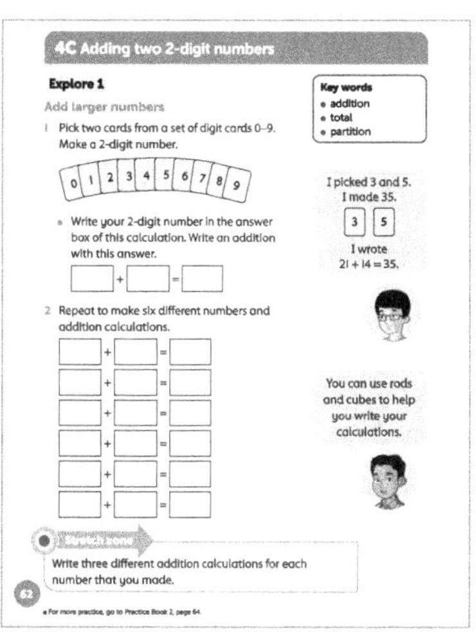

Differentiation

Supporting: Ask students to take their time completing the examples of adding two 2-digit numbers. Support them by modelling with ones-cubes and tens-rods.

Consolidating: Ask students to explain to you how they formed the numbers and added them.

Extending: Ask students to form calculations where the smaller number is written first, for example 7 + 18, to see whether they can use the commutative principle and work it out as 18 + 7.

Stretch zone: *Write three different addition calculations for each number that you made.*

Check that students have written addition number sentences correctly, using the number either as the addend or the sum.

Reflection time

Ask students to share their strategies for forming additions and finding the totals using tens-rods and ones-cubes. Try adding the numbers to the first example in a different order to remind students that they can add a list of numbers in any order.

Practice Book: Students complete page 64 of the Practice Book. They can do this directly after the main activity, as homework, or as the focus of a separate mathematics session to help students consolidate their learning and build fluency. Encourage them to record all their workings on a separate sheet of paper and use ones-cubes and tens-rods to add the pairs of numbers (or use them to check their answers).

Differentiated outcomes	
All students	should make 2-digit numbers and partition them into tens and ones.
Most students	should make 2-digit numbers and partition them in more than one way.
Some students	may solve addition of 2-digit numbers mentally and use the commutative principle to add, for example, 9 + 34.

Answers

Student Book page 62

Check that students have calculated correctly when adding the numbers they made with the digit cards.

Practice Book page 64

1 41, 30, 17, 12, 24   **4** 54, 43, 30, 25, 37

2 62, 51, 38, 33, 45   **5** 55, 44, 31, 26, 38

3 70, 59, 46, 41, 53

Stretch zone: Near-doubles could have been used for 19 + 18, 20 + 24, 20 + 18.

4C Adding two 2-digit numbers

Explore 2 — Student Book page 63 · Practice Book page 65

Specific learning focus

- Solve addition word problems involving addition of two 2-digit numbers.

Global skills

- **Creative skills:** exploring/problem solving
- **Real-world skills:** presenting information
- **Interpersonal skills:** communication

Key vocabulary

- addition, add, total, partition, altogether, +, =

Resources

- base-10 equipment: tens-rods and ones-cubes

Language support

Help students to identify the key words in the word problems that tell them the numbers to use and the calculation to perform and to define any words that they are unfamiliar with.

 **Introductory activity**

Display and read aloud the following story: *A shop sells 43 newspapers on Monday and 29 newspapers on Tuesday. Altogether, the shop sells 72 newspapers over these two days.*

Discuss with students which parts of the story they need to form a calculation. *How do we know it is an addition calculation?* (The word 'altogether'; 43 and 29 make 72 so it is an adding calculation.) *How could we represent the story using a part-whole diagram?* Invite a student to draw the diagram on the board.

 **Main activity**

Write the following word problem on the board and read it aloud: *I pick 24 apples on Saturday and another 17 on Sunday. How many apples did I pick over the weekend in total?* Ask students to think about a part-whole diagram to represent the problem and what the parts would be and what the whole would be. Agree that the numbers of apples on Saturday and Sunday will be the parts and the total (answer) will be the whole. *How many apples for each part?* (24 and 17) Ask *Is it an addition or a subtraction problem?* (Addition) *How do you know?* Focus on key vocabulary, that is, 'in total'. Ask students to represent the numbers using tens-rods and ones-cubes and then partition and re-group to find the total. (41)

Ask students to put the answer into context – *What does the 41 represent?* (The total number of apples picked over the weekend)

Students should work through the problems on page 63 of the Student Book independently. Remind them to look carefully at the text in the word problems to find out what numbers to use and what to do with them to find the answer. They can model the problems with tens-rods and ones-cubes, if necessary.

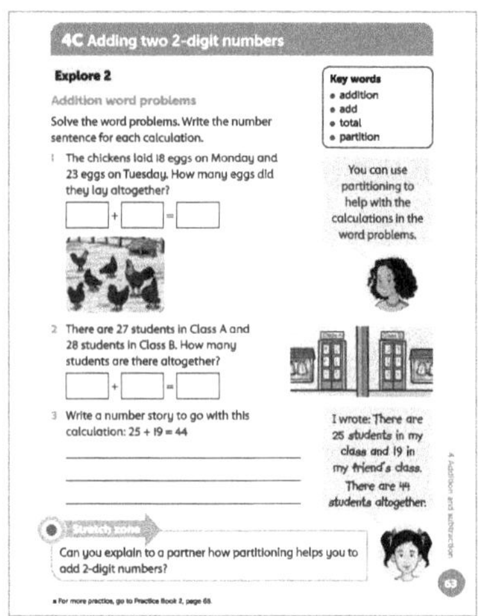

Differentiation

Supporting: Read the word problems together and help students to pick out key vocabulary and information needed to solve the problem. Model the problem with tens-rods and ones-cubes.

Consolidating: Ask students to check their answers by adding using a different calculation strategy.

Extending: Encourage students to try to write two different word problems involving 2-digit numbers for the same calculation.

Stretch zone: *Can you explain to a partner how partitioning helps you to add 2-digit numbers?*

Students should be able to talk through how they found the necessary calculation and then used partitioning to help find the answer. Listen to see whether they can describe partitioning and re-grouping.

 **Reflection time**

Ask students to discuss with a partner what they did to solve the word problems. Can they share with the whole class what they did and see whether there were similar approaches or any differences?

Finally, ask them to share their number stories for question 3.

Practice Book: Students complete page 65 of the Practice Book. They can do this directly after the main activity, as homework, or as the focus of a separate mathematics session to help students consolidate their learning and build fluency. If the activity was done as homework, set time aside in class for students to share their problems in pairs and have their partner solve them.

Differentiated outcomes	
All students	should recognise the parts in each word problem and understand that they need to be added to solve the problem.
Most students	will use an appropriate strategy to add the two 2-digit numbers in a word problem.
Some students	may use another calculation strategy to check that their answers are correct.

Answers

Student Book page 63

1 $18 + 23 = 41$ eggs

2 $27 + 28 = 55$ students

3 Check the number story for $25 + 19 = 44$

Practice Book page 65

Check that students have formed three sensible word problems from the picture, written them as calculations and found the correct totals. For example, The greengrocer has 21 oranges and 12 carrots. How many is this in total? $21 + 12 = 33$.

Stretch zone: Answers will vary.

4D Finding the difference

Discover Student Book page 64 • Practice Book page 66

Specific learning foci

- Find a small difference between pairs of 2-digit numbers.
- Understand subtraction as both difference and 'take away'.

Global skills

- **Creative skills:** exploring
- **Real-world skills:** presenting information
- **Interpersonal skills:** communication/teamwork

Key vocabulary

- subtraction, difference, counting on, take away

Resources

- interlocking cubes
- base-10 equipment

Language support

Display an illustrated example of how to find the difference (similar to those in the Student Book) using tens-rods and ones-cubes.

 **Introductory activity**

Show students two towers of identical interlocking cubes, one with 5 cubes and one with 7 cubes. Ask students to talk to their partner about what is the same and what is different about the two towers. Share ideas, focusing on what is the same first. Explain that when we compare two numbers, we look to see how many more to make them the same, so we count the **difference**.

Take two cubes off the taller tower, show that the remaining tower is the same height as the smaller tower, and then replace the two cubes. asking students to find the difference. When students are confident, link to the number sentence, saying, *The difference between 7 and 5 is 2, 7 − 5 = 2.*

Main activity

Display on the board the following word problem and read it aloud: *Rachael has 33 postcards. Sami has 24. How many fewer cards does Sami have?*

Who has more postcards, Rachael or Sami? (Rachael)
What do we need to find out? (how many fewer cards Sami has)
What do we know that can help us? (Rachael has 33 cards, Sami has 24.)

Model how many cards Rachael has using three tens-rods and three ones-cubes. Then model how many cards Sami has using two tens-rods and four ones-cubes.

What can we do with the rods and cubes to find out how many fewer cards Sami has? (Count the difference between the numbers of cubes in the two models.) *How many cubes is that?* (9)

Demonstrate this by swapping one rod for ten cubes in Rachael's model and then removing 9 cubes.

What is the difference between the number of cards that Rachael has and the number Sami has? (The difference between 33 and 24 is 9.) Record the answer in a number sentence: $33 − 9 = 24$.

Ask students to work in pairs on the activities on Student Book page 64. Explain that they will be finding the difference between numbers, and that they should use tens-rods and ones-cubes to help them.

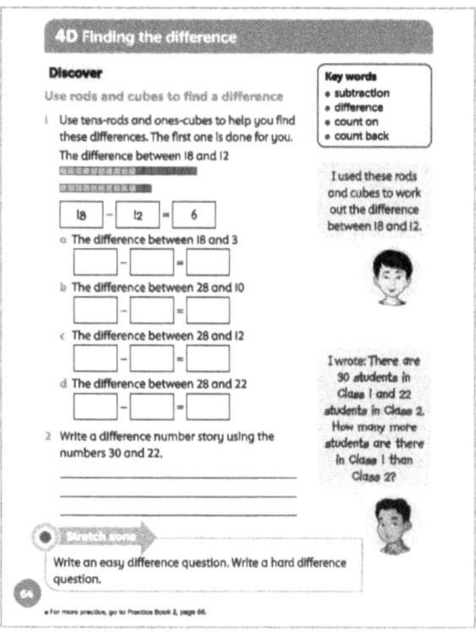

Practice Book: Students complete page 66 of the Practice Book. They can do this directly after the main activity, as homework, or as the focus of a separate mathematics session to help students consolidate their learning and build fluency. Give students interlocking cubes or tens-rods and ones-cubes to build each number and then compare to find the difference as well as showing the calculation on the number lines.

Differentiated outcomes	
All students	should find the difference using concrete resources to make each number and then comparing.
Most students	will find the difference using concrete resources to compare numbers and may link this with a number line representation.
Some students	may find the difference by counting on or counting back and show this on a number line.

Answers

Student Book page 64

1 **a** $18 - 3 = 15$ **c** $28 - 12 = 16$

   b $28 - 10 = 18$ **d** $28 - 22 = 6$

2 Check that students have written a sensible number story for $30 - 22 = 8$

Practice Book page 66

1 $38 - 31 = 7$ **4** $35 - 25 = 10$

2 $42 - 36 = 6$ **5** $14 - 8 = 6$

3 $29 - 21 = 8$

Stretch zone: Check that students have drawn tens-rods and ones-cubes correctly for question 5.

Differentiation

Supporting: Work with students to use ones-cubes and tens-rods to find the difference.

Consolidating: Ask students to explain how they found the difference using ones-cubes and tens-rods.

Extending: Ask students to explain how they found the difference and to model finding the difference on a number line.

Stretch zone: *Write an easy difference question. Write a hard difference question.*

Ask students why they chose the questions they did, and what makes a question easy or hard. For example, $27 - 2$ might be an easy example because you only need to count back 2 to find the difference, but $31 - 17$ is harder as it crosses two tens boundaries.

 **Reflection time**

Model the word problem in the second speech bubble on page 64 of the Student Book with the whole class. Ask students to share the number stories they created for the same calculation. For each one ask, *Do we find the difference to solve this problem or are we finding out how many are left?*

4D Finding the difference

Explore Student Book page 65 • Practice Book page 67

Specific learning foci

- Find a small difference between pairs of 2-digit numbers.
- Understand subtraction as difference.

Global skills

- **Creative skills:** exploring
- **Real-world skills:** presenting information
- **Interpersonal skills:** communication/teamwork

Key vocabulary

- subtraction, difference, counting on, counting back

Resources

- large 0–30 number line for front of class
- 0–100 number lines, one per pair
- counting sticks labelled 0–10, 0–100 (in tens)

Language support

Display a number line showing counting on to find the difference and counting back to find the difference. Add a caption or speech bubble saying, 'We can count on or count back to find the difference.'

Show students a counting stick marked in ones from 0 to 10. Count together up and back in ones, not always starting or finishing at one end of the stick.

Then use a counting stick marked in tens from 0 to 100 and repeat, counting up and back in tens, not always starting or finishing at one end of the stick.

Explain to students that they are going to learn how to find the difference between two numbers in another way. Write on the board 23 – 17. Display a number line from 0 to 30 and ask whether students can show you how to find the difference between 23 and 17. They might say, for example, that the answer is how big the space is between 23 and 17 on the line.

Circle 23 and 17. Start at 23 and ask students to count back with you until you reach 17: 22, 21, 20, 19, 18, 17. *How many numbers were jumped?* (6) Complete the number sentence on the board: 23 – 17 = 6.

Can you think of another way to find the difference on the number line? Students may suggest (if not, show them) starting at 17 and jumping forward to 23, counting: *18, 19, 20, 21, 22, 23. Is the number of jumps you made the same or different?* (the same, 6) Write this as a missing-number addition sentence 17 + [ ] = 23 and then fill in the missing number box with 6.

Say, *You can find the difference by counting on or back between two numbers. True or false?* Ask students to work in pairs to give you examples to support their answer.

Take feedback from the class and agree that it is true. Ask students to share some of their examples to support this or use examples to convince any students who thought that the statement was false.

Students now complete the questions on page 65 of the Student Book in pairs.

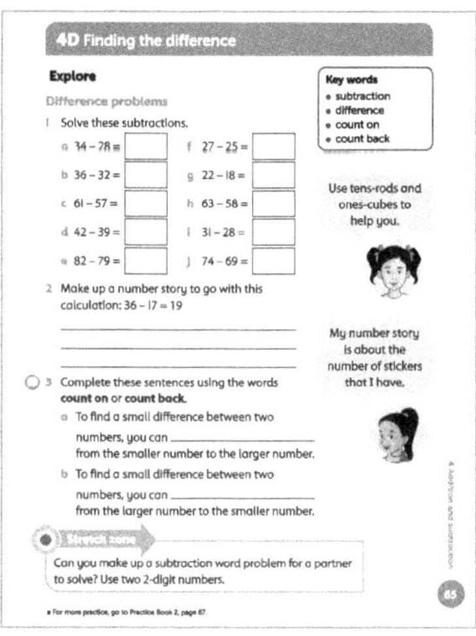

Differentiation

Supporting: Give students tens-rods and ones-cubes to help them find the difference. Model the use of a number line with students.

Consolidating: Ask students to explain how they found the difference. Did they count on or back? Can they explain why?

Extending: Encourage students to make their own difference questions that cross more than one tens boundary (e.g. 46 – 28) and to solve them using a number line.

Stretch zone: *Can you make up a subtraction word problem for a partner to solve? Use two 2-digit numbers.*

Check that students have made suitable word problems for a partner and that they can solve them themselves.

Reflection time

Discuss with students how they solved some of the problems. Ask them to explain which number they started on and how they counted on or back to get to the other number. *Did you find any strategies that helped you?* For example, in doing 34 – 28, they could count back 4 from 34, then count back 2 more to get to 28.

Ask students to share the word problems they made on page 65 of the Student Book.

Practice Book: Students can complete page 67 of the Practice Book. They can do this directly after the main activity, as homework, or as the focus of a separate mathematics session to help students consolidate their learning and build fluency. Explain that to make a number sentence with a given difference, they can choose a 2-digit number, then count back by that difference to find the other number in their sentence. For example, for a difference of 7, they could choose 15, then count back 7 from 15 to get to 8, making the number sentence 15 – 7 = 8. Alternatively, they could choose a number and then count on 7 from that number. Show this by starting with 8 and counting on 7 to 15. Go through a couple of examples together.

Differentiated outcomes	
All students	should use ones-cubes and tens-rods to find a difference.
Most students	will count on or back to find the difference between two numbers on a number line.
Some students	may find the difference using different strategies appropriate to the calculation, for example jumping to the nearest 10 first on a number line.

Student Book page 65

1
  a $34 - 28 = 6$
  b $36 - 32 = 4$
  c $61 - 57 = 4$
  d $42 - 39 = 3$
  e $82 - 79 = 3$
  f $27 - 25 = 2$
  g $22 - 18 = 4$
  h $63 - 58 = 5$
  i $31 - 28 = 3$
  j $74 - 69 = 5$

2 Check that students have made an accurate number story.

3 **a** count on   **b** count back

Practice Book page 67

Check that students have completed correct number sentences.

Stretch zone: Students may notice that two numbers with a difference of 10 have the same ones digit, for example 35 and 25. What is different about the numbers is that the tens digit is always 1 ten more or 1 ten less.

4E Missing numbers

Discover   Student Book page 66 · Practice Book page 68

Specific learning focus

- Solve missing number sentences such as $27 + \boxed{} = 30$.

Global skills

- **Creative skills:** exploring
- **Interpersonal skills:** communication/teamwork

Key vocabulary

- addition, subtraction, difference, total, missing number

Resources

- counters
- large 0–30 number line for display
- base-10 equipment

Language support

Read a missing-number question with students and model what the sentence is asking you to find out. This will help them to understand how to solve the calculation.

Introductory activity

Give each pair of students 20 counters and explain that they are going to play a game. With the 20 counters on the table in front of them, one student places a hand on top of some of the counters and slides them away. They could slide them off the table into their other hand or keep them on the table. Their partner then works out how many counters are hidden. Give students around five minutes to play the game.

Main activity

Ask students to describe how they worked out how many counters were hidden in the introductory activity. Students may talk about, counting up, counting back, finding the difference. Remind students about parts and wholes. Draw a part-whole diagram on the board, showing 20 as the whole, the remaining number of counters (e.g. 10) as a part and then a question mark to show the missing part that they are looking for. Now use this to help students write different number sentences such as $20 - 10 = 10$, $20 - 13 = 7$ and so on.

Write a missing-number problem such as $16 + \boxed{} = 20$ on the board. *Which is a part? Which is the whole?* Agree that 20 is the whole, 16 is a part. *Is the **missing number** a part or a whole?* (a part) *Can you work out the missing number?*

Some students will be able to use number bonds for 20 to tell you that 4 must be the missing number. Model counting on from 16 to 20 on a number line to find the number. Tell students that they can use whichever strategy they prefer and that they can use concrete resources to help if they wish, for example, making the whole and the part with cubes and rods and comparing to find the difference.

Students then complete the questions in the Student Book page 66 either individually or in pairs. Encourage them to tailor their strategy to each question, for example, using known number bonds to solve questions 3a and 3g. If students complete the questions individually, leave time for them to compare and discuss strategies in pairs for working out the missing numbers.

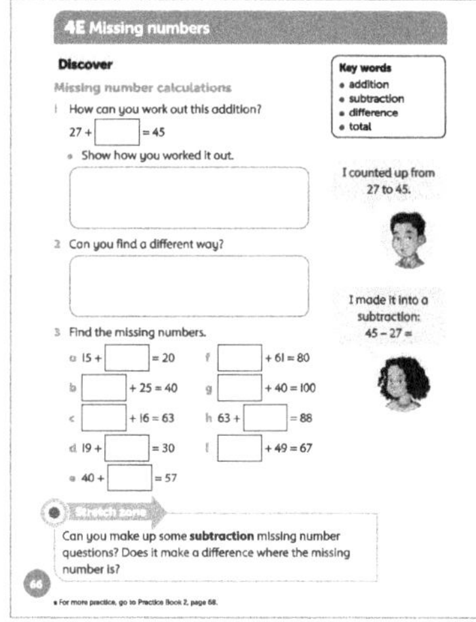

Differentiation

Supporting: Give students ones-cubes and tens-rods to find differences. Model the process using a number line as well so that they begin to have more than one strategy for finding differences. They could continue with the counters game rather than work on the Student Book activities.

Consolidating: Ask students to explain their strategy for finding the missing number.

Extending: Ask students to find a different strategy and then compare strategies to see which was more efficient.

Stretch zone: *Can you make up some subtraction missing-number questions? Does it make a difference where the missing number is?*

Check that the subtraction missing-number problems are accurate and appropriate: for example, students should be able to explain whether the missing number is the whole or one of the parts.

 **Reflection time**

Ask students to share and compare their strategies. Ask questions such as: *What makes this strategy a good one? Would this one work with bigger numbers? Do you need to use this strategy with smaller numbers or is there an easier one?*

Practice Book: Students complete Practice Book page 68. They can do this directly after the main activity, as homework, or as the focus of a separate mathematics session to help students consolidate their learning and build fluency. Tell students to ask for support from an adult to record their answers if necessary, or to record their answers in a diagram.

Differentiated outcomes	
All students	should identify the whole and the parts in a missing-number problem.
Most students	will find an efficient strategy to solve missing-number problems.
Some students	may use more than one strategy and be able to choose the most efficient one for solving missing-number problems.

Answers

Student Book page 66

1 $27 + 18 = 45$. Check students' calculation method.

2 Check that the alternative method is correct.

3 **a** $15 + 5 = 20$ **f** $19 + 61 = 80$

   b $15 + 25 = 40$ **g** $60 + 40 = 100$

   c $47 + 16 = 63$ **h** $63 + 25 = 88$

   d $19 + 11 = 30$ **i** $18 + 49 = 67$

   e $40 + 17 = 57$

Practice Book page 68

1 $9 + 16 = 25$ **3** $19 + 17 = 36$

2 $28 + 11 = 39$ **4** $22 + 50 = 72$

Check that the explanations are correct.

Stretch zone: Ask students to explain why they think their choices are easy and hard.

4E Missing numbers

Explore   Student Book page 67 · Practice Book page 69

Specific learning focus

- Find the missing number so that two sides of an equation balance.

Global skills

- **Creative skills:** exploring/investigating
- **Interpersonal skills:** communication/teamwork

Key vocabulary

- addition, subtraction, difference, total, balance, equal, number bond

Resources

- balance scales, sets of weights or objects to balance
- Numicon shapes, interlocking cubes, base-10 equipment

Language support

Display a poster showing simple balance scales with $17 + 3$ on one side and the question 'Which numbers could we use to make the scales balance?' on the other side. Pin a large piece of paper next to the scales for students to write their own ideas. You could write one yourself to get students started. If no one writes a subtraction calculation such as $27 - 7$, write one yourself and encourage students to think of some more.

 **Introductory activity**

Show students a pan balance. Ask them to explain what it means when the scales (pans) **balance**. They might say, for example, that when the pans balance, the objects on both sides weigh the same. Ask two students to put some weights on the scales until they balance. *How do you know the weights on each side have the same total weight even though they may not look the same?* (The scales are balanced.) Ask other students to try different objects in the pans, adjusting them so that they balance.

Agree that no matter what students put in the pans, if the scales are balanced the contents of the pans are **equal** in weight.

Main activity

Draw a simple set of balance scales on the board with an equals sign somewhere in the middle. Explain that we can balance numbers too. Write a number bond for 10 on one side (e.g. 7 + 3) and ask students to tell you what would need to be on the other side to make the scales balance. They should choose another number bond for 10 to write on the other side (e.g. 8 + 2). You can also show students the equivalences pictorially, for example, by using Numicon shapes 8 and 2, two towers of interlocking cubes, one of 8 cubes and one of 2 cubes. Ask students to write the matching number sentence (e.g. 8 + 2 = 7 + 3) and model it pictorially as well.

Write another balancing number sentence, but this time with a missing number, for example 6 + 4 = ☐ + 9. Ask students to discuss in pairs how they might find the missing number. Ask pairs to share their strategies. Reinforce the idea that one side is equal to the other side, linking back to the balance scales. Ask students to write the compelted number sentence. Repeat with a number sentence where the total of each side is a number between 11 and 20.

Ask students to continue to work in pairs to complete page 67 of the Student Book. Give students concrete resources to help them make equivalent expressions. Model how they could use base-10 equipment to make the total on one side and then take away the number of rods and cubes that represent the number on the other side of the balance to find the missing number.

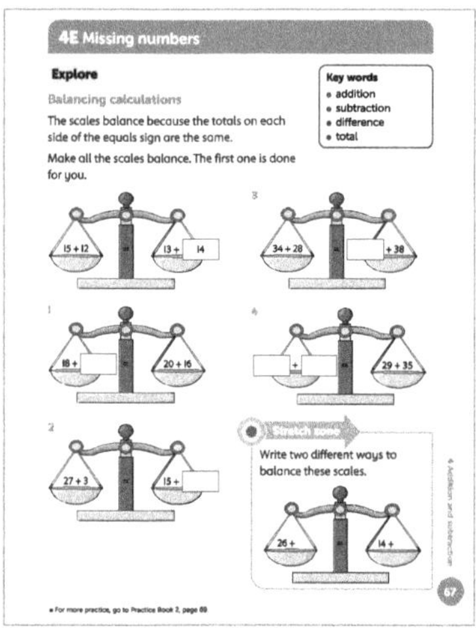

Differentiation

Supporting: Help students to use number bonds for 10 to total each side of the balances and use concrete resources to check. Work with students rather than have them working in pairs.

Consolidating: Ask students to explain how they balanced pairs of expressions.

Extending: Ask students to find more than one strategy to balance the pans.

Stretch zone: *Write two different ways to balance these scales.* (26 + ☐ = 14 + ☐)

Check that students understand that there are many ways to balance the scales, for example: 26 + 1 = 14 + 13, or 26 + 5 = 14 + 17 and so on.

Reflection time

Students share how they balanced the scales. Ask them to explain the strategies they used. Ask them to describe which strategies they found most helpful and efficient, for example, did they look for bonds for 10?

Practice Book: Students then complete page 69 of the Practice Book. They can do this directly after the main activity, as homework, or as the focus of a separate mathematics session to help students consolidate their learning and build fluency. Students can continue to use concrete resources to find totals.

Differentiated outcomes	
All students	should find the total for one side of the balances.
Most students	will find the missing number to make the pans balance.
Some students	may use different strategies to make the pans balance.

Answers

Student Book page 67

1 18

2 15

3 24

4 Any number pair that adds to make 64

Practice Book page 69

1 15

2 14

3 40

4 25

5 Any pair where the second number is 10 more than the first.

Stretch zone: Answers will vary as students choose their own numbers. Check that their examples of easy and hard calculations fit the description. For example, 24 + 6 = 20 + 10 would be easy; 14 + 19 = 25 + 8 would be more difficult.

4 Addition and subtraction

Connect Student Book page 68

Big idea

There are lots of strategies for adding and subtracting numbers. I can use different strategies depending on the numbers.

Global skills

- **Creative skills:** exploring/investigating
- **Interpersonal skills:** communication/teamwork
- **Self-development skills:** reflecting on learning

Key vocabulary

- addition, subtraction, difference, counting on, counting back

Resources

- 100-square for each pair

Language support

Encourage students to talk about the patterns they notice and to listen to each other as they describe what they see. This will help them to develop their use of mathematical vocabulary.

 **Introductory activity**

Write several different calculations on the board – each should lend themselves to different calculation methods, for example:

$17 + 3$ $87 - 37$ $43 - 30$ $29 + 40$ $46 - 7$ $33 + 49$

Ask students to discuss in small groups how they would find the answer. As they discuss, ask questions such as, *Which strategy will you use for this one? Can you think of a more efficient strategy?* Then come together as a class to share what they have discussed.

 **Main activity**

Ask students to look at page 68 in the Student Book. Tell them that they are going to explore what happens when you take a 2-digit number, swap the digits round and add the numbers together. Give pairs plenty of time to explore this in question 1. They can colour the answers on a 100-square to spot the pattern. Then look at the pattern of numbers created together. What have they found?

Now ask pairs to work on question 2, looking at the two numbers they have made and subtracting the smaller from the larger. Again, give time to explore this (colouring the answers on a 100-square) before discussing the results.

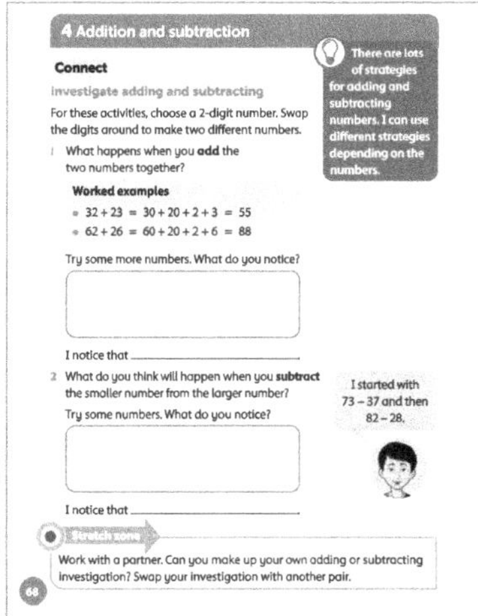

Differentiation

Supporting: Help students to identify the differences between the numbers using a strategy they know.

Consolidating: Ask students to describe the patterns they see.

Extending: Ask students to join with another pair who are struggling to see the pattern and, using the pair's chosen numbers, support them to discover the pattern.

Stretch zone: *Work with a partner. Can you make up your own adding or subtracting investigation? Swap your investigation with another pair.*

You can give students suggestions to help them here, for example, let them choose three digits, then make a 2-digit number with two of them, for example, 2, 4, 7 and make 27. Then explore what happens when they add and subtract the numbers. $27 + 4 = 31$, $27 - 4 = 23$. Now add the answers: $31 + 23 = 54$, which is the same as $27 + 27$. Does this always work? Check that students have written and calculated correctly.

 **Reflection time**

Discuss as a class what students noticed when they added the numbers together. The pattern is easier to see for addition than for subtraction. For addition, if the answer is a 2-digit number, both digits in the answer are the same and each is the sum of the original two digits. For subtraction, students may recognise that the answers are a multiple of 9 (9, 18, 27, 36 and so on) and that the two digits in the answer add to 9. Where both digits are already the same, e.g. 44, the pattern is the same for addition, i.e. the digits in the answer are both the same ($44 + 44 = 88$). However, for subtraction, the answer will always be zero ($44 - 44 = 0$; $55 - 55 = 0$, and so on). Students could discuss whether this fits the pattern.

Differentiated outcomes	
All students	should explore several calculations.
Most students	will see the pattern described above for addition.
Some students	may explore subtraction as well as addition.

4 Addition and subtraction

Review Student Book page 69 · Practice Book page 70

Global skills

- **Creative skills:** problem solving/exploring
- **Interpersonal skills:** communication
- **Self-development skills:** reflecting on learning

Student Book

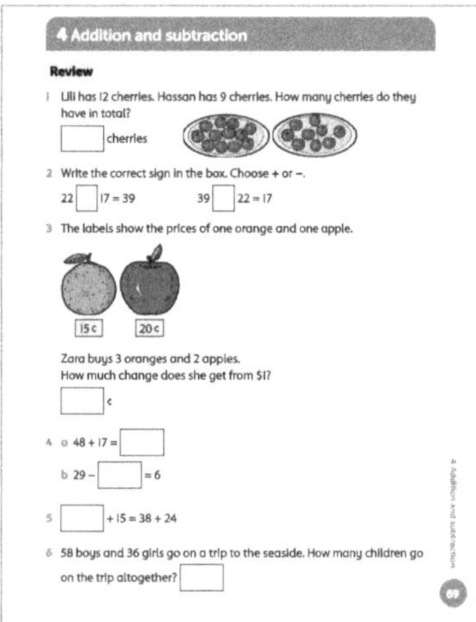

With young children, assessment activities are most effective when carried out as an everyday classroom activity. Students should be able to explore numbers in the context of addition and subtraction, developing strategies for calculation and an understanding that addition and subtraction are inverse operations. Students can use number lines or tens-rods and ones-cubes to help them solve the problems. Ask questions to assess their understanding, for example, *How did you know which operation to use? How did you set out the numbers using the equipment? How can you check your answer?*

Answers

Student Book page 69

1   21

2   +, −

3   15 ¢

4   **a**   48 + 17 = 65

    b   29 − 23 = 6

5   47 + 15 = 38 + 24

6   94

Practice Book

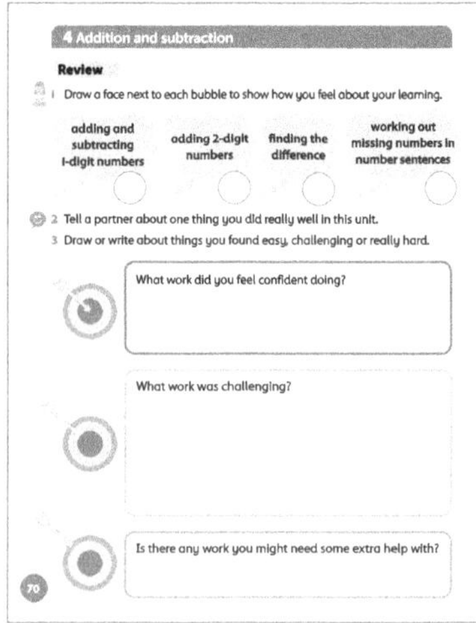

It is appropriate to complete this Practice Book review as a whole-class discussion. You may choose to keep a record of the class discussion or a copy of the review page for your own records. The review provides an opportunity for students to reflect on their learning from the unit, to discuss any areas of mathematics that they feel went particularly well, and any areas that they feel less confident about. Ensure all students have a copy of the Student Book as a reminder of the areas of mathematics that they have worked on in this unit.

Allow students plenty of time for discussion before asking them to complete the Practice Book page individually, and then, if appropriate, to share their responses with the rest of the class. If students complete this self-assessment at home, encourage them to discuss this with adults. Make a note of areas that students still feel unsure about.

Additional material

There are additional end-of-unit assessments available on the *Oxford Owl* website.

5 Multiplication and division

Big idea

This unit focuses on multiplication and division. The Big idea includes the concepts that: multiplication is commutative but that division is not, patterns in numbers help us calculate because division is about equal groups. Multiplication is both repeated addition (3×5 is $5 + 5 + 5$) and grouping (five groups of three). In both cases, writing this as 3×5 is shorthand for what we mean. When solving problems, students will sometimes find it useful to think about repeated addition and sometimes to think about groups, so it is useful for them to experience both ideas as this will support their understanding.

A secure understanding and recall of number facts, and how to use them, is essential. It is not enough to learn times tables as an ordered list. Students need to be more flexible in their thinking, recognising that if we know that $5 \times 3 = 15$, we also know that $3 \times 5 = 15$, $15 \div 3 = 5$ and $15 \div 5 = 3$. Students need this flexibility to be able to calculate mentally and to communicate their methods. Developing this understanding will ensure that students recognise the meaning of formal methods of recording.

Students must understand that division is about equal groups. The groups must be exactly the same size and anything left over is a remainder. This equality must be stressed with students. Ask students to imagine sharing sweets or something similar and to think about fairness. For the sharing and grouping to be fair, everyone must get exactly the same amount. In other words, the groups must be equal.

Look out for

- **Students who have difficulties with some of the multiplication tables.** This may be because the numbers lack an obvious pattern, for example the 3 times table pattern is less obvious than that of the 5 times table. Spend more time practising and building on their understanding of multiplication as repeated addition and the facts they can recall. For example, if you know that $3 \times 3 = 9$ and that this is the same as $3 + 3 + 3$, then 4×3 will be one more 3 (12).

- **Students who persist with counting in ones** but can count in twos, fives or tens when prompted. Help students to count in twos, fives or tens by using number lines or a 100-square.

- **Students who may not be fluent or accurate in counting on in twos, fives and tens.** Practise with students to count in different multiples using concrete resources as necessary.

Possible misconceptions

- **Students may confuse the concept of odd and even numbers, not understanding the connection to multiples of 2.**

- **Students may think that multiplication makes numbers bigger and division makes them smaller.** Since this does not always work (for example, when multiplying or dividing by fractions and decimals), it is better not to make that generalisation.

- **Students may assume that division is commutative because multiplication is.** Students quickly realise that multiplication is commutative, which means that it can be done in any order, just like addition. However, they need to understand that division is not commutative, just like subtraction. Take time to model with arrays their incorrect calculations to show them how division is not commutative.

Key vocabulary

- multiplication, multiply, divide, division
- count in …, groups of …, odd, even
- Carroll diagram, criteria, double, halving
- array, row, column, times table
- multiplication sign ($\times$), multiplication fact, multiples of …, division sign ($\div$), division fact
- equal groups, grouping, fact family
- remainder, left over, calculation

Learning focus	Learning outcomes (the ENC objectives)
Odd and even	Recognise odd and even numbers.
Doubles	Recall and use multiplication and division facts for the 2 multiplication table.
Twos, fives, tens, threes and fours	Recall and use multiplication and division facts for the 2, [3, 4] 5 and 10 multiplication tables. Solve problems involving multiplication using repeated addition.
Arrays	Solve problems involving multiplication using materials and arrays. Show that multiplication of two numbers can be done in any order (commutative).
Division	Solve problems involving multiplication and division using materials and arrays. Calculate mathematical statements for multiplication and division within the multiplication tables and write them using the multiplication ($\times$), division ($\div$) and equals ($=$) signs. Show that multiplication of two numbers can be done in any order (commutative) and division of one number by another cannot.
Remainders	Solve problems involving multiplication and division, including problems in contexts.

5 Multiplication and Division

Engage Student Book page 70

Big question

What strategies help us multiply and divide?

Global skills

- **Creative skills:** exploring/investigating
- **Real-world skills:** interpreting information/financial literacy
- **Interpersonal skills:** communication/teamwork

Key vocabulary

- count in twos, threes, four, fives, tens, groups of …

Resources

- concrete resources such as tens-rods and ones-cubes, interlocking cubes, counters and number rods
- 0–100 number lines, 100-squares

Language support

Display some empty packaging labelled with an arrow pointing to the number of items contained inside (e.g. 12 pencils). Next to the box, you could display a speech bubble saying 'Two boxes of pencils is 24. 24 = 12 + 12, 24 = 2 × 12.'

 **Introductory activity**

Look together at page 70 of the Student Book. Display it on the IWB, if possible. Ask students what they notice. Ask questions to focus their attention and familiarise them with the vocabulary, such as: *What are in the blue boxes? How many in a pair? How many rows of packs of pencils? How many pencils in each pack? How much for two packs of cakes? What is the man in the back right-hand corner doing? Which is more expensive, a pack of cakes or a pack of pencils?*

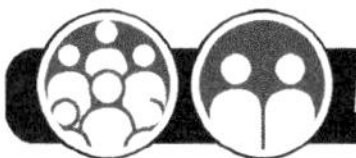 **Main activity**

Give students access to concrete resources and 0–100 number lines to support their calculating. Explain that you would like them to work out, in pairs, how many shoes, cakes and pencils they can see in the drawing of the supermarket. Each pair should think of the most effective way to count the objects and explain why they are using that strategy.

Differentiation

Supporting: Use 100-squares, number lines and concrete resources to reinforce counting in twos, threes, fours and fives.

Consolidating: Ask students to try different methods of counting to check answers. Can they explain why one was more efficient than another?

Extending: Ask students to make their own questions for a partner to answer using the supermarket drawing in the Student Book.

 **Reflection time**

Discuss which strategies students used to find the total number of items in each picture. Did they count in twos, fives or tens? Did they find some part totals along the way? As students explain their ideas, talk about which method was more efficient.

5A Odd and even

Specific learning focus

- Understand even and odd numbers and recognise these up to at least 20.

Global skills

- **Creative skills:** exploring
- **Real-world skills:** presenting information
- **Interpersonal skills:** communication/teamwork

Key vocabulary

- odd, even

Resources

- number cards 1–10 (one set per pair)
- interlocking cubes or counters, number lines, 100-squares

Language support

Display and organise quantities of up to 10 or 20 cubes or counters in patterns of 2 so that students can quickly see what we mean by odd and even, for example:

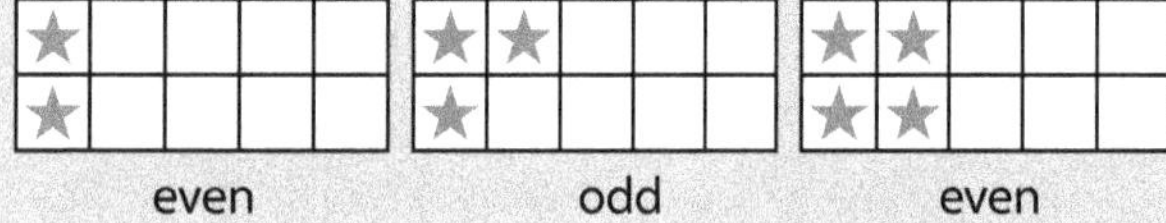

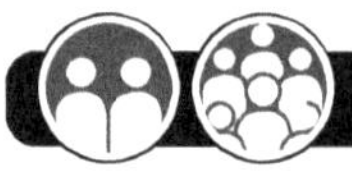 **Introductory activity**

Give each pair of students a set of number cards 1–10 and some interlocking cubes. Ask students to order their cards and match each card with the correct number of cubes. *Now, for each number, put your cubes in pairs. How many in a pair?* (2) The students should click their cubes together to make pairs.

Ask: *Which numbers make pairs with no single cubes left over?* (2, 4, 6, 8 and 10). Write the numbers on the board and say, T*hese are **even** numbers.* Write 'even' above the numbers. *Which numbers have cubes left over?* (1, 3, 5, 7 and 9). These are **odd** numbers.

Write 1–10 in a row on the board and then invite a student or a pair of students to mark the even and the odd numbers in some way. Ask them whether they can see the pattern and describe it.

 **Main activity**

Students should work through page 71 of the Student Book in pairs, talking with each other about what they are doing and continuing to use concrete resources to check their work. For question 1, explain that they should

continue to count on from the numbers at the top of the columns to complete the table. For question 4, tell students to look at which 2-digit numbers are odd and even to help them complete the sentences. Suggest that they make some other 2- and 3-digit numbers to check. As you walk around the room, listen for the correct use of odd and even. *How do you know that this is an even number? How do you know that this is an odd number? What do you think the next number will be? Why?*

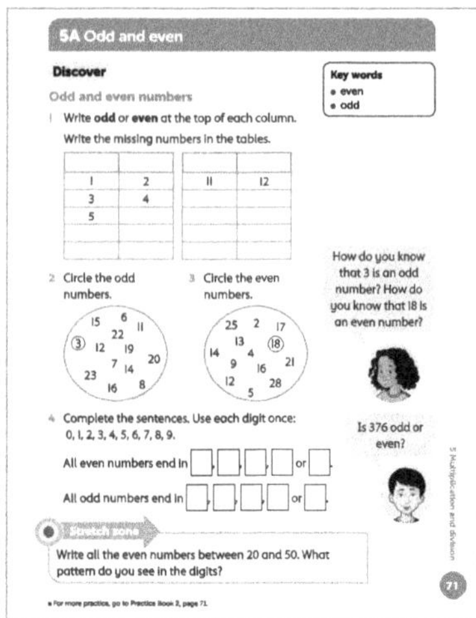

Differentiation

Supporting: Give students concrete resources and a number line or 100-square for them to mark the odd and even numbers as they go along.

Consolidating: Ask students to predict whether larger numbers will be odd or even and then check using cubes. *Can they explain how they made their prediction?*

Extending: Ask students to define what they mean by an odd and an even number. Ask questions, for example: *Can you tell me an odd number greater than 82 but less than 90? Is there more than one answer?*

Stretch zone: *Write all the even numbers between 20 and 50. What pattern do you see in the digits?*

Check that students are noticing that all the even numbers end in 0, 2, 4, 6, 8.

 **Reflection time**

Take feedback from students on their answers to the Student Book activities and agree correct answers.

Ask students in pairs to write down any numbers from 20 to 30 that they know are odd or even. Then repeat from 30 to 40, then 40 to 50. *Can they describe any patterns in the ones digits? What can we say about odd and even numbers?* Thinking back to question 4 in the Student Book students should say that all even numbers end in 0, 2, 4, 6 or 8; all odd numbers end in 3, 5, 7 or 9. Reinforce this by looking back at examples that students have shared.

Practice Book: Students complete page 71 of the Practice Book. They can do this directly after the main activity, as homework, or as the focus of a separate mathematics session to help students consolidate their learning and build fluency. Talk them through the rules, perhaps modelling how to make two 2-digit numbers with only two digit cards.

Differentiated outcomes	
All students	should understand the meaning of even and odd and relate this to making pairs.
Most students	will recognise most odd and even numbers up to 20 and start to see the pattern beyond 20.
Some students	may extend the pattern of odd and even numbers beyond 20.

Answers

Student Book page 71

1

odd	even	odd	even
1	2	11	12
3	4	13	14
5	6	15	16
7	8	17	18
9	10	19	20

2 3, 7, 11, 15, 19, 23

3 2, 4, 12, 14, 16, 18, 28

4 All even numbers end in 0, 2, 4, 6 or 8.

All odd numbers end in 1, 3, 5, 7 or 9.

Practice Book page 71

Answers will vary as students can make their own numbers by choosing number cards. Check that students have written numbers in the correct column (odd or even).

Stretch zone: Students might say they can make an odd number as long as one of the digits is odd.

5A Odd and even

Explore Student Book page 72 · Practice Book page 72

Specific learning foci

- Understand even and odd numbers and recognise these up to at least 20.
- Use Carroll diagrams to sort odd and even numbers.

Global skills

- **Creative skills:** exploring/investigating
- **Real-world skills:** presenting information
- **Interpersonal skills:** communication/teamwork

Key vocabulary

- odd, even, Carroll diagram, criteria

Resources

- number cards 0–30, enough for one card per student
- interlocking cubes

Language support

Display a large Carroll diagram with labels such as 'odd', 'not odd', 'less than 20', 'not less than 20' and some numbers for students to add to the diagram. Make a speech bubble label for each area of the diagram, for example: 'Odd numbers less than 20 belong here'. Ask questions such as: *Does 13 belong here?*

 **Introductory activity**

Give out the number cards randomly, one for each student. Call out *Odd!* or *Even!* and ask students to stand up, depending on what you call out. Start with odd, then even, making sure that everyone has taken their turn at standing. Ask students who are standing to show their number so that everyone can check. Then ask them to explain how they know, encouraging them to say that they can tell by the number in the ones column (e.g. all numbers that end in 3 are odd). Students then sit down and wait for the next call.

The student with the 0 card may be unsure whether it is an odd or even number. Discuss as a class; 0 can be split into two equal-sized groups (0 and 0) so it is even.

Go on to call out instructions with two **criteria** such as: odd numbers over 10, even numbers less than 20, odd numbers less than 15, even numbers with no tens.

 **Main activity**

Ask students to look at page 72 of the Student Book (display on an IWB if possible) and check that they understand how to complete the **Carroll diagram**. Do they recall, from their work in the previous year, what they use a Carroll diagram for? Agree that it is for sorting. Read all the labels aloud and ask for a number that is an example for each category (e.g. less than 12). *Who can tell me a number less than 12? And another?* Point to each box in turn and say, *What numbers could go in this box? Can you tell me a number that could not?* Agree, for example, that the top left box is for numbers that are odd and less than 12 – you could write '7', but not '19'.

Ask students to work in pairs to complete the Carroll diagram in their books and then answer the related questions. As you walk around the classroom, listen out for the correct use of odd and even. Ask questions such as, *Why have you put that number there? How do you know that is the correct place?*

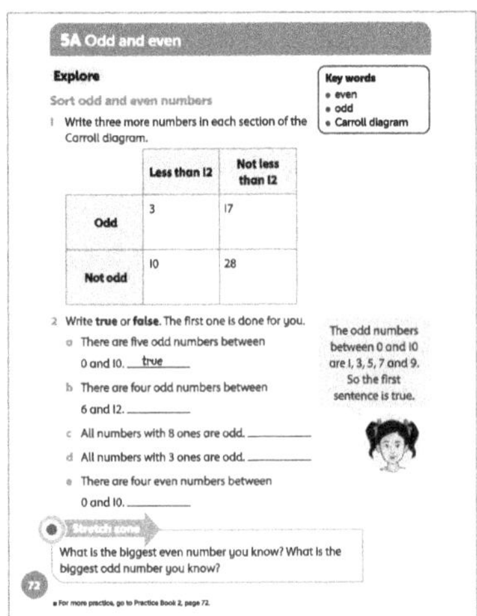

Differentiation

Supporting: Revise numbers that are even between 0 and 9 with students (building them with pairs of cubes and labelling each if necessary), ask them to recall the rule for odd and even numbers and help them to apply this.

Consolidating: Ask students to explore numbers between 30 and 40, predicting which will be odd or even and then checking with cubes and justifying their answers.

Extending: Ask students to explore numbers between 100 and 130, saying which will be odd or even and justifying their answers.

Stretch zone: *What is the biggest even number you know? What is the biggest odd number you know?*

This activity will help students reinforce the idea that odd and even numbers continually increase in steps of 2, and always end in a digit from one of the sets seen previously. Check that students can say the number correctly as this will help to ensure that they choose their number correctly.

 Reflection time

Copy the Carroll diagram from the Student Book onto the board and write in some of the students' numbers, checking that they are correct. Share some of the larger numbers that students have suggested. Ask them to only suggest numbers they are fairly confident they can read aloud. For numbers less than 1000 that are suggested but said incorrectly, model how to read them correctly. For each number, discuss how students know that the number was odd or even.

Practice Book: Students complete page 72 of the Practice Book. They can do this directly after the main activity, as homework, or as the focus of a separate mathematics session to help students consolidate their learning and build fluency. Ask them, *Which column do you look at to decide whether a 2-digit number is odd or even?* (ones) *Which digit may be in this column if a number is odd?* (1, 3, 5, 7, 9) *What if it is even?* (0, 2, 4, 6, 8)

Differentiated outcomes	
All students	should identify whether a number is odd or even, using concrete resources for support.
Most students	will identify whether a number is odd or even independently.
Some students	may extend the pattern of odd and even numbers beyond 100.

Answers

Student Book page 72

1 Check that students have entered numbers correctly into the parts of the Carrol diagram.

2 a true

   b false

   c false

   d true

   e true

Practice Book page 72

Check that the numbers made are placed into the Carroll diagram correctly.

Stretch zone: Students may say that, to make a number larger than 30, one of the two digits must be 3 or larger.

5B Doubles

Specific learning focus

- Find and learn doubles for all numbers up to 10 and also 15, 20, 25 and 50.

Global skills

- **Creative skills:** exploring/investigating
- **Interpersonal skills:** communication

Key vocabulary

- double

Resources

- dominoes (up to double 8), one set per pair, or domino cards cut out from Resource Sheet 2, copied onto card
- interlocking cubes

Language support

Display pairs of items (or pictures of them) such as socks, gloves or shoes with labels such as double 1 is 2, double 2 is 4, double 3 is 6 and so on as appropriate.

 ## Introductory activity

Show students a set of dominoes. Give each student a domino. Ask them to look closely at their domino and those given to the students around them. *What do they notice?* They might say that there are two sets of spots on each domino, and some of the dominoes have matching sets of spots.

Explain that, to start a game of dominoes, you need a **double**, that is, a domino with the same number on each part. Write the word 'double' on the board. *Does anyone have a double domino?*

Ask students with doubles to come to the front. Order the students from double 1 to double 6. Use ordering the dominoes to practise the vocabulary of position, by asking, for example, *Which domino should be first? In what position is the double 5 domino?* Then ask each student in turn to tell you what double their number is and how they know. Some students will be able to add the two numbers, others will need to count the spots.

 ## Main activity

Walk around the classroom and observe students as they complete the Student Book page 73. Ask individuals: *How did you work out this double? What is double 4? How do you know? What do you notice about all the double numbers?* (They are all even numbers, even if the start number was odd.)

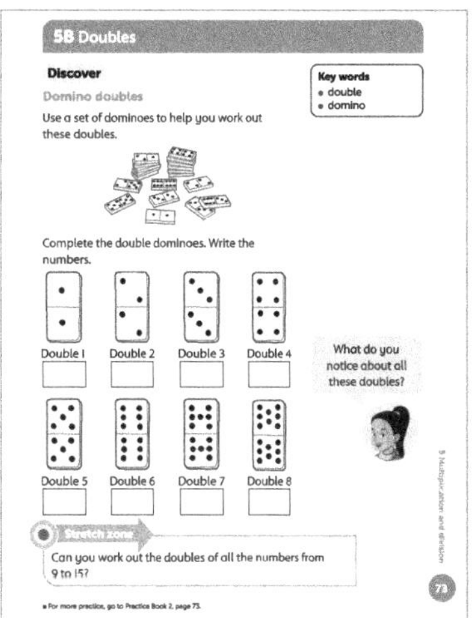

Differentiation

Supporting: Model counting in twos to find doubles. Give students concrete resources to build 'doubles'.

Consolidating: Ask students to count up in twos stopping at a given double.

Extending: Ask students to predict how many spots would be on a double 18 domino.

Stretch zone: *Can you work out the doubles of all the numbers from 9 to 15?*

This activity will help students see the connections between numbers and their doubles as they work with larger numbers. Encourage students to try to think of an efficient strategy to work out all the doubles, for example, partitioning and then re-combining.

 ## Reflection time

Ask students what they noticed about all the doubles. They might have noticed that all doubles have an even number of spots.

Call out random numbers to 10 and ask the students to answer with the double. Call out a number (e.g. 12) and say, *12 is double which number?* Repeat for other numbers up to 24. Include some numbers that are not doubles (e.g. 13 and 21). Can they explain why these numbers are not doubles?

Practice Book: Students complete page 73 in the Practice Book. They can do this directly after the main activity, as homework, or as the focus of a separate mathematics session to help students consolidate their learning and build fluency. Explain that they can ask for help writing their answers for questions 2, 3 and 4, or talk to someone about their answers, but they should still record the pairs of doubles in their book.

Differentiated outcomes	
All students	should understand the meaning of 'doubles' and find doubles by counting spots.
Most students	will count in twos to find doubles.
Some students	may extend the pattern of doubling beyond double 8.

Student Book page 73

2, 4, 6, 8, 10, 12, 14, 16

Stretch zone: 18, 20, 22, 24, 26, 28, 30

Practice Book page 73

1 2, 4, 6, 8, 10, 12   14, 16, 18, 20, 22, 24

2 Double 1 = 2, double 10 = 20

3 Double 2 = 4, double 20 = 40

4 Double 5 = 10, double 50 = 100

Students should notice that the tens digit doubles in the multiples of 10.

Stretch zone: 30, 60, 150, 300

5B Doubles

Explore 1   Student Book page 74 · Practice Book page 74

Specific learning focus

- Calculate doubles for numbers up to 50.

Global skills

- **Creative skills:** exploring
- **Interpersonal skills:** communication/teamwork

Key vocabulary

- double, multiple, even, odd

Resources

- base-10 equipment

Language support

Display number sentences such as '3 + 3 = 6' alongside written sentences such as 'double 3 is 6' to help students make the link between the number sentence and the word 'double'. When you read a number sentence, ask whether it is a double number sentence. Ask students how they know.

 ### Introductory activity

Draw a picture of a phone keypad on the board.

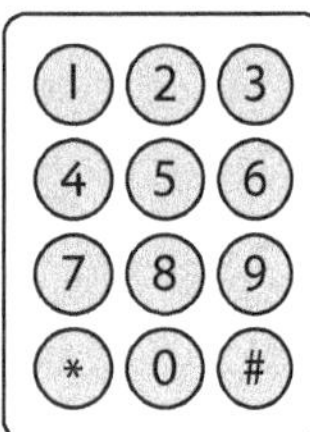

Ask students to look at the layout and explain that you are going to ask them to 'read' a row, but say the double of each of the numbers. For example, the first row would be 2, 4, 6. Give students a few moments to work out the double of each number, then point out various lines for them to say. Include diagonals and up and down the columns.

 **Main activity**

Explain that now that students know doubles to 10, they can work out any double. *What do you think I mean? How do you think knowing double 3 and double 10 might help me work out double 13?*

Give students time to discuss with a partner. Share ideas as a class. Use these as a basis for talking students through partitioning a number such as 13 into 10 and 3, doubling the ones and the tens and then re-combining to find the total. Demonstrate this on the board and with concrete resources such as tens-rods and ones-cubes.

Ask students to look at the worked example on page 74 of the Student Book. Display on the IWB, if possible. Choose a student to explain what the diagram shows. They should notice that it shows double 12 and that 12 is partitioned into one ten (one tens-rod) and two ones (2 ones-cubes). The double is shown as double 10, which is 20 (2 tens-rods), and double 2, which is 4 (4 ones-cubes). *What is double 12?* (24)

Ask students work in pairs on questions 1 and 2 in the Student Book to find double 14 and 17. Provide them with ones-cubes and tens-rods to model their doubling. Students should then move on to doubling the numbers on the number line. As students work, circulate and ask questions such as: *How did you split 17 into tens and ones? How is finding double 17 different from finding double 14?* (You need to swap 10 ones-cubes for a tens-rod before recombining.) *How did you know that double 25 is 50?* Remind students to check the doubles as well as following the pattern.

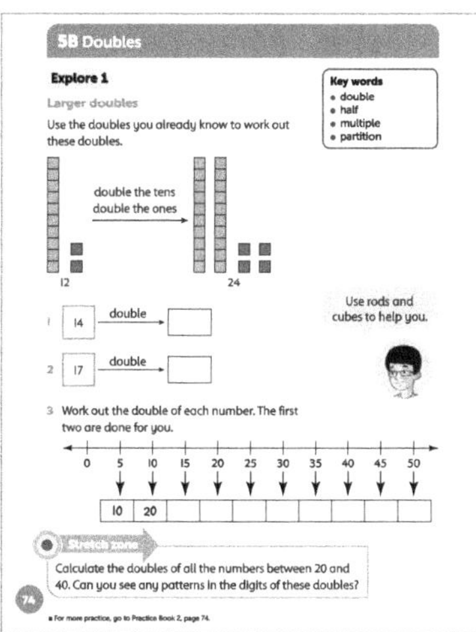

Differentiation

Supporting: Model how to find doubles up to 20 using tens-rods and ones-cubes.

Consolidating: Ask students to use tens-rods and ones-cubes to find doubles of any number up to 50 and explain to you how they found them.

Extending: Ask students to calculate doubles mentally, explaining how they worked them out.

Stretch zone: *Calculate the doubles of all the numbers between 20 and 40. Can you see any patterns in the digits of these doubles?*

Check that students are finding the correct doubles. They should notice the even number ending on each double.

 **Reflection time**

Ask students to discuss in pairs what they noticed about doubling **multiples** of 5. Ask pairs to share their ideas with the whole class. They may say, for example, that every double of a multiple of 5 gives a multiple of 10. Pairs should then work out the biggest double – they can using base-10 equipment for support. Discuss each pair's answers, checking that the doubles are correct.

Practice Book: Students complete page 74 of the Practice Book. They can do this directly after the main activity, as homework, or as the focus of a separate mathematics session to help students consolidate their learning and build fluency. Look at the worked example together, building both 13 and 20 with tens-rods and ones-cubes as you show students how to record the partitioning using the place-value card frames.

Differentiated outcomes	
All students	should double numbers up to 20 using base-10 equipment for support.
Most students	will double numbers up to 50 by partitioning and recombining.
Some students	may double mentally multiples of 5 using partitioning.

Answers

Student Book page 74

1 28

2 34

3 10, 20, 30, 40, 50, 60, 70, 80, 90, 100

Practice Book page 74

1 16 = 10 + 6. Double each part: 20 + 12, 20 + 12 = 32, so double 16 = 32.

2 18 = 10 + 8. Double each part: 20 + 16, 20 + 16 = 36, so double 18 = 36.

3 19 = 10 + 9. Double each part: 20 + 18, 20 + 18 = 38, so double 19 = 38.

4 11 = 10 + 1. Double each part: 20 + 2, 20 + 2 = 22, so double 11 = 22.

Stretch zone: Check that students have found the doubles of each 2-digit number by doubling each part and then adding the answers together.

5B Doubles

Explore 2 Student Book page 75 • Practice Book page 75

Specific learning focus

- Double a single-digit number more than once and halve even numbers.

Global skills

- **Creative skills:** exploring/investigating
- **Real-world skills:** presenting information
- **Interpersonal skills:** communication/teamwork

Key vocabulary

- double, half, halving

Resources

- 100-squares
- base-10 equipment
- digit cards 1–9 (one set for each pair)

Language support

Support students with the language of 'double and double again', helping them to understand that it means repeat the doubling process, doubling a number and then doubling that answer.

Start with the number 1. Ask a student to double it (2). Ask another student to double 2 (4), then another student to double 4 (8), and one more to double 8 (16).

Ask students whether they think numbers can be doubled forever? Students may say doubles go on forever because all numbers go on forever. Ask: *How far can we go?* Continue to double again and again. Give students more time and access to base-10 equipment as the numbers get larger.

Start with number 5 and ask students to double it (10), then double 10 (20), double 20 (40) and finally double 40 (80). *Do you notice anything about the answers?* Students might notice that, from 10 onwards, all the answers end in 0. *Will the numbers always end in 0 if we kept on doubling? Can anyone explain why?* Students may notice that double 0 in the ones will always give 0.

Ask students whether they can explain what **halving** means. Ask them, in pairs, to use several numbers to explain, modelling using tens-rods and ones-cubes. Ask: *Which numbers can you halve and which can you not? Why?* Students should suggest that they can only halve even numbers; odd numbers will always mean that there is one left over.

Now start with number 80 and ask students to find half of 80 (40), then find half of 40 (20), half of 20 (10) and half of 10 (5). Ask students to look at the answers to the doubling and halving sequences. Can they notice anything? Can they see a link between doubling and halving?

Students now complete the activities on page 75 of the Student Book in pairs. Work through an example as a class first to show students how to complete the table. Once they have discussed question 2 with their partner, they can join with another pair to share their thinking.

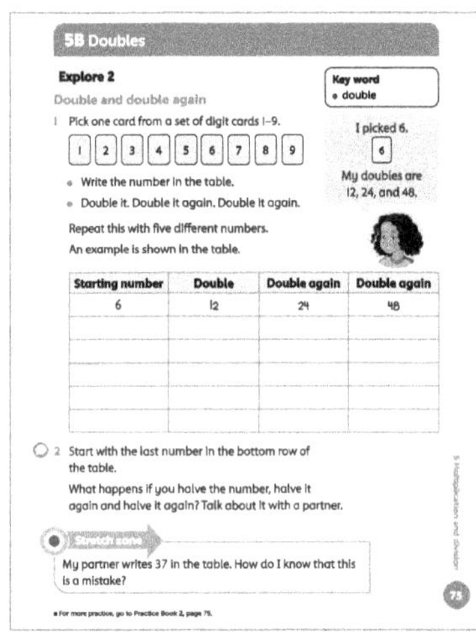

Differentiation

Supporting: Focus on helping students halve numbers as they are likely to be less comfortable halving than doubling. Model this with tens-rods and ones-cubes.

Consolidating: Ask students to double until they pass 100 and halve until they reach an odd number. Ask what might happen if they halved an odd number.

Extending: Ask students to look for patterns when they double numbers such as 2, 4 and 8, or 3 and 6. Can they make a connection between the sequences?

Stretch zone: *My partner writes 37 in the table. How do I know that this is a mistake?*

Check that the students understand that an odd number cannot be the double of a whole number.

Reflection time

Ask some students to describe what happened when they kept doubling a number and to share an example. How many times do they think they could keep doubling?

Then ask students about halving. Can they explain that halving can done to even numbers, but halving odd numbers leaves one left over?

Practice Book: Students complete page 75 of the Practice Book. They can do this directly after the main activity, as homework, or as the focus of a separate mathematics session to help students consolidate their learning and build fluency. Students can share their answers for questions 2 and 3 with an adult orally rather than writing if they wish.

Differentiated outcomes	
All students	will double small numbers (under 10).
Most students	will double any number up to 50 or halve any even number to 100, using partitioning where necessary.
Some students	will understand the inverse relationship between doubling and halving.

Answers

Student Book page 75

1

Starting number	Double	Double again	Double again
6	12	24	48
1	2	4	8
2	4	8	16
4	8	16	32
7	14	28	56
8	16	32	64

The numbers shown are examples; students may start with any number from 1 to 9.

2 64, 32, 16, 8. The numbers are the reverse of the doubling sequence.

Practice Book page 75

1 22, 24, 26, 28, 42, 44, 46, 48

32, 34, 36, 38, 52, 54, 56, 58

0, 10, 20, 30, 40, 50, 60, 70

2 Check to see whether students have noticed patterns in the starting numbers, for example, 'they go up one at a time' or 'they go up in fives'.

3 Check to see whether students have identified patterns in the answers, for example, 'they go up in twos', or 'they go up in tens'.

Stretch zone: Answers will vary.

5C Twos

Discover Student Book page 76 · Practice Book page 76

Specific learning focus

- Use arrays to represent multiplication facts in the 2 times table.

Global skills

- **Creative skills:** exploring
- **Real-world skills:** presenting information
- **Interpersonal skills:** communication

Key vocabulary

- array, row, column, multiplication, multiplicaton fact, times table, multiples of 2

Resources

- counters or cubes

Language support

Focus on using the terms 'row' and 'column' in an array and link the numbers in the 2 **times table** fact to the numbers of rows and columns.

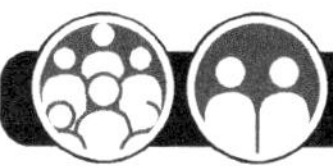 **Introductory activity**

Give each pair of students eight counters. Ask whether they can arrange them into equal rows of 2. They should find four **rows** of 2. *How can you describe your arrangement? Think of several ways.* Students might come up with a number of descriptions, such as four groups of 2, two **columns** of 4, a 4 by 2 **array**. Make a list of all the descriptions on the board, adding any students may have missed.

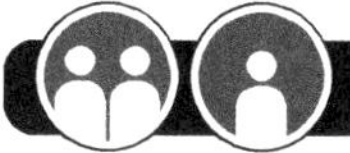 **Main activity**

Ask students to make 14 counters into equal rows of 2 counters in an array. Ask: *How many lots of 2 have you made? Can you write that as an addition of lots of 2?*

*Now can you write a **multiplication fact** to show how many lots of 2 you have?* Students should be able to tell you that there are 7 lots of 2 counters. This can be written as 2 + 2 + 2 + 2 + 2 + 2 + 2, or as a multiplication fact, 7 × 2.

Students complete the activities on page 76 of the Student Book individually. For each multiplication, they should make an array showing rows of 2 and then record it as a repeated addition. Once they have completed the questions, they can join with another student to discuss and investigate the two speech bubbles.

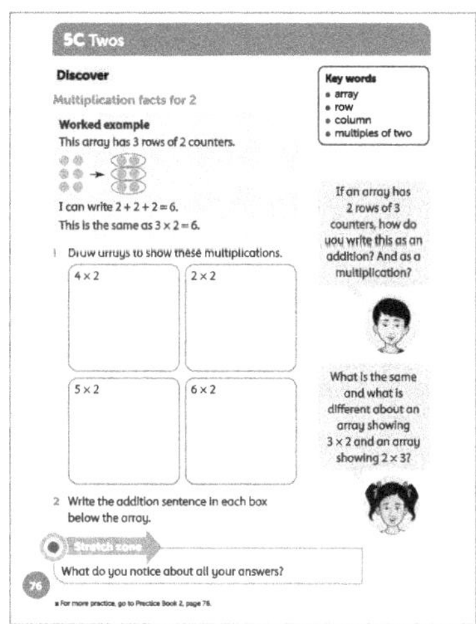

Differentiation

Supporting: Model counting the number of rows and counters in the array by touching the two counters in the first row, then touching each row in turn.

Consolidating: Ask students to count the rows and then how many counters in a row to reinforce 'lots of'.

Extending: Ask students to extend their arrays to more than 10 lots of 2 and write the repeated addition and multiplication sentences.

Stretch zone: *What do you notice about all your answers?*

Students may notice that their answers are all double the amount being multiplied by 2 and that their answers are all even numbers.

 **Reflection time**

Ask students to share their arrays and talk about how they formed them. Can they describe how the rows and columns relate to the parts of the **multiplication** sentence?

Practice Book: Students complete page 76 of the Practice Book. They can do this directly after the main activity, as homework, or as the focus of a separate mathematics session to help students consolidate their learning and build fluency. Ask students to also record each multiplication as an array.

Answers

Student Book page 76

1 4 × 2 [4 rows of 2]; 2 × 2 [2 rows of 2]; 5 × 2 [5 rows of 2]; 6 × 2 [6 rows of 2]

2 2 + 2 + 2 + 2 = 8

2 + 2 = 4

2 + 2 + 2 + 2 + 2 = 10

2 + 2 + 2 + 2 + 2 + 2 = 12

Practice Book page 76

1 4 × 2 = 8

2 8 × 2 = 16

3 9 × 2 = 18

4 11 × 2 = 22

5 12 × 2 = 24

Check that students have drawn the correct array in each case.

Stretch zone: **Multiples of 2** have a 0, 2, 4, 6 or 8 in the ones place because they are all even numbers.

5C Twos

Explore Student Book page 77 • Practice Book page 77

Specific learning foci

- Learn and recognise multiples of 2 and derive the related division facts.
- Understand multiplication as repeated addition and use the × sign.

Global skills

- **Creative skills:** exploring/investigating
- **Real-world skills:** presenting information
- **Interpersonal skills:** communication/teamwork

Key vocabulary

- multiples of 2, count in twos, multiplication sign (×)

Resources

- counters or cubes
- poster-sized pieces of paper
- large 0–20 number line for front of class, small 0–20 number lines for student use

Language support

Add the 2 times table to the class display, including a speech bubble saying: 'All the numbers in the 2 times table are even numbers' to support students with their discussions about the 2 times table. You could also put up two large poster-sized pieces of paper, one with the heading 'These numbers are in the 2 times table. They end with 0, 2, 4, 6 or 8' and one with the heading 'These numbers are not in the 2 times table. They end in 1, 3, 5, 7 or 9'. Encourage students to write numbers over 20 on the correct poster.

 **Introductory activity**

Count around the class in twos from zero to 40 and then back down to zero. If possible, ask the students to stand in a circle and count round the circle. If students are confident, replace every other count with a clap. So the count will be two, clap, six, clap, ten, …

 **Main activity**

Draw or display a 0–20 number line. Ask students how many jumps of 2 are needed to make 6 and record this as a repeated addition sentence: 2 + 2 + 2 = 6.

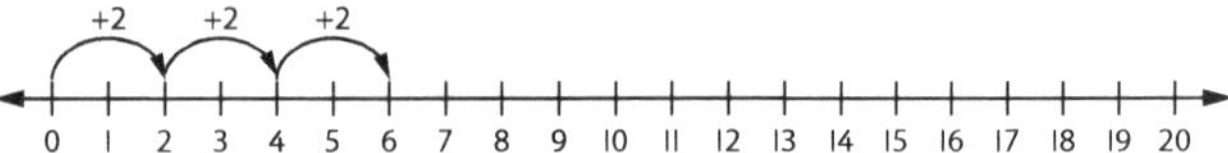

Repeat for 4 and 8. Represent each calculation as an array and build up a staircase as an additional representation.

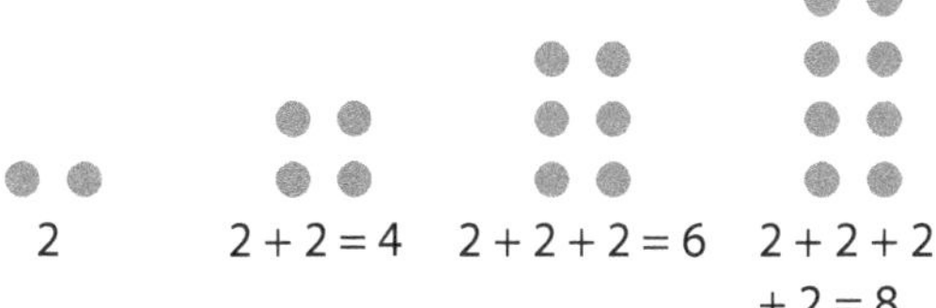

$$2 \qquad 2+2=4 \qquad 2+2+2=6 \qquad 2+2+2+2=8$$

Formally introduce the **multiplication sign (×)** explaining that we can use 'multiplication' as a way of recording repeated additions, which represent 'groups of'. Add to the staircase:

$2 + 2 = 4$	2×2	2 groups of 2
$2 + 2 + 2 = 6$	3×2	3 groups of 2
$2 + 2 + 2 + 2 = 8$	4×2	4 groups of 2

Ask students to complete page 77 in the Student Book leaving the first line blank (0×2). To help them they can continue to build a staircase, use a number line and make jumps of 2, or use cubes or counters to make multiple groups of 2 to work out the calculations and record their answers in the table. Discuss the questions in the speech bubbles as a class.

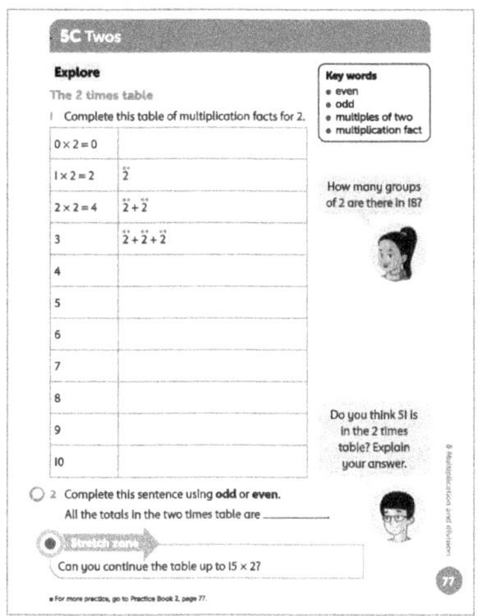

Differentiation

Supporting: Use jumps of 2 on number lines to model groups of 2 for students.

Consolidating: Say a 2-digit number and ask students to predict whether or not the number will be in the 2 times table.

Extending: Ask students to explain how they know a number is in the 2 times table.

Stretch zone: *Can you continue the table up to 15×2?*

Students should be able to show how they use repeated addition to find the multiples of 2 up to 30.

 ## Reflection time

Ask pairs to discuss why $0 \times 2 = 0$ has been written on the first row. Why is there is no picture for 0×2? Share students' reasons after the discussion, for example, 'There are no groups so there are zero things. If I start at zero and count on zero twos, I stay at zero.'

Ask students to explain their answer to question 2. *Why do you think the two times table has even answers?*

Give students a double to represent, say, double 6. Ask, *Can you draw an array to show this? What addition can you write?* (6 + 6) *What multiplication?* (2×6)

Practice Book: Students complete page 77 of the Practice Book. They can do this directly after the main activity, as homework, or as the focus of a separate mathematics session to help students consolidate their learning and build fluency. Look at the worked example. *How is the picture of the rods and cubes different from the array of a multiplication? What do they both show that is the same?*

Differentiated outcomes	
All students	should be able to count in twos.
Most students	will recognise that counting in twos is the same as groups of 2.
Some students	may recall the facts in the 2 times table up to 24.

Answers

Student Book page 77

1

$0 \times 2 = 0$	0
$1 \times 2 = 2$	2
$2 \times 2 = 4$	$2 + 2 = 4$
$3 \times 2 = 6$	$2 + 2 + 2 = 6$
$4 \times 2 = 8$	$2 + 2 + 2 + 2 = 8$
$5 \times 2 = 10$	$2 + 2 + 2 + 2 + 2 = 10$
$6 \times 2 = 12$	$2 + 2 + 2 + 2 + 2 + 2 = 12$
$7 \times 2 = 14$	$2 + 2 + 2 + 2 + 2 + 2 + 2 = 14$
$8 \times 2 = 16$	$2 + 2 + 2 + 2 + 2 + 2 + 2 + 2 = 16$
$9 \times 2 = 18$	$2 + 2 + 2 + 2 + 2 + 2 + 2 + 2 + 2 = 18$
$10 \times 2 = 20$	$2 + 2 + 2 + 2 + 2 + 2 + 2 + 2 + 2 + 2 = 20$

2 All the totals in the 2 times table are even.

Practice Book page 77

1 $11 + 11 = 22$ $\qquad$ $2 \times 11 = 22$

2 $14 + 14 = 28$ $\qquad$ $2 \times 14 = 28$

3 $12 + 12 = 24$ $\qquad$ $2 \times 12 = 24$

4 $17 + 17 = 34$ $\qquad$ $2 \times 17 = 34$

Stretch zone: Students should show 1 rod and 8 cubes twice, then show these recombined (3 rods and 6 cubes).

Discover — Student Book page 78 · Practice Book page 78

Specific learning focus

- Use arrays to represent multiplication facts in the 5 times table

Global skills

- **Creative skills:** exploring
- **Real-world skills:** presenting information
- **Interpersonal skills:** communication

Key vocabulary

- array, row, column, multiples of five

Resources

- counters or cubes (at least 30 for each pair)

Language support

Focus on using the terms 'row' and 'column' in an array and link the numbers in the 5 times table fact to the numbers of rows and columns.

 Introductory activity

Give each pair of students 15 counters. Ask them to try to arrange them in equal rows of 5. They should make an array with 3 rows of 5. Ask students to describe their array in as many ways as they can, for example, 5 + 5 + 5, 3 × 5, 3 lots of 5, 3 groups of 5, 3 × 5. Record their suggestions on the board, adding any they may have missed.

 Main activity

Look together at the Student Book page 78 and work on the first part of question 1 together as a whole class. Give each pair of students 20 counters and ask them to arrange them in equal rows of 5 counters in an array. Ask: *How many lots of 5 have you made? Can you write that as an addition of lots of 5? Now can you write a multiplication fact to show how many lots of 5 you have?* Students should be able to tell you that there are 4 groups of 5 counters, and write it as 5 + 5 + 5 + 5, or as a multiplication fact, 4 × 5.

Students complete the rest of the questions on page 78 of the Student Book in pairs. Each pair needs 30 counters. For each multiplication, they should make an array showing rows of 5. Ask them to discuss the two questions in the speech bubbles in their pairs.

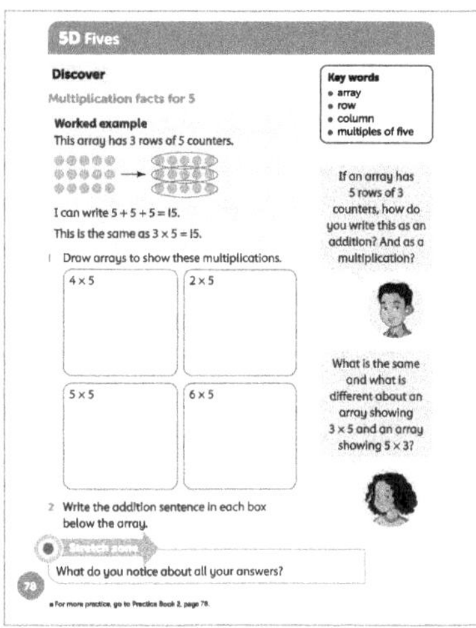

Differentiation

Supporting: Model counting the numbers of rows and counters in the array by touching the five counters in the first row, then touching each row in turn and counting up in fives.

Consolidating: Ask students to count the rows and then how many counters in a row to reinforce the idea of 'lots of'.

Extending: Ask students to extend their arrays to more than 10 lots of 5 and write the multiplication sentences.

Stretch zone: *What do you notice about all your answers?*

Students should notice that all answers end in 0 or 5.

 Reflection time

Ask students to share their arrays and talk about how they formed them. Can they describe how the rows and counters relate to the addition and multiplication sentences?

Practice Book: Students complete page 78 of the Practice Book. They can do this directly after the main activity, as homework, or as the focus of a separate mathematics session to help students consolidate their learning and build fluency. Students can also record each multiplication as a repeated addition. The stretch zone question is quite challenging so students may prefer to give their answer orally.

Differentiated outcomes	
All students	should make arrays for multiples of 5 showing rows of 5.
Most students	will make an array in rows of 5 and record the corresponding multiplication fact.
Some students	may understand the commutative relationship that is expressed in an array.

Student Book page 78

1 Check that students have drawn the correct arrays to match the multiplication sentences.

2 $5 + 5 + 5 + 5 = 20$

 $5 + 5 = 10$

 $5 + 5 + 5 + 5 + 5 = 25$

 $5 + 5 + 5 + 5 + 5 + 5 = 30$

Practice Book page 78

1 $8 \times 5 = 40$

2 $9 \times 5 = 45$

3 $12 \times 5 = 60$

Check that students have drawn the correct array in each case.

Stretch zone: Students might say that all multiples of 5 have a 0 or 5 in the ones place because odd multiples of 5 end in 5 but even multiples of 5 are also multiples of 10, which end in 0. Alternatively, they may say that when you count up in fives from zero, you first land on 5 and then a second jump lands on 10, which ends in 0. You repeat this pattern each time you make a jump so it is always a 5 or a 0.

5D Fives

Explore Student Book page 79 · Practice Book page 79

Specific learning focus

- Understand multiplication as repeated addition and use the × sign for the 5 times table.

Global skills

- **Creative skills:** exploring/investigating
- **Real-world skills:** presenting information
- **Interpersonal skills:** communication

Key vocabulary

- multiples, multiplication fact, multiple of 5

Resources

- 0–50 number line for display
- counter or cubes
- poster-sized piece of paper

Language support

Add the 5 times table to the classroom display to support students in their discussions about the 5 times table. You could also put up two large poster-sized pieces of paper, one with the heading 'These numbers are the 5 times table. They have 0 or 5 in the ones place' and one with the heading 'These numbers are not in the 5 times table. They do not have 0 or 5 in the ones place'. Encourage students to write numbers over 50 on the correct poster.

Introductory activity

Count around the class in fives from zero to 50 and then back down to zero. Say, *We are counting in multiples of 5.* If possible, ask students to stand in a circle so that you can count round the circle. If students are confident, replace every other count with a clap. So the count will be: 5, clap, 15, clap, 25,…

Repeat counting in fives to 100 and back.

 **Main activity**

Display or draw a 0–50 number line on the board. Ask students how many jumps of five are needed to make 15 and record this as $5 + 5 + 5 = 15$. Continue as you did with twos, building an array 'staircase'.

Ask students whether they can remember the shorter way of writing these. Write the matching multiplication fact next to each addition.

Ask students to look at page 79 in the Student Book. Leave the first line ($0 \times 5 = 0$) until reflection time. Talk through with students the pattern that is forming for 1×5, 2×5 and 3×5. *How does each line change from the previous one?* Students should be able to say that each line has another group of 5 and so the addition sentence has an extra 5 added. After they complete the multiplication for $5 + 5 + 5$ (3×5), students should complete the remaining rows independently or in pairs. They should have access to counters or cubes to support them if needed.

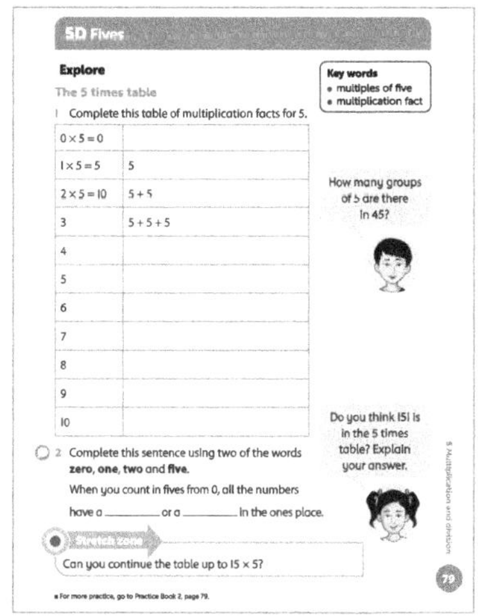

Differentiation

Supporting: Give students counters and number lines to support them with multiplication facts for fives.

Consolidating: Ask students to predict whether a number will be included when counting on in fives from a range of numbers.

Extending: Ask students to work out **multiples of 5** beyond 50.

Stretch zone: *Can you continue the table up to 15 × 5?*

Students should be able to show how they use repeated addition to find the multiples of 5 up to 75. They can also use the pattern of 0 and 5 in the ones place to help them.

 ### Reflection time

Ask students: *What should we write in the column next to 0 × 5 = 0?* (nothing) *Why is this?* Students might say that no groups of 5 means there is nothing to add.

Ask students in pairs to look at the multiplication facts (times tables) for twos and fives. *Which numbers appear in both sets of facts? Which only appear in one set of facts?* Ask students to discuss this in pairs and try to explain the why some numbers appear in both sets of multiplication facts. Students may notice that the numbers in both times tables end in 0: numbers ending in 0 are the only even numbers in the 5 times table and all the numbers in the 2 times table are even.

Practice Book: Students complete page 79 of the Practice Book. They can do this directly after the main activity, as homework, or as the focus of a separate mathematics session to help students consolidate their learning and build fluency. Play a round of the game together. *How will you decide who is correct?* They may suggest counting up in fives, using a 100-square or checking their answers from page 79 of the Student Book.

Differentiated outcomes	
All students	should count in fives to 100.
Most students	will recognise that counting in fives is the same as groups of 5 and record the number of groups of 5 as a multiplication or repeated addition sentence.
Some students	may recall the facts in the 5 times table up to 60.

Answers

Student Book page 79

1

$0 \times 5 = 0$	0
$1 \times 5 = 5$	5
$2 \times 5 = 10$	$5 + 5 = 10$
$3 \times 5 = 15$	$5 + 5 + 5 = 15$
$4 \times 5 = 20$	$5 + 5 + 5 + 5 = 20$
$5 \times 5 = 25$	$5 + 5 + 5 + 5 + 5 = 25$
$6 \times 5 = 30$	$5 + 5 + 5 + 5 + 5 + 5 = 30$
$7 \times 5 = 35$	$5 + 5 + 5 + 5 + 5 + 5 + 5 = 35$
$8 \times 5 = 40$	$5 + 5 + 5 + 5 + 5 + 5 + 5 + 5 = 40$
$9 \times 5 = 45$	$5 + 5 + 5 + 5 + 5 + 5 + 5 + 5 + 5 = 45$
$10 \times 5 = 50$	$5 + 5 + 5 + 5 + 5 + 5 + 5 + 5 + 5 + 5 = 50$

2 When you count in fives from 0, all the numbers have a 5 or a 0 in the ones place.

Practice Book page 79

Check that students have multiplied their number by 5 correctly each time.

Stretch zone: Students answers will vary. Ask them to explain their choices.

5E Tens

Discover Student Book page 80 • Practice Book page 80

Specific learning focus

- Use arrays to represent multiplication facts in the 10 times table.

Global skills

- **Creative skills:** exploring/investigating
- **Real-world skills:** presenting information
- **Interpersonal skills:** communication/teamwork

Key vocabulary

- array, row, column, multiples of 10, groups of

Resources

- counters or cubes (at least 60 per pair)
- tens-rods

Language support

Focus on using the terms 'row' and 'column' in an array and link the numbers in the 10 times table to the numbers of rows and columns.

 ### Introductory activity

Begin by counting in groups of 10. You can hold up some tens-rods and ask, *How many groups of 10?* Students count the rods and tell you how many groups of 10 there are, and the total. For example, hold up 4 rods. Students say that there are 4 groups of 10, which is 40. Repeat with other numbers of rods. Students can complete Practice Book page 80 individually, if appropriate.

 **Main activity**

Give each pair of students 30 counters. Ask them to try to arrange them into equal rows of 10. They should find 3 rows of 10. *Look at your rows. How can we describe this array as a repeated addition, as a 'groups of' statement and as a multiplication?* Draw the array on the board and then use students' answers to say and write on the board $10 + 10 + 10 = 30$, 3 groups of $10 = 30$ and $3 \times 10 = 30$.

Ask students to make 40 counters into equal rows of 10 counters in an array and record the repeated addition and mutliplication fact.

Ask students to look at the worked example on page 80 of the Student Book. Display it on the IWB, if possible. *Does your array match this picture?* Students can now complete the questions on page 80 of the Student Book.

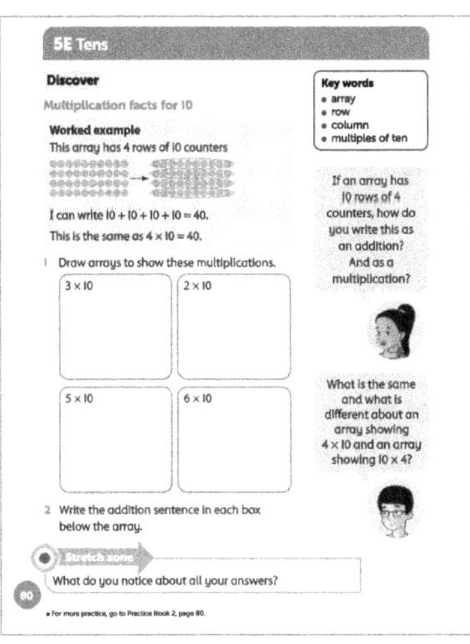

Differentiation

Supporting: Model counting the numbers of rows and counters in the array by touching the ten counters in the first row, then touching each row in turn.

Consolidating: Ask students to describe their arrays using 'groups of'.

Extending: Ask students to extend their arrays to more than 6 groups of 10 and write the multiplication sentences.

Stretch zone: *What do you notice about all your answers?*

Check what students have written. They may have described the answers as groups of 10 or said that they all end in 0 or have 0 in the ones place. You can suggest that they call them multiples of 10 to embed that vocabulary.

 **Reflection time**

Ask students to share their arrays and talk about how they formed them. Can they describe how the rows and counters relate to the addition and multiplication sentences? Ask students who completed the stretch zone question to share their answers. Reinforce that when you add groups of 10 your total will be a **multiple of 10**.

Read the top speech bubble on page 80 of the Student Book. Ask pairs of students to make the array and write the matching repeated addition and multiplication fact $(10 \times 4 = 40)$. *What is the same and what is different about these facts and the ones in the worked example?* (The total is the same but the size of group and the number of groups are different.)

Practice Book: Students can complete page 80 of the Practice Book if not already completed in the introduction (the activity works well as a quick follow-on written activity from the introductory activity). They could also do this directly after the main activity, as homework, or as the focus of a separate mathematics session to help students consolidate their learning and build fluency.

Differentiated outcomes	
All students	should make arrays showing rows of 10 and say how many groups of 10 and how many altogether.
Most students	will make an array in rows of 10 and record the corresponding repeated addition and multiplication fact.
Some students	may understand the commutative relationship that is expressed in an array.

Answers

Student Book page 80

1 Check that students have drawn the correct arrays to match the multiplication sentences.

2 $10 + 10 + 10 = 30$

$10 + 10 = 20$

$10 + 10 + 10 + 10 + 10 = 50$

$10 + 10 + 10 + 10 + 10 + 10 = 60$

Practice Book page 80

1 30

2 60

3 80

4 70

5 100

Stretch zone: 110. Answers to the second part will vary.

5E Tens

Explore Student Book page 81 • Practice Book page 81

Specific learning foci

- Learn and recognise multiples of 2, 5 and 10 and derive the related division facts.
- Understand multiplication as repeated addition and use the × sign.

Global skills

- **Creative skills:** exploring/investigating
- **Real-world skills:** presenting information
- **Interpersonal skills:** communication/teamwork

Key vocabulary

- multiples, multiplication facts

Resources

- base-10 equipment, 0–100 number lines
- poster-sized pieces of paper
- digit cards 0–9

Language support

Add the 10 times table to the classroom display. You could also put up two large poster-sized pieces of paper, with the headings 'These numbers are in the 10 times table. They have 0 in the ones place.' and 'These numbers are not in the 10 times table. They do not have 0 in the ones place.' Encourage students to write numbers over 100 on the correct poster.

 **Introductory activity**

Count around the class in tens from zero to 100 and then back down to zero. If possible, ask students to stand in a circle so that you can count round the circle. If students are confident, replace every other count with a clap. So the count will be: 10, clap, 30, clap, 50, ….

Repeat counting in tens to 100 and back.

 **Main activity**

Ask students to complete the table on page 81 in the Student Book individually. *Why is there nothing in the first line?* (The first line is left blank as there is nothing to write for 0 × 10 because zero groups means there are zero things.) Talk through with the students the pattern that is forming for 1 × 10, 2 × 10 and 3 × 10. *How does each line change from the previous one? Can you use the pattern to predict what the next line in the table would be after 10? And the next?* They should be able to say that the next line will have another group of 10 and so the

addition sentence has an extra 10 added, it will be a multiple of 10 and will have a zero in the ones place. Ask students to discuss the questions in the speech bubbles first in pairs and then come together as class to discuss answers.

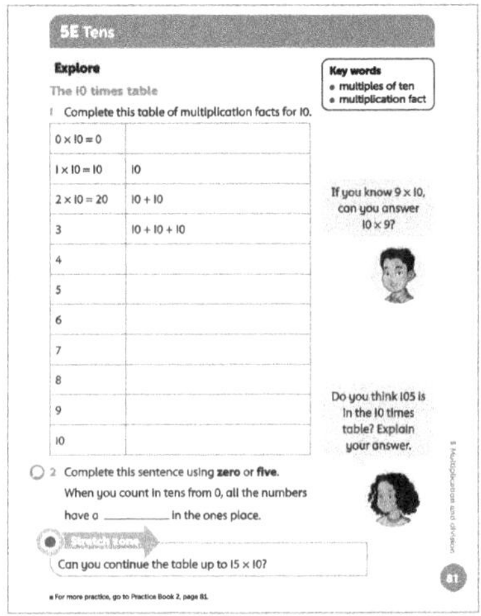

Differentiation

Supporting: Give students base-10 equipment and number lines to support counting in tens.

Consolidating: Ask questions to connect thinking about groups of 10, repeated addition and multiplication facts. *Can you show me 8 groups of 10 in your table? How do you know?*

Extending: Ask students to work out multiples of 10 beyond 100.

Stretch zone: *Can you continue the table up to 15 × 10?*

Students should be able to show how they use repeated addition to find the multiples of 10 up to 150.

 **Reflection time**

Ask students, in pairs, to look at the multiplication facts for twos, fives and tens referring back to previous pages in their Student Books. *Which numbers appear in all three sets of facts? Which appear in only one set of facts?* They should discuss this in pairs and try to explain why some numbers appear in several sets of multiplication facts.

Discuss as a class. Students should notice that the numbers ending in 0 that appear in the 2 times table will also appear in the 5 and 10 times tables and that all numbers ending in 0 in the 5 times table are also in the 10 times table. Draw out, if necessary, that these are all multiples of 10 and that multiples of 10 are even.

Practice Book: Students can complete Practice Book page 81. They can do this directly after the main activity, as homework, or as the focus of a separate mathematics session to help students consolidate their learning and build fluency. Discuss an example with students to ensure that they understand how to fill in the table.

<table>
<tr><td colspan="2">Differentiated outcomes</td></tr>
<tr><td>All students</td><td>should repeat the multiplication facts up to 10 × 10 in order.</td></tr>
<tr><td>Most students</td><td>will work out the multiplication facts to 10 × 10 in any order.</td></tr>
<tr><td>Some students</td><td>may see the pattern and predict the 'next answer' when doing the 10 times table in order.</td></tr>
</table>

Answers

Student Book page 81

1

$0 \times 10 = 0$ 0

$1 \times 10 = 10$ 10

$2 \times 10 = 20$ $10 + 10 = 20$

$3 \times 10 = 30$ $10 + 10 + 10 = 30$

$4 \times 10 = 40$ $10 + 10 + 10 + 10 = 40$

$5 \times 10 = 50$ $10 + 10 + 10 + 10 + 10 = 50$

$6 \times 10 = 60$ $10 + 10 + 10 + 10 + 10 + 10 = 60$

$7 \times 10 = 70$ $10 + 10 + 10 + 10 + 10 + 10 + 10 = 70$

$8 \times 10 = 80$ $10 + 10 + 10 + 10 + 10 + 10 + 10 + 10 = 80$

$9 \times 10 = 90$ $10 + 10 + 10 + 10 + 10 + 10 + 10 + 10 + 10 = 90$

$10 \times 10 = 100$ $10 + 10 + 10 + 10 + 10 + 10 + 10 + 10 + 10 + 10 = 100$

2   When you count in tens from 0, all the numbers have a 0 in the ones place.

Practice Book page 81

The order of the answers will vary but should include five answers shown in the table below.

My number	× 2	× 5	× 10
1	2	5	10
2	4	10	20
3	6	15	30
4	8	20	40
5	10	25	50
6	12	30	60
7	14	35	70
8	16	40	80
9	18	45	90

Stretch zone: The answers in the ×5 column all end in 0 or 5. The answers in the ×10 column all end in 0.

5F Threes and fours

Discover   Student Book page 82 • Practice Book page 82

Specific learning focus

- Represent multiplication facts in the 3 times tables and 4 times tables

Global skills

- **Creative skills:** exploring
- **Interpersonal skills:** communication/teamwork

Key vocabulary

- array, row, column, multiples of 3, multiples of 4

Resources

- paper or plastic rectangles and triangles (4–8 per pair)
- counters or cubes

Language support

Display the 3 and 4 times tables and key vocabulary 'multiplied by', 'times', 'groups of', '×', 'multiples of 3', 'multiples of 4' and refer to these during the class activities.

 **Introductory activity**

Give each pair of students a small number of triangles. Ask them to count how many sides on one triangle (3). Now ask them to count how many triangles they have, and to work out how many sides in total there are in their set of triangles.

How did you find how many sides there are in total? Some pairs may say they counted all the sides to find the total. If no students suggest it, ask *Did anyone count in groups of 3? Why?* (Yes, because each triangle has 3 sides.) Work through an example and record it as a repeated addition and a multiplication fact. For example, if they have 6 triangles, they will have 6 groups of 3 sides, a total of 18 sides. Write this as $3 + 3 + 3 + 3 + 3 + 3 = 18$ and $6 \times 3 = 18$.

Now repeat with rectangles, asking pairs to work out how many groups of 4 they have.

Display on the board the following word problem: *A stool has 3 legs. How many legs are there in total on 5 stools?* Show a picture of (or draw) a stool with three clearly visible legs. Ask students to identify the key numbers in the word problem. Ask them to decide which number represents one 'group' and which number shows how many 'groups' are being counted. They should recognise that the calculation is 5 groups of 3 legs. Using counters or cubes, they can work out that 5 groups of 3 (5×3) = 15.

Students complete the activities on page 82 of the Student Book individually or in pairs. Explain that they need to look at the pictures and write a word problem. Read the example in the speech bubble and show how it relates to the picture of the tables. Explain and write any key vocabulary on the board to help students write their word problem. You can scribe students' word problems if they are struggling to write them.

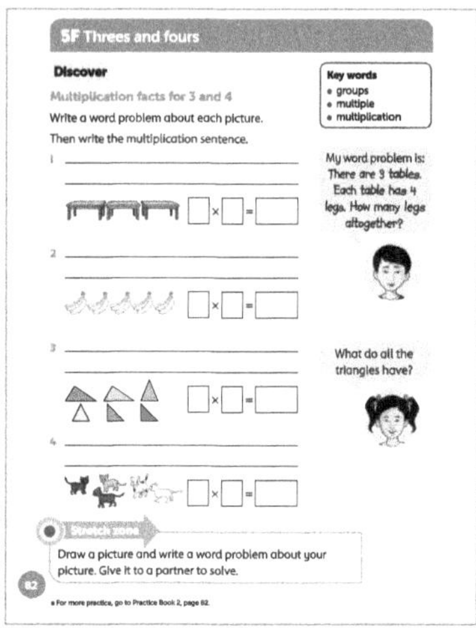

Differentiation

Supporting: Model counting the numbers of items of each object by touching them, then touching each object in turn (e.g. counting legs on each table and how many tables). Record each amount as you count to build a multiplication sentence. Model with counters arranged in an array.

Consolidating: Ask students to say their word problem aloud using 'groups of', for example, the monkey ate 5 groups of 3 bananas.

Extending: Ask students questions such as, *I add two more tables. What problem can you make up now? What multiplication sentence shows it?*

Stretch zone: *Draw a picture and write a word problem about your picture. Give it to a partner to solve.*

Check that students have drawn and written correct problems for their partners. You can also share these problems as part of the reflection time.

Ask some students to share their word problems that involve the 3 times or 4 times table from the Stretch zone question. Choose one student to try to write the multiplication sentence for the problem and then the rest of the class can try to solve it.

Practice Book: Students complete page 82 of the Practice Book. They can do this directly after the main activity, as homework, or as the focus of a separate mathematics session to help students consolidate their learning and build fluency. Encourage students to draw pictures of counters rather than the actual objects to help them solve the problems so they are not distracted by trying to draw the objects accurately.

Differentiated outcomes	
All students	should record a multiplication sentence for each problem with support.
Most students	will make up a simple multiplication word problem and write a corresponding multiplication sentence to match each picture.
Some students	may write their own word problems using their understanding of groups of 3 and 4.

Answers

Student Book page 82

1   There are 3 tables. Each table has 4 legs. How many legs altogether? $3 \times 4 = 12$

2   There are 5 bunches of bananas. Each bunch has 3 bananas? How many bananas altogether? $5 \times 3 = 15$

3   There are 6 triangles. Each triangle has 3 sides. How many sides altogether? $6 \times 3 = 18$

4   There are 5 cats. Each cat has 4 legs. How many legs altogether? $5 \times 4 = 20$

Practice Book page 82

1   16

2   18

3   20

4   15

5   9

Check that students have drawn a correct picture in each case.

Stretch zone: Check that students have created a multiplication word problem with an answer of 24 that they have correctly represented with a drawing

Explore — Student Book page 83 • Practice Book page 83

Specific learning focus

- Work out multiplication and division facts for the 3 times and 4 times tables.

Global skills

- **Creative skills:** exploring/investigating
- **Real-world skills:** interpreting information
- **Interpersonal skills:** communication/teamwork

Key vocabulary

- multiple, multiplication fact

Resources

- sticky notes, counting stick
- counters or cubes
- 100-squares (one per pair)
- a strip of squared paper, 3 squares wide and 10 squares long (one per student)

Language support

Add to the 3 and 4 times tables display two speech bubbles saying, *Which numbers are in both the 3 and 4 times tables? Why are they in both?* Include some blank speech bubbles for students to record their answers.

 **Introductory activity**

Use sticky notes to attach multiples-of-3 number labels to each of the lines on a counting stick. Practise counting forward along the counting stick, then remove two of the labels (e.g. 12 and 21).

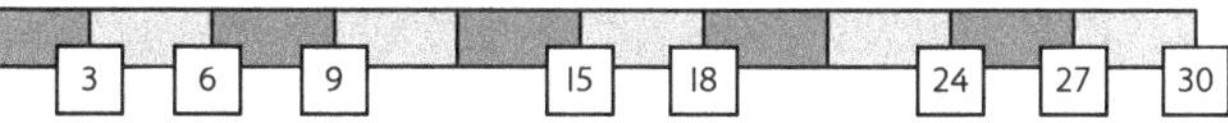

Count along again, helping students to fill in the missing numbers. Take two more of the labels away and repeat. Point to the lines where you have removed the labels and ask students which number should be there and how they know.

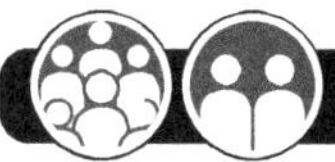 **Main activity**

Repeat the following activity from lesson 1E: Ask students to raise three fingers (or their thumb and two fingers) on their left hand and count along them in ones repeatedly to 30. They whisper or think the numbers for the first two fingers and only say the third one aloud, so that they are counting in threes. Ask students to do this in pairs to keep a check on each other. Remind them that each number they say aloud is a **multiple of 3**.

Give each pair a 100-square. Ask them to repeat their count and shade each number they say aloud. *What do you notice about the pattern?* (The shaded numbers make a line on the diagonal.) *Can you use the pattern to tell me what the next two multiples of 3 are after 30?* (33 and 36)

Repeat, for **multiples of 4** counting to 40.

Ask students to complete the questions on page 83 of the Student Book. *How can you use our shaded 100-squares to help you?* Point to '9' on the 100-square. *How many groups of 3 does this show?* Point to and count the shaded numbers up to and including 9 (3, 6, 9). *Can you write 3 groups of 3 as a multiplication and a repeated addition?* (3×3 and $3 + 3 + 3$)

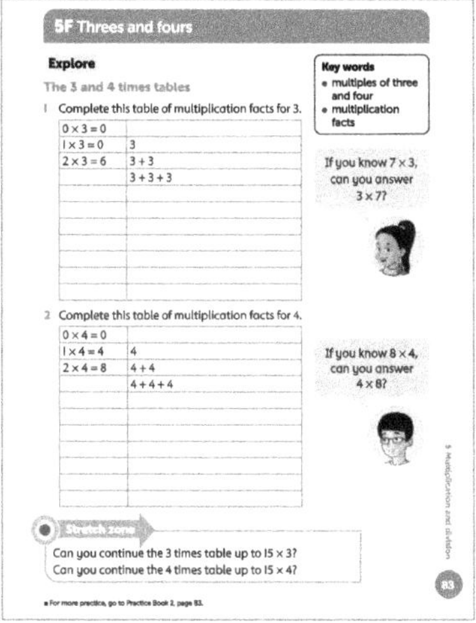

Differentiation

Supporting: Ask students to group and count counters or cubes to be secure in their answers.

Consolidating: Ask students some multiplication facts for 3 and 4 out of sequence.

Extending: Ask students to make up a word problem for facts for 7, 8 or 9 groups of 3 and 4.

Stretch zone: *Can you continue the 3 times table up to 15×3? Can you continue the 4 times table up to 15×4?*

Students should be able to show how they use repeated addition to find the multiples of 3 and 4 up to 60, for example, by building up in threes from 30.

 **Reflection time**

Give students a strip of squared paper, 3 squares wide and 10 squares long. Show them which way up to place the paper (the edge with 3 squares along the top), then ask them to number the squares, starting with 1 in the top left-hand corner, continuing along the row, carrying on with the next number on the next row.

1	2	3
4	5	6
7	8	9
10	11	12
13	14	15
16	17	18
19	20	21
22	23	24
25	26	27
28	29	30

What do you notice about the numbers in the right-hand column? What would happen if the paper were 4 squares wide?

Practice Book: Students complete page 83 of the Practice Book. They can do this directly after the main activity, as homework, or as the focus of a separate mathematics session to help students consolidate their learning and build fluency.

Differentiated outcomes	
All students	should complete the multiplication facts for 3 and 4 with support.
Most students	will work out all the multiplication facts for 3 and 4.
Some students	may use these facts to work out new facts beyond $\times$ 10.

Answers

Student Book page 83

1

$0 \times 3 = 0$ 0

$1 \times 3 = 3$ 3

$2 \times 3 = 6$ $3 + 3 = 6$

$3 \times 3 = 9$ $3 + 3 + 3 = 9$

$4 \times 3 = 12$ $3 + 3 + 3 + 3 = 12$

$5 \times 3 = 15$ $3 + 3 + 3 + 3 + 3 = 15$

$6 \times 3 = 18$ $3 + 3 + 3 + 3 + 3 + 3 = 18$

$7 \times 3 = 21$ $3 + 3 + 3 + 3 + 3 + 3 + 3 = 21$

$8 \times 3 = 24$ $3 + 3 + 3 + 3 + 3 + 3 + 3 + 3 = 24$

$9 \times 3 = 27$ $3 + 3 + 3 + 3 + 3 + 3 + 3 + 3 + 3 = 27$

$10 \times 3 = 30$ $3 + 3 + 3 + 3 + 3 + 3 + 3 + 3 + 3 + 3 = 30$

2

$0 \times 4 = 0$ 0

$1 \times 4 = 4$ 4

$2 \times 4 = 8$ $4 + 4 = 8$

$3 \times 4 = 12$ $4 + 4 + 4 = 12$

$4 \times 4 = 16$ $4 + 4 + 4 + 4 = 16$

$5 \times 4 = 20$ $4 + 4 + 4 + 4 + 4 = 20$

$6 \times 4 = 24$ $4 + 4 + 4 + 4 + 4 + 4 = 24$

$7 \times 4 = 28$ $4 + 4 + 4 + 4 + 4 + 4 + 4 = 28$

$8 \times 4 = 32$ $4 + 4 + 4 + 4 + 4 + 4 + 4 + 4 = 32$

$9 \times 4 = 36$ $4 + 4 + 4 + 4 + 4 + 4 + 4 + 4 + 4 = 36$

$10 \times 4 = 40$ $4 + 4 + 4 + 4 + 4 + 4 + 4 + 4 + 4 + 4 = 40$

Practice Book page 83

My number	$\times 2$	$\times 3$	$\times 4$
1	2	3	4
2	4	6	8
3	6	9	12
4	8	12	16
5	10	15	20
6	12	18	24
7	14	21	28
8	16	24	32
9	18	27	36

Stretch zone: The answers in the '$\times$ 4' column are double the answers in the '$\times$ 2' column.

5G Arrays

Discover
Student Book page 84 • Practice Book page 84

Specific learning focus
- Understand multiplication as describing an array.

Global skills
- **Creative skills:** exploring
- **Real-world skills:** presenting information/ interpreting information
- **Interpersonal skills:** communication/teamwork

Key vocabulary
- array, row, column, multiplication fact

Resources
- mini whiteboards and markers
- counters or cubes (16 per pair)
- ink stamps

Language support
Add an array to the classroom display. Ask students to help you decide how to label it. For example: This is an array; an array must have the same number of items in each row; an array must have the same number of items in each column; the number of items in each row and each column do not have to be the same.

 **Introductory activity**

Ask 12 students to come to the front of the class and sit in a 3 by 4 array. Ask the class whether they can think of another way to arrange the students in an array. *How many in each row and column?* Try out students' ideas, making sure that they include 1×12, 12×1, 2×6, 6×2, and 4×3. Record the matching multiplication number sentences.

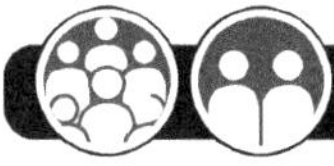 **Main activity**

Give each pair of students 16 counters. Ask them to arrange the counters in one long line and write the two matching multiplication sentences on their mini whiteboard ($16 \times 1 = 16$ and $1 \times 16 = 16$).

Ask students to move the end counter to start a second row under the first.

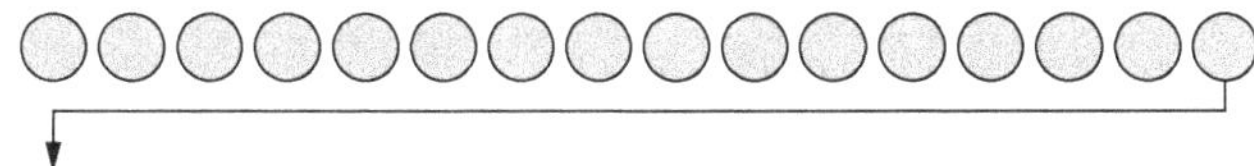

They should then keep moving counters until they have two equal rows. Then ask them to write the two matching multiplication sentences on their mini whiteboard ($2 \times 8 = 16$ and $8 \times 2 = 16$). Ask students to do the same to make 3 equal rows. Give them time to do this, then ask them what the problem is. Agree that they cannot make 3 equal rows using 16 counters. Now ask them to make 4 equal rows. This time there is no problem and they can record $4 \times 4 = 16$. Give each pair 8 more counters so that they have 24 counters altogether and ask them to explore arrays for 24, recording their arrays in the Student Book page 84. Remind them that they must use all the counters and have the same number in each row.

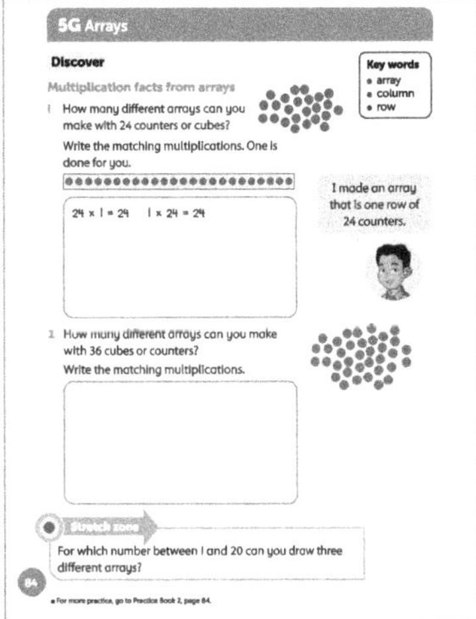

Differentiation

Supporting: Support students in making arrays using counters. Rather than replacing arrays, keep a record of the different arrays for each number.

Consolidating: Ask students how they know they have all the possible arrays. Ask them to record their answers and link to factors.

Extending: Ask students to record division facts as well as multiplication facts.

Stretch zone: *For which number between 1 and 20 can you draw three different arrays?*

Number 9 has three different arrays (1×9, 3×3, 9×1) Can students explain why? They should recognise that when you multiply a number by itself (3×3) it means instead of two possible arrays only one is possible.

 **Reflection time**

Ask students how many different arrays they found for 24. Draw a 6 by 4 array on a large piece of paper. Ask students whether they drew an array similar to yours. Turn the array a quarter of a turn so that it now shows a 4 by 6 array. Ask students what is the same and what is different. Establish that it is the same array and it does not matter which way up students drew it.

Practice Book: Students can complete page 84 of the Practice Book. They can do this directly after the main activity, as homework, or as the focus of a separate mathematics session to help students consolidate their learning and build fluency. If students have access to ink stamps, you may suggest they use these to create their arrays so they can record them more easily.

Differentiated outcomes	
All students	should create some arrays for each number.
Most students	will create all possible arrays for each number.
Some students	will work systematically and be able to explain how they know they have found all possible arrays.

Student Book page 84

1 $24 \times 1 = 24, 1 \times 24 = 24, 12 \times 2 = 24, 2 \times 12 = 24,$ $8 \times 3 = 24, 3 \times 8 = 24, 6 \times 4 = 24, 4 \times 6 = 24$

2 $36 \times 1 = 36, 1 \times 36 = 36, 18 \times 2 = 36, 2 \times 18 = 36,$ $12 \times 3 = 36, 3 \times 12 = 36, 9 \times 4 = 36, 4 \times 9 = 36,$ $6 \times 6 = 36$

Practice Book page 84

$20 \times 1 = 20, 1 \times 20 = 20, 10 \times 2 = 20, 2 \times 10 = 20,$ $5 \times 4 = 20, 4 \times 5 = 20$

Stretch zone: 8

5G Arrays

Explore Student Book page 85 · Practice Book page 85

Specific learning focus

- Understand multiplication as describing an array.

Global skills

- **Creative skills:** exploring/investigating
- **Real-world skills:** presenting information/interpreting information
- **Interpersonal skills:** communication/teamwork

Key vocabulary

- array, row, column, multiplication fact

Resources

- images of arrays taken from magazines or the internet, for example, aerial views of car parks, seed trays, boxes of candles, paint sets
- counters
- mini whiteboards and markers

Language support

Add example arrays and their matching number sentences to the classroom display. Include a written description of each array, e.g. 'This array has 6 rows of 5 stars.'

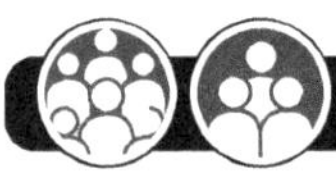 Introductory activity

Give each group a selection of images of arrays. Ask them to discuss what is the same and what is different about the images. Take feedback from the groups and use this feedback to focus on the key vocabulary of rows and columns.

 Main activity

Ask students to look at the worked example on page 85 of the Student Book. Display on the IWB, if possible. *How many multiplication facts can you see?* (2) *What does each describe?* Students should say that that 3×5 relates to 3 rows of 5 and 5×3 describes 5 columns of 3.

Ask pairs to look at one or two of the images from the introductory activity. They should work out the total number in their image and decide what multiplication facts are linked with it. Show some examples on the board and record the matching multiplication facts.

Students should then complete the activity on page 85 in the Student Book in pairs so they can discuss and answer at least the first question in the speech bubbles.

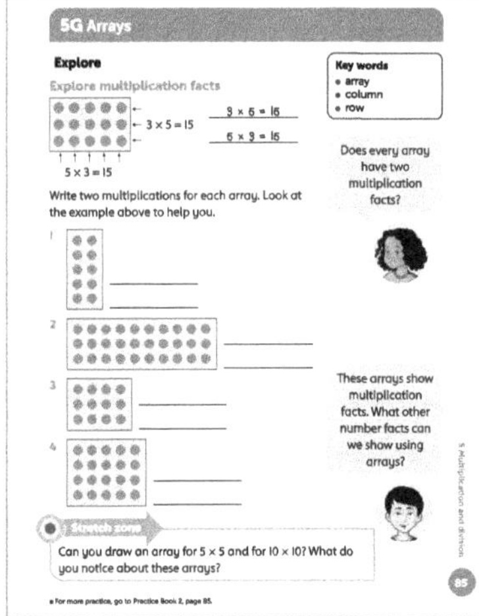

Differentiation

Supporting: Ask, *How many rows? How many in each row? How many altogether?* Use each of the student's answers to make a multiplication sentence in steps.

Consolidating: Ask students to identify which arrays represent particular facts.

Extending: Ask students the question in the second speech bubble on page 85 of the Student Book. They should make the connection to repeated addition but may also describe how an array represents division.

Stretch zone: *Can you draw an array for 5 × 5 and for 10 × 10? What do you notice about these arrays?*

Students should notice that if they rotate them the multiplication sentence stays the same.

 **Reflection time**

Choose an array from the Student Book or draw one and ask students to tell you a story to match the array. For example, for a 3 by 8 array: 3 rows of 8 cakes in a box. That's 24 cakes in the box.

Practice Book: Students complete page 85 of the Practice Book. They can do this directly after the main activity, as homework, or as the focus of a separate mathematics session to help students consolidate their learning and build fluency. Tell students to record their array as part of the stretch zone question on a separate piece of paper.

Differentiated outcomes	
All students	should find the total for each array by counting rows and columns.
Most students	will use their knowledge of multiplication facts to find totals.
Some students	will be able to find all possible arrays for a given number.

Answers

Student Book page 85

1   5 × 2 = 10, 2 × 5 = 10

2   3 × 10 = 30, 10 × 3 = 30

3   4 × 3 = 12, 3 × 4 = 12

4   5 × 4 = 20, 4 × 5 = 20

Practice Book page 85

1   6 × 3 = 18, 3 × 6 = 18

2   6 × 4 = 24, 4 × 6 = 24

3   7 × 5 = 35, 5 × 7 = 35

4   8 × 2 = 16, 2 × 8 = 16

Stretch zone: Check that students have drawn an array and written two matching multiplication sentences.

5H Division

Discover   Student Book page 86 • Practice Book page 86

Specific learning foci

- Understand division using arrays.
- Understand division as grouping and use the ÷ sign.

Global skills

- **Creative skills:** exploring/investigating
- **Real-world skills:** interpreting information
- **Interpersonal skills:** communication/teamwork

Key vocabulary

- array, equal groups, division, division sign (÷), divide

Resources

- counters or cubes for grouping
- poster-sized paper

Language support

Make a poster showing a large division sign to display in the classroom. Label it with, 'This is the division sign. It means arranging a quantity into equal groups of the same size.'

 **Introductory activity**

Ask 12 students to come to the front of the class and sit in a 3 by 4 array. Tell students that this time, you would like them to look at the whole array, 12 students. Ask: *If 12 students sit in rows of 3, how many rows are there?* Record on the board: 12 students in rows of 3 makes 4 rows. Under it, write 12 ÷ 3 = 4. Say, *12 shared into in groups of 3 makes 4 groups.* Explain that '÷' is the **division sign**. Slowly write it on its own on the board. Explain that it means 'shared into equal groups' or 'put into groups of' and we can say the division sentence as: *12 **divided** by 3 equals 4. We have divided 12 by 3.* Repeat for a 2 × 5 array of students.

 **Main activity**

Draw a 2 × 6 array on the board. Show students how to circle each row (or column) to help them see how many is in each row (or column) and how many rows (or columns). Say as you do this, for example, *I am circling a group of 2. And a second group of 2… that's 6 groups of 2 altogether. 12 arranged in groups of 2 is how many groups?* (6) Check that students have understood the two interpretations of the array as 12 ÷ 2 = 6 and 12 ÷ 6 = 2. Record both **division** sentences on the board and work through the example again, circling the relevant groups in the array and linking each part of the division sentence to the array.

Repeat with a 3 × 6 array.

Ask students to work in pairs to complete page 86 in the Student Book. Encourage them to continue to use counters or cubes to model the arrays or to draw rings around groups in each array.

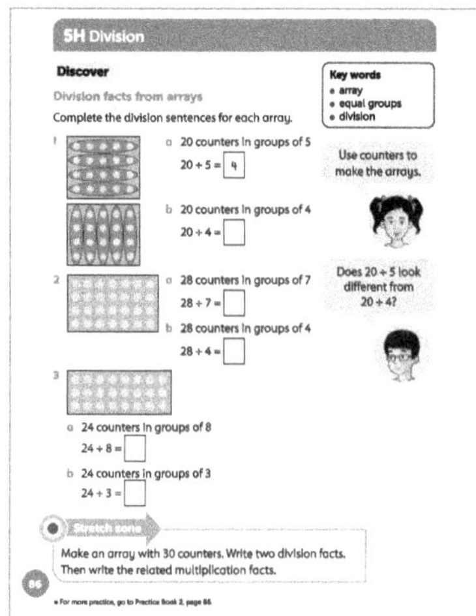

Differentiation

Supporting: Help write the answers for the correct division sentences for students who can explain division as equal grouping.

Consolidating: Encourage students to form their own arrays and write related division sentences.

Extending: Take a handful of cubes and ask students to arrange them in an array. Students use the array to write multiplication and division facts (e.g. $3 \times 6 = 18$; $6 \times 3 = 18$; $18 \div 3 = 6$; $18 \div 6 = 3$).

Stretch zone: *Make an array with 30 counters. Write two division facts. Then write the related multiplication facts.*

Students should be able to write, for example, $30 \div 6 = 5$, $30 \div 5 = 6$, or one of the other possible facts for 30, and the related multiplication facts.

 Reflection time

Discuss with students the questions they worked on in the Student Book. Make the link between the rows and columns of the arrays and equal groups. For example, in a 4×5 array, 20 can be seen as 4 equal groups of 5 or 5 equal groups of 4. Ask pairs of students to make up a matching story to go with one of the arrays shown in the Student Book. For example, for $20 \div 4 = 5$ the story could be: 20 students are going on a trip by car, 4 students can go in each car, so they take 5 cars.

Practice Book: Students complete page 86 of the Practice Book. Tell students that they may arrange the objects in an array to show and solve each calculation They can do this directly after the main activity, as homework, or as the focus of a separate mathematics session to help students consolidate their learning and build fluency. Students will need 40 counters or similar small household objects for this activity.

Differentiated outcomes	
All students	should relate arrays to division and division to grouping.
Most students	will represent division sentences as arrays and then record and solve them.
Some students	may understand that multiplication and division are inverse operations.

Answers

Student Book page 86

1 $20 \div 4 = 5$

2 $28 \div 7 = 4$
$28 \div 4 = 7$

3 $24 \div 8 = 3$
$24 \div 3 = 8$

Practice Book page 86

1 8

2 9

3 4

4 7

5 9

6 8

Stretch zone: Students should answer two of the following:

$16 \div 8 = 2$, $16 \div 2 = 8$, $16 \div 4 = 4$

5H Division

Explore Student Book page 87 • Practice Book page 87

Specific learning foci

- Understand multiplication as describing an array.
- Understand division as grouping and use the ÷ sign.

Global skills

- **Creative skills:** exploring
- **Interpersonal skills:** communication/teamwork

Key vocabulary

- array, equal groups, grouping, division, multiplication facts, division facts, fact family

Resources

- mini whiteboards and markers
- counters or cubes

Language support

Add the matching division sentences to the arrays already on display in the classroom.

 Introductory activity

Ask students to find at least two different arrays they can make using 18 counters. They should draw these on a mini whiteboard and write down as many facts as they can using this image. Encourage students to use repeated addition as well as multiplication and division. Share the results as a whole class and model how you record all the multiplication facts and division facts for each array. Include the 1 × 18 array.

 Main activity

Ask pairs to write a multiplication fact that they know from the 2, 5 or 10 times tables. They could draw the array or make it with counters or cubes. They then record the multiplication and **division facts** for that array. Explain that there is a **fact family** of multiplication and division sentences for every fact in the 2, 5 and 10 times tables.

Ask students to work in pairs to complete Student Book page 87. Check that they understand the relationship between the three numbers in each triangle. *Look at the bottom two numbers. What number would I get if I multiplied them?* (the top number) *What number would I get if I divided the top number with the number on the left?* (the number on the bottom right) Suggest that students continue to use counters or cubes to build the arrays to model their calculations.

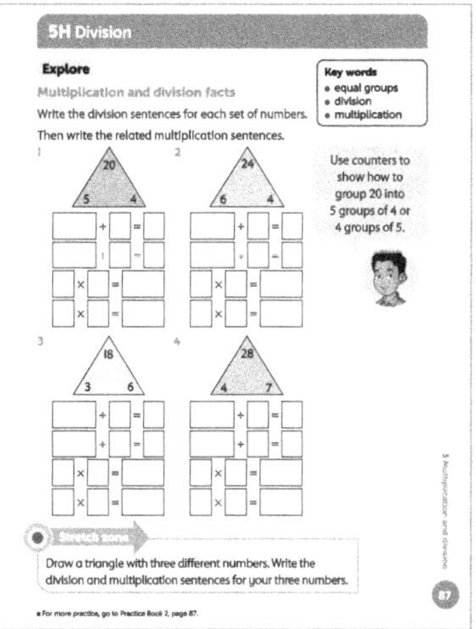

Differentiation

Supporting: Encourage students to use concrete resources to carry out calculations by creating arrays. Discuss the link between each row or column and the missing part of the multiplication and division sentences. You can help write the answers for students if necessary.

Consolidating: Ask students to explore patterns that they notice in the fact families and describe them to you.

Extending: Ask students to explore patterns that they notice in the fact families and explain why these patterns occur.

Stretch zone: *Draw a triangle with three different numbers. Write the division and multiplication sentences for your three numbers.*

Check that the numbers used and the facts derived are correct, and that the numbers are recorded in the correct positions in the triangle.

 Reflection time

Discuss with students the fact families they completed. Ask them to describe any patterns they noticed, perhaps that for each fact there was a reverse one, for example, 5 × 4 = 20 and 4 × 5 = 20, also 20 ÷ 5 = 4 and 20 ÷ 4 = 5.

Practice Book: Students complete the activities in the Practice Book, page 87. They can do this directly after the main activity, as homework, or as the focus of a separate mathematics session to help students consolidate their learning and build fluency. They make fact families for the 3 and 4 times tables. Ask, *Do you think there will be a fact family for all the facts in the 3 and 4 times tables?* Agree that if they can write a multiplication fact with 3 or 4, then there will also be a division fact that uses the same numbers as well.

Differentiated outcomes	
All students	should complete fact families using concrete resources for support.
Most students	will complete the fact families without support.
Some students	may describe and explain patterns in the fact families.

Answers

Student Book page 87

1 $20 \div 5 = 4, 20 \div 4 = 5, 5 \times 4 = 20, 4 \times 5 = 20$

2 $24 \div 6 = 4, 24 \div 4 = 6, 6 \times 4 = 24, 4 \times 6 = 24$

3 $18 \div 6 = 3, 18 \div 3 = 6, 6 \times 3 = 18, 3 \times 6 = 18$

4 $28 \div 7 = 4, 28 \div 4 = 7, 7 \times 4 = 28, 4 \times 7 = 28$

Practice Book page 87

1 $2 \times 3 = 6, 3 \times 2 = 6, 6 \div 3 = 2, 6 \div 2 = 3$

2 $4 \times 3 = 12, 3 \times 4 = 12, 12 \div 4 = 3, 12 \div 3 = 4$

3 $8 \times 3 = 24, 3 \times 8 = 24, 24 \div 8 = 3, 24 \div 3 = 8$

4 $10 \times 3 = 30, 3 \times 10 = 30, 30 \div 10 = 3, 30 \div 3 = 10$

5 $2 \times 4 = 8, 4 \times 2 = 8, 8 \div 2 = 4, 8 \div 4 = 2$

6 $4 \times 4 = 16, 4 \times 4 = 16, 16 \div 4 = 4, 16 \div 4 = 4$

7 $8 \times 4 = 32, 4 \times 8 = 32, 32 \div 8 = 4, 32 \div 4 = 8$

8 $10 \times 4 = 40, 4 \times 10 = 40, 40 \div 10 = 4, 40 \div 4 = 10$

Stretch zone: For example, for a 10×8 array, the fact family is: $10 \times 8 = 80, 8 \times 10 = 80, 80 \div 10 = 8, 80 \div 8 = 10$.

---

51 Remainders

Discover Student Book page 88 • Practice Book page 88

Specific learning foci

- Understand division as grouping and use the ÷ sign.
- Understand that division can leave some 'left over' and describe this as a remainder.

Global skills

- **Creative skills:** exploring/investigating
- **Interpersonal skills:** communication/teamwork

Key vocabulary

- division, divide, equal groups, grouping, remainder, left over

Resources

- counters or cubes
- mini-whiteboards and markers

Language support

Add to your classroom display a division calculation that has a remainder. Draw a matching illustration. Also display a sign saying: *'r' means remainder. The remainder is what is left over after dividing.*

 Introductory activity

Ask 16 students to come to the front of the class. Ask, *Can you group yourselves in pairs? Are there any students* **left over**? Repeat for groups of 4. Then ask the same students to arrange themselves in groups of 3, and then groups of 5. *Why is there a student left over?* (Because we can't divide 16 into equal groups of 3 or 5.)

Ask pairs to explore how many equal groups they can make with 24 counters and which **groupings** leave remainders. Introduce the word **remainder** for 'left over'.

 Main activity

Give each pair of students 10 cubes and ask them to use them to make groups of 2, and then groups of 3, and record the division sentence for each. They should then write the answer, saying how many are left over.

Go through each of the examples (e.g. 10 cubes in groups of 3). *How many altogether?* (10) *What size of group?* (groups of 3) *How many groups of 3?* (3) *How many are left over?* (1) *What is the remainder?* (1) Model writing the result as: $10 \div 3 = 3 \text{ r1}$

Explain that the 'r' stands for 'remainder', which is the same as saying 'left over'.

Students now complete page 88 of the Student Book. They should use concrete resources such as cubes to help them complete the questions.

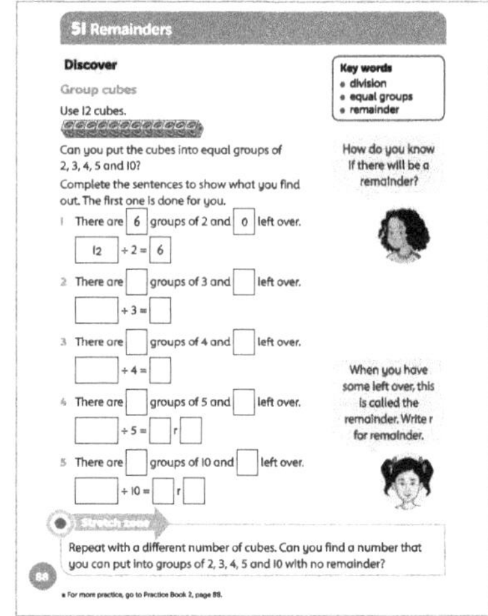

Differentiation

Supporting: Encourage students to use concrete resources to carry out calculations. You can help write the answers for these students if necessary.

Consolidating: Ask students to explain why we sometimes have some 'left over' and other times not.

Extending: Ask students to create a word problem that includes a remainder to match one of the questions.

Stretch zone: *Repeat with a different number of cubes. Can you find a number that you can put into groups of 2, 3, 4, 5 and 10 with no remainder?*

Students may use trial and error to find a number, using cubes. Encourage them to consider what they know about the 2, 5 and 10 times tables to help them narrow down the possibilities. For example, all numbers (products) in the 2 times table are even and all the numbers in the 10 times table are multiples of 10 so they end in 0. This means the answer can only be a number that ends in 0. Students can look back at their shaded number squares for the 3 and 4 times tables (lesson 5F Explore) for support.

 **Reflection time**

Make up a story about putting 25 cakes (or something similar) onto 2, 3, 4 and 5 plates ready for a party, talking about how many cakes are left over each time. The story ends with 5 cakes on each plate and none left over. Model this with pretend or real cakes.

Practice Book: Students can complete page 88 of the Practice Book. This can be done directly after the main activity, as homework, or as the focus of a separate mathematics session to help students consolidate their learning and build fluency. Encourage students to count out the amounts using counters or similar and arrange them in groups before drawing each picture.

Differentiated outcomes

Differentiated outcomes	
All students	should carry out the calculations using concrete resources and understand that when there are objects left over the number cannot be divided into equal groups.
Most students	will carry out the calculations using concrete resources and record remainders as 'r'.
Some students	may move away from using concrete resources and refer to their knowledge of multiplication and division facts.

Answers

Student Book page 88

2 4, 0, $12 \div 3 = 4$

3 3, 0, $12 \div 4 = 3$

4 2, 2, $12 \div 5 = 2$ r2

5 1, 2, $12 \div 10 = 1$ r2

Practice Book page 88

1 $13 \div 4 = 3$ r1

2 $18 \div 5 = 3$ r3

3 $17 \div 3 = 5$ r2

4 $9 \div 2 = 4$ r1

Stretch zone: Students will choose their own numbers to make a division problem. For example, Put 16 counters in groups of 5. $16 \div 5 = 3$ r 1.

5I Remainders

Explore Student Book page 89 • Practice Book page 89

Specific learning foci

- Understand division as grouping and use the ÷ sign.
- Understand that division can leave some left over.

Global skills

- **Creative skills:** problem solving/investigating
- **Interpersonal skills:** communication/teamwork

Key vocabulary

- division, equal groups, grouping, remainder, calculation

Resources

- counters or cubes
- mini whiteboards and markers

Language support

Refer to the classroom display of division calculations with remainders, and the sign saying: 'r means remainder. The remainder is what is left over after dividing.' Use these for support as students complete the activities.

 **Introductory activity**

Write '71' on the board and ask students whether it can be divided by 2 with no remainder. They should discuss this in pairs and then share their discussion. Students should notice that 71 is not an even number. Therefore, it cannot be split into two equal groups so it will have a remainder. Repeat for 76. Students may draw on their previous discussion to support them in solving this problem, that is, recognising that 76 is an even number because it ends in 6.

Repeat for 32 divided by 3, 41 divided by 4 and 57 divided by 5. Students should use their knowledge of the multiplication facts they have learned to support them in this discussion, as well as having cubes or counters available.

Give each pair 16 counters and ask them to find out whether 16 can be shared exactly into equal groups of 2, 3, 4, 5 and 10. They should record each result as a division **calculation**.

Next, ask them to choose two of their calculations to use to write word problems for another pair to solve.

They should stay in their pairs to complete the questions on page 89 of the Student Book. If some students may be distracted by drawing or find it hard to draw each group, you could suggest that they circle each group instead.

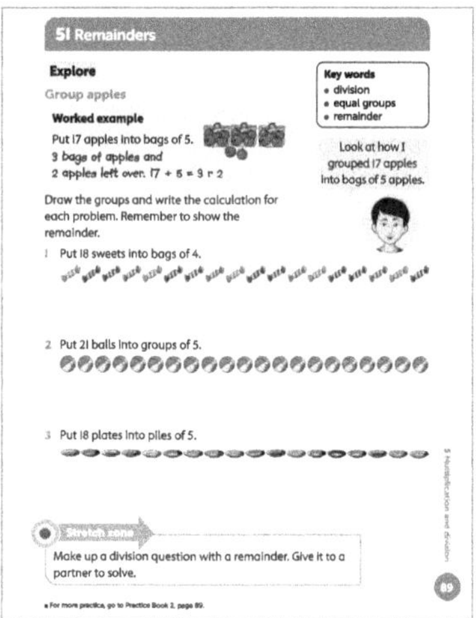

Differentiation

Supporting: Encourage students to use concrete resources to carry out calculations. You can help write the calculations for these students.

Consolidating: Choose one of the questions and ask students to explain how they know that they have calculated and recorded their answer correctly.

Extending: Ask students to create their own word problems that include remainders.

Stretch zone: *Make up a division question with a remainder. Give it to a partner to solve.*

Check that students' questions are sensible and that their partner can solve it.

 Reflection time

Ask pairs of students to come to the front of the class and explain how they carried out their calculations. Use these presentations to model the appropriate mathematical language focusing on the key words.

Practice Book: Students complete page 89 of the Practice Book. They can do this directly after the main activity, as homework, or as the focus of a separate mathematics session to help students consolidate their learning and build fluency. Encourage students to use counters or similar to support their calculations. This activity may suit being worked on over more than one lesson, as arranging and re-arranging seven arrays may be time consuming.

Differentiated outcomes	
All students	should carry out the calculations using concrete resources.
Most students	will use knowledge of multiplication and division facts to predict answers before putting objects into groups.
Some students	may create their own story problems involving remainders.

Answers

Student Book page 89

1 $18 \div 4 = 4$ r2

2 $21 \div 5 = 4$ r1

3 $18 \div 5 = 3$ r3

Practice Book page 89

		÷ 3	÷ 4	÷ 5	÷ 10
1	18	$18 \div 3 = 6$	$18 \div 4 = 4$ r2	$18 \div 5 = 3$ r3	$18 \div 10 = 1$
2	22	$22 \div 3 = 7$ r1	$22 \div 4 = 5$ r2	$22 \div 5 = 4$ r2	$22 \div 10 = 2$
3	26	$26 \div 3 = 8$ r2	$26 \div 4 = 6$ r2	$26 \div 5 = 5$ r1	$26 \div 10 = 2$
4	16	$16 \div 3 = 5$ r1	$16 \div 4 = 4$	$16 \div 5 = 3$ r1	$16 \div 10 = 1$
5	20	$20 \div 3 = 6$ r2	$20 \div 4 = 5$	$20 \div 5 = 4$	$20 \div 10 = 2$
6	23	$23 \div 3 = 7$ r2	$23 \div 4 = 5$ r3	$23 \div 5 = 4$r3	$23 \div 10 = 2$
7	25	$25 \div 3 = 8$ r1	$25 \div 4 = 6$ r1	$25 \div 5 = 5$	$25 \div 10 = 2$

Stretch zone: Answers will vary, for example, 'I know that 3×7 is 21 so I know that $22 \div 3$ would have a remainder.'

5 Multiplication and division

Connect Student Book page 90

Big idea

We can use repeated addition, grouping, equal sharing and arrays to help us multiply and divide.

Global skills

- **Creative skills:** investigating
- **Real-world skills:** research/presenting information
- **Interpersonal skills:** communication/teamwork
- **Self-development skills:** reflecting on learning

Key vocabulary

- multiply, multiplication, divide, division, equal grouping, remainder

Resources

- counters or cubes

Language support

Record sentence frames on the board to support students with the main activity. For example,
There are ___ students in the class today.
___ does not have a partner. Everyone is in a group.
___ is not in a group.

 **Introductory activity**

Tell students you would like them to count how many students are in the class today. Ask *What is a sensible way to work out how many students there are?* Take and discuss their suggestions. Students might suggest standing in a line to count each student one at a time, or counting how many students at each table and how many tables. Choose two strategies to count how many. *Did we get the same number both times? Was one way faster than another? Was one way easier than the other?*

Record the final number of students in the class on the board.

 **Main activity**

Ask students to look at page 90 of the Student book. Display it on the IWB, if possible. Read through the questions together.

Ask students to predict whether the class could be put into groups of 2, 3, 4, 5 and 10 without anyone being left out. Students can use cubes or counters to support them in checking whether their predictions are correct and drawing pictures to show their groups of 2, 3, 4, 5 and 10.

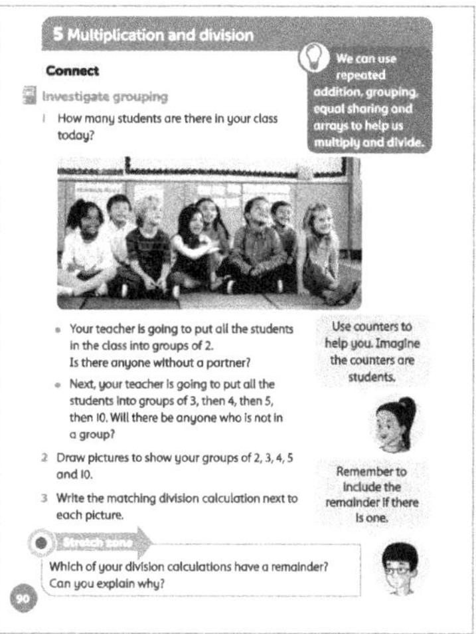

Differentiation

Supporting: Help students to structure their work, for example, recording their work in a table.

Consolidating: Ask students, for example, what array they would draw to show all the students in groups of 4. Do they think that there will be a remainder? Why or why not?

Extending: Ask students what they notice about making groups of 2 and 4, and groups of 5 and 10.

Stretch zone: *Which of your division calculations have a remainder? Can you explain why?*

This activity will help students show whether they have understood the concept of division and how it can be impossible to share some amounts equally when whole numbers are involved.

 **Reflection time**

Play a grouping game in an open space. Call out a number such as 2, 3, 4, 5 or 10. Students get into groups of that size. Students in groups that are too small or too big must sit out for the rest of that game. Keep calling out numbers until only two students are left. They are the winners.

Differentiated outcomes	
All students	should arrange counters in equal-sized groups of 2, 3, 4, 5 and 10 with support.
Most students	will arrange counters in equal-sized groups of 2, 3, 4, 5 and 10 and record it as a division sentence, including any remainders.
Some students	may use their knowledge of the relationship between numbers to adapt their groups rather than create new ones each time (e.g. make groups of 5 by halving existing groups of 10).

5 Multiplication and division

Review   Student Book page 91 · Practice Book page 90

Global skills

- **Creative skills:** problem solving/exploring
- **Interpersonal skills:** communication
- **Self-development skills:** reflecting on learning

Student Book

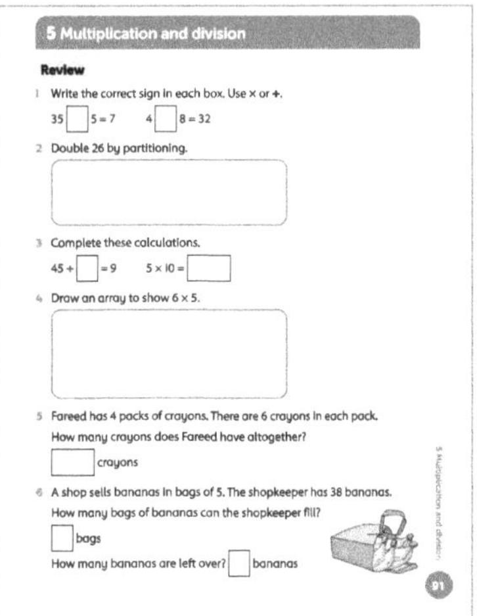

With young children, assessment activities are most effective when carried out as an everyday classroom activity. Students could have base-10 equipment, cubes and counters available to support them, if appropriate.

You may choose to ask students to explain their thinking after they have completed the questions. Asking them to explain what they did will help you gain a better understanding of their skills in this area. Ask questions such as: *How did you get started? What did you do next? How did that help you? What were you looking out for?*

Watch as students complete the sentences that represent multiplication and division situations. Check that they understand the language of multiplication, division, arrays and equal groups.

Answers

Student Book page 91

1 $35 \div 5 = 7$ $4 \times 8 = 32$

2 $26 = 20 + 6$, Double $20 + 6 = 40 + 12$, and $40 + 12 = 52$ so double 26 is 52.

3 $45 \div 5 = 9$ $5 \times 10 = 50$

4   Check for an array 6×5.

5   24 crayons

6   7 bags, and 3 bananas are left over

Practice Book

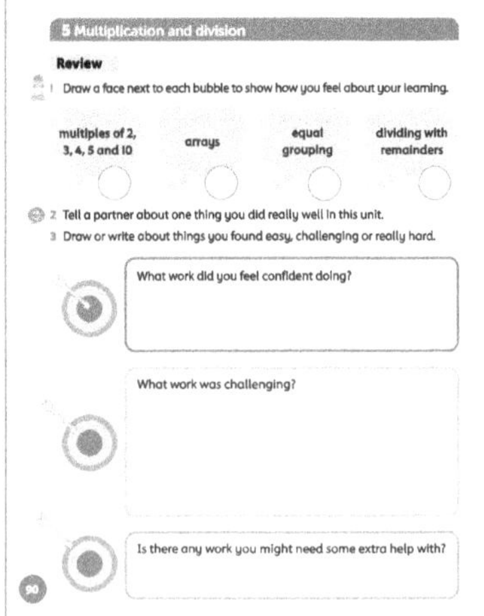

It is appropriate to complete this Practice Book review as a whole-class discussion. You may choose to keep a record of the class discussion or a copy of the review page for your own records. The review provides an opportunity for students to reflect on their learning from the unit, to discuss any areas of mathematics that they feel went particularly well, and any areas that they feel less confident about. Ensure all students have a copy of the Student Book as a reminder of the areas of mathematics that they have worked on in this unit.

Allow students plenty of time for discussion before asking them to complete the Practice Book page individually, and then, if appropriate, to share their responses with the rest of the class. If students complete this self-assessment at home, encourage them to discuss this with adults. Make a note of areas that students still feel unsure about.

Additional material

There are additional end-of-unit assessments available on the *Oxford Owl* website.

"

6 Fractions

Overview

Big Idea

The Big idea for this unit is that both shapes and quantities can be split into halves and quarters. Each part of a whole must be of equal size.

Finding half of a quantity can begin with finding half of two circles. Students find half of each circle and put them together to make one circle, discovering that half of 2 is 1. The idea of being fair is very useful when sharing a quantity between two to find halves and then between four to find quarters, as this emphasises that the pieces must be equal.

It is easy to spend a lot of time on fractions of shapes and then expect students to transfer their understanding to quantities with very little further experience. Students need plenty of practice of finding fractions of a whole when the whole is more than one object.

In this unit, this is done by arranging numbers as arrays to help find the fractional parts. We can then move onto fractions of a quantity from a secure level of understanding. There is a strong link between fractions and division; finding half of 6 is the same as dividing 6 by 2. We help students to start to make the link between division and fractions.

Look out for

- **Students whose everyday experience of the language of 'half' interferes with their understanding of fractions being made of equal parts.** Most students will have some experience of cutting something in half, but may have also heard people talk about, for example, the 'big' half and the 'small' half. They may accept both parts of a shape cut into two pieces as halves, even when the pieces are not the same size. Students need to recognise that they must check that the parts are equivalent, starting with two halves and then moving on to four quarters.

Possible misconceptions

- **Students may think that fractions with larger numbers are bigger,** e.g. $\frac{6}{8}$ must be bigger than $\frac{3}{4}$. This can be overcome by showing representations of a whole, marked to show the different fractions.
- **Students may not believe that fractions can be equivalent because the numbers are different,** e.g. that $\frac{1}{2}$ cannot be equivalent to $\frac{2}{4}$.
- **Students may make the false generalisation that a fraction has two numbers, 'a small one on top and a big one on the bottom'.** Use precise language when talking about fractions, referring to the denominator and numerator from early on in students' fraction learning and emphasising that the denominator is the number of equal parts and that the numerator is the number of those equal parts.

Key vocabulary

- fraction, half, halves, quarter, three-quarters, share, equal, part, whole, equivalent
- equal-sized, split, divide, same
- unequal, thirds, array
- numerator, denominator, row, column
- eighth, equivalent

Coverage in lessons

Learning focus	Learning outcomes (the ENC objectives)
Fractions of shapes	Recognise, find, name and write fractions $\frac{1}{3}$, $\frac{1}{4}$, $\frac{2}{4}$ and $\frac{3}{4}$ of a shape.
Fractions of numbers and amounts	Recognise, find, name and write fractions $\frac{1}{3}$, $\frac{1}{4}$, $\frac{2}{4}$ and $\frac{3}{4}$ of a length, set of objects or quantity.
Equivalent fractions	Recognise the equivalence of $\frac{2}{4}$ and $\frac{1}{2}$.

6 Fractions

Big question

- What sorts of fractions can we find?

Global skills

- **Interpersonal skills:** communication

Key vocabulary

- half, quarter, equal-sized, part, split, divide, fraction, equal, same, share, whole

Resources

- mini whiteboards and markers
- orange, string, jug of water, two glasses, modelling clay, small cake to cut in half and something to cut it with

Language support

Start a fraction display. Cut a large circle in half. Label one '$\frac{1}{2}$' and the other 'half'. Display the two halves together to show one whole circle. Add a label saying 'Two halves make one whole. Both halves must be the same size or they are not halves.'

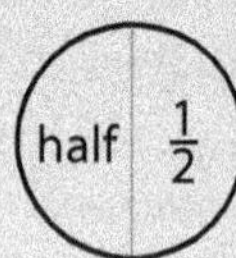

Introductory activity

Ask students to talk to their partner and draw any pictures that represent one **half**. Show students a collection of objects: an orange, a length of string, a jug of water and two glasses, a ball of modelling clay and a cake. Ask: *How do you find half of an orange, half of a piece of string, half of a jug of water, half of a piece of modelling clay and half of a cake?* As a class, share ideas and pictures. Talk through how to find half of each item, following the students' instructions (e.g. **splitting** the modelling clay into two balls of roughly the **same** size, placing them on a balance scale and adding or taking away clay from one and adding or taking away from the other until they balance; folding the piece of string so the ends touch and marking where the bend is to show half). Use this as a chance to introduce and model the key fraction vocabulary, for example, 'same', 'equal parts', 'half', 'halves', 'quarter', 'divide' and 'split'. Draw out that half means one of two **equal-sized** parts.

 **Main activity**

Ask students to look at page 92 of the Student Book. Display on the IWB, if possible *What are these children doing?* (Having a picnic) Ask students what they notice about the foods in the picture. Ask a series of questions to draw out students' understanding of fractions such as: *Can you see anything that has been cut into equal-sized pieces? How many **equal** pieces? If I ate one **part** of one of the sandwiches, how many parts would be left on the plate? Can you describe this as a **fraction**? If you ate half of the pizza, how many slices would that be? Can each child at the picnic have a **whole** sandwich? How many parts are in a whole sandwich? So how many **halves** is that?*

Read each of the speech bubbles. Students then work in pairs to answer each of the questions. Ask whether they can find any other ways to **share** the food equally.

Draw students' attention to the first bullet point on page 92. Point to the fractions and say, *What do we call these?* (fractions) Model how to read them, over-enunciating each part of the fractions: *one-third, one-quarter, two-quarters, three-quarters.* Explain that you will look more closely at these in the next lesson.

Differentiation

Supporting: Encourage students to use concrete resources to find halves and quarters.

Consolidating: Ask students to explain how they know that the parts are equal.

Extending: Ask students to come up with fraction stories relating to the picture (e.g. Toby ate two slices of pizza so only half a pizza was left for the other children).

 **Reflection time**

Ask pairs of students to share their solutions. Different pairs may suggest different ways, so make sure that the class can understand the different methods. Compare these methods and ask what is the same and what is different. For example, some students may suggest splitting each apple into four equal parts to share between the four children and others may suggest splitting each apple into two and giving one part to each child. Model with circles to show that in both cases children would get the same amount of apple.

6A Fractions of shapes

Discover
Student Book page 93 • Practice Book page 91

Specific learning foci
- Recognise when shapes are divided into equal and unequal parts.
- Recognise simple fractions of shapes such as halves, thirds and quarters.
- Shade simple fractions on shapes.

Global skills
- **Creative skills:** exploring/investigating
- **Real-world skills:** presenting information
- **Interpersonal skills:** communication/teamwork

Key vocabulary
- parts, equal, unequal, fraction, half, quarters, thirds

Resources
- assorted paper shapes (e.g. triangles, rectangles, diamond shapes) that are easy to cut in half
- scissors
- examples of a shape (e.g. rectangle, triangle – not a circle) split into halves; examples of a shape split into two unequal parts
- paper squares (at least four per small group)
- mini whiteboards and markers

Language support
Model the vocabulary of fractions for students to copy, for example 'half', 'third', 'quarter'. Make posters with students to show the meanings of the words. Use students' examples of halves from the introductory activity to make a 'halves' display.

 **Introductory activity**

Give students a range of paper shapes (e.g. a circle, rectangle, triangle, but not a square) to fold and cut in half. Remind them that a shape has not been cut in half unless it is cut into two equal parts, that is, both pieces are the same size. Explain that they may need to flip one half over or turn the shape round to check, or to put one piece on top of the other to see whether they match. Show students a few shapes (e.g. a rectangle or triangle) split into halves in more than one way, and also a shape split into two unequal parts. Use these to model how to check whether the two parts are equal or **unequal**. When students have halved a few shapes, take feedback and show examples on the board of ways to split each shape into halves.

Model how to record a fraction using digits by saying, *We have now cut up whole shapes. Into how many parts did you cut them to show halves?* (2) Record '$\frac{}{2}$' on the board. Hold up two parts of a shape and then lift one part higher. Point to it and say, *How many parts is this?* (1) Add '1' to '$\frac{}{2}$' to make the fraction $\frac{1}{2}$. *This is another way of writing one half.*

 **Main activity**

Arrange students in small groups and give each group several paper squares. Ask students to work together to find as many ways as they can of cutting a square in half. The shapes could be stuck into a workbook or on a sheet of paper. After exploring simple straight-line cuts through opposite sides or corners, students can explore more complex ways of cutting the square into two halves. Students can then move on to explore ways of finding thirds and quarters of the square of paper.

Talk to students about finding other fractions of a square. Build on the method of finding half of a square. *You split this square into 2 equal parts. Is it possible to split it into 4 equal parts?* Explain that a shape split into four equal parts is split into **quarters**. Students should try this, and then do the same for **thirds**. Model how to write these fractions with an emphasis on the number of parts in the whole and link to what each part is of that whole.

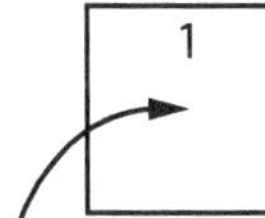 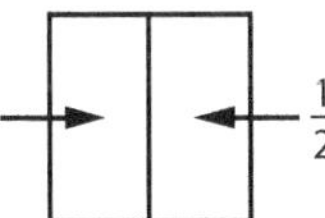 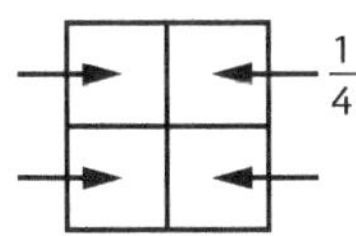

Students then complete the activities on Student Book page 93 in pairs. Encourage them to discuss their answer choices. Check that they are saying each fraction correctly and remind them of key vocabulary they need to describe fractions (e.g. parts, equal parts, whole).

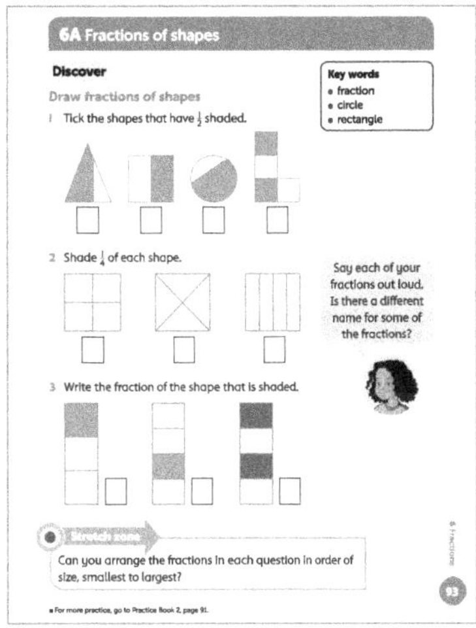

Differentiation

Supporting: Encourage students to describe what they see and then help them to make links to written fractions (e.g. *The shape has two parts so we need to record 2 here below the line*).

Consolidating: Ask students to explain how they know they have found different fractions of shapes.

Extending: Ask students to draw their own shapes for a partner to colour a half or other fractions.

Stretch zone: *Can you arrange the fractions in question 3 in order of size, smallest to largest?*

Students should identify that $\frac{1}{4}$ is the smallest, then $\frac{1}{3}$, then $\frac{2}{4}\left(\frac{1}{2}\right)$.

Ask students to share their ways of dividing a square into halves. Ask them to explain how they knew they had made two equal parts. Discuss how to divide a square into equal parts to show quarters and thirds. *What did you notice about the different parts? Which were the bigger parts, halves or thirds? Which is smaller, a quarter or a third? Which is bigger, a half or a quarter? Can you say how much bigger a half is than a quarter?* (double or twice as big)

Practice Book: Students complete page 91 of the Practice Book. They can do this directly after the main activity, as homework, or as the focus of a separate mathematics session to help students consolidate their learning and build fluency. Look at the first question together. Ask students to imagine a circle and think about how they would fold it to make quarters; or give them a paper circle to fold into quarters. Agree that they should fold into half and then fold that half into half again. Look at the circles in question 1. *How many of these parts match your quarter?* (2) Ask students to shade two different parts for each circle. *Does it matter which two parts?* Agree that it does not matter but that it must be two of these parts to show one quarter.

Differentiated outcomes	
All students	should find $\frac{1}{2}, \frac{1}{3}, \frac{1}{4}$ of a square.
Most students	will explain how they know they have found $\frac{1}{2}, \frac{1}{3}, \frac{1}{4}$ of a square.
Some students	will find $\frac{1}{2}, \frac{1}{3}, \frac{1}{4}$ and $\frac{3}{4}$ of other shapes and explain their method.

Answers

Student Book page 93

1   Students should have ticked the square and the L shape.

For question 1, students should be able to see that the shaded fraction of the square and the L-shape is the same.

2   Check that students have shaded $\frac{1}{4}$ of each shape.

3   $\frac{1}{3}, \frac{1}{4}, \frac{2}{4} \left(\frac{1}{2}\right)$

Practice Book page 91

Check that the shapes are coloured correctly to represent each fraction.

Stretch zone: Check that students have divided the square into equal parts and shaded it in two different ways to show, for example, $\frac{1}{2}$ and $\frac{1}{4}$.

6A Fractions of shapes

Explore   Student Book page 94 • Practice Book page 92

Specific learning focus

- Recognise and shade fractions of rectangles.

Global Skills

- **Creative skills:** exploring/investigating
- **Real-world skills:** presenting information
- **Interpersonal skills:** communication/teamwork

Key vocabulary

- half, quarter, three quarters, third, array

Resources

- squared paper
- counters or cubes

Language support

Use the language of 'half', 'quarter' and 'three quarters' regularly to remind students of it. Create a display of these words with images for each fraction.

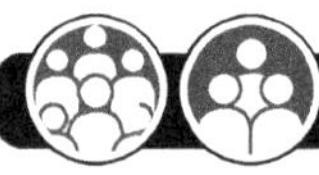 **Introductory activity**

Draw three grids on the board: 5×2, 8×4, 10×6. Ask students to think about how they might colour half of each grid. *How can you work out how much a half is in each case?'* Students might notice visually how many squares on each grid will be half. Or they may say they need to work out how many squares in total on each grid and then find half that number. For example, on a 5×2 grid, there are 10 squares, so half will be 5 squares. *Is there more than one way of doing it for each grid?* Once students recognise that half of 10 squares is 5 squares, they can find several ways of shading 5 squares. Ask students to work in small groups and give them squared paper. Allow some time for discussion and then share a solution for each grid.

 **Main activity**

Draw a 10×4 grid on the board. Ask students to say how many squares are in the grid. (40) Compare the grid to an **array**. *How many rows and columns does the grid have?* (10 rows and 4 columns, or 4 rows and 10 columns depending on which way it is drawn) Ask students how you could colour half of the grid. They may suggest you could colour half of the rows or half of the columns. If students don't suggest it, ask if you could colour half of the 40 squares in another way. *Would any 20 shaded squares be a half?* Agree that it would.

Repeat for a quarter of the grid. Show students the 10×4 grid with half shaded, say two columns, labelled $\frac{1}{2}$. Then show them $\frac{1}{4}$ shaded (one column), remind them that this is $\frac{1}{4}$ and ask them what they notice. They should see that it is half the number of shaded squares (or columns) for $\frac{1}{2}$.

What if I wanted to shade two quarters of the grid? How many squares would that be? How can I use what I know about $\frac{1}{4}$ of the total grid to help me? Give students some time to consider these questions in pairs. Then ask, *If one quarter is one column, how many columns will two quarters be?* Agree that it will be two columns or 20 squares. *Does it matter which 20 squares you shade to show two quarters?* (No)

Show students the grids on page 94 of the Student Book. *How many squares are in each of these grids?* (24) *How do you know?* Ask students to share their strategies for finding how many.

Ask students to work on the activities in pairs.

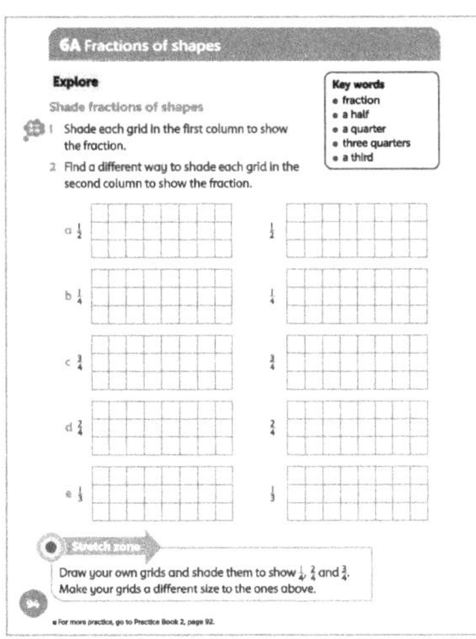

Differentiation

Supporting: Students could use counters or cubes for finding half or a quarter of a grid.

Consolidating: Ask students to explain how they found unit fractions of a grid.

Extending: Ask students to make up 'backward' fraction shading problems for another student solve (e.g. I shade 12 squares. What fraction is that?)

Stretch zone: *Draw your own grids and shade them to show $\frac{1}{4}$, $\frac{2}{4}$ and $\frac{3}{4}$. Make your grids a different size to the ones above.*

This activity will enable students to show their understanding of finding different numbers of quarters of each grid. Before they begin, suggest a size of grid to draw for each fraction, if appropriate, and explain that each grid needs to have an even number of squares.

Reflection time

Ask students to share how they shaded different fractions on the grids. Ask them to explain their method and talk about any similarities or differences with others' grids.

Practice Book: Students complete page 92 of the Practice Book. They can do this directly after the main activity, as homework, or as the focus of a separate mathematics session to help students consolidate their learning and build fluency. Remind students to think about how many squares for each grid.

Differentiated outcomes	
All students	should find a half or quarter of each grid and show it in one way.
Most students	will find different ways to show a half or quarter of each grid and at least one way to show a third, two quarters and three quarters of each grid.
Some students	may say that shading $\frac{1}{2}$ and $\frac{2}{4}$ of the same-sized grid means that the same number of squares will be shaded.

Answers

Student Book page 94

There will be various possible answers here – check that each colouring solution is correct.

The numbers of squares coloured should be as follows:

a 12 **b** 6 **c** 18 **d** 12 **e** 8

Practice Book page 92

There will be various possible answers here – check that each colouring solution is correct.

The numbers of squares coloured should be as follows:

1 6

2 12

3 18

4 8

5 6

Stretch zone: Students should see that $\frac{1}{4}$ of 24 squares is the same number of squares as $\frac{1}{2}$ of 12 squares.

6B Fractions of numbers and amounts

Discover
Student Book page 95 • Practice Book page 93

Specific learning foci

- Understand and use fraction notation, recognising that fractions are several parts of a whole.
- Begin to relate a fraction of a number to division.

Global Skills

- **Creative skills:** exploring/investigating
- **Interpersonal skills:** communication/teamwork

Key vocabulary

- fraction, half, quarter, three quarters, numerator, denominator, array

Resources

- cubes or counters
- strips of paper 12 cm and 24 cm long (one per pair)
- tens-rods and ones-cubes or 30 cm rulers

Language support

Emphasise the vocabulary of 'array', 'fraction', 'half', 'quarter' and 'three quarters' as students work on finding each number, and link the spoken fractions to $\frac{1}{2}$, $\frac{1}{4}$ and $\frac{3}{4}$.

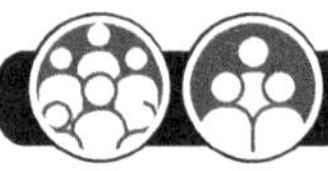 **Introductory activity**

Put four chairs in a row at the front of the class and ask four students to stand behind them. Now ask half of these students to sit down. Ask them to talk to each other to decide how many of them need to sit and who will sit. Ask *How did you work it out?* Emphasise the vocabulary and say that *half of 4 is 2.* Ask the class, *Can half of them sit down in a different way? How many different ways can you think of?* Ask students to explore this in small groups. (They may be able to find 6 different ways for half or them to sit down.) Then repeat for six chairs. *Is there more than one way to seat half the students? How many different ways can you find?* There are more possibilities here, 20 in all, so suggest that they find 8 ways. Again, emphasise the language by saying *half of 6 is 3.*

 **Main activity**

Give each pair of students counters or cubes to work with. Ask them to set out a pile of 12 counters. Explain that they are going to find half of the pile. Ask, *How can you do this?* Students might suggest sharing the counters into two equal groups, or making an array and splitting it into two equal parts. If appropriate to your students' progress, link the sharing of 12 counters into two groups to division ($12 \div 2 = 6$).

Now ask them to find a quarter of the pile of 12 counters. *How can you do this?* Students may say that a quarter is one of 4 equal parts, and suggest that sharing the counters into 4 equal groups will give quarters. They can do this with their counters to find that a quarter of 12 is 3 counters. Using this, ask them to find two quarters (6 counters) and **three quarters** (9 counters).

Next, give each pair a strip of paper 12 cm long. *How long do you think this strip of paper is? How can you find out?* Agree that they can use a ruler to measure its length. Alternatively, explain that a ones-cube is 1 centimetre long. They can line up a tens-rod and ones-cubes to find the total length. *How can I find half the length?* Students may fold the strip in half and then measure this with ones-cubes or may work out that half of 12 is 6 so one half would be 6 cm. Ask them to find $\frac{1}{4}$, $\frac{2}{4}$, $\frac{1}{3}$ and $\frac{3}{4}$ of 12 cm. Repeat with a strip of paper 24 cm long.

Students can then work on the activities on Student Book page 95 in their pairs. They may want to work with cubes or counters to model each problem or alternatively they could draw shapes and split them into the same number of parts as in the problem and then share the objects into each of the parts to find out what the fraction is for each amount.

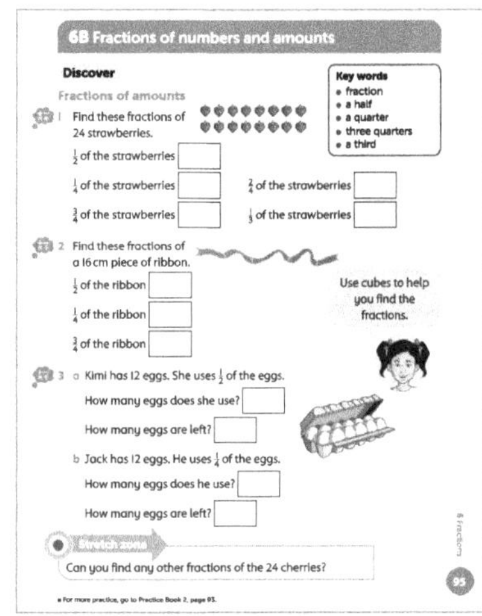

Differentiation

Supporting: Ask students to say fractions aloud as they work. Students should use concrete resources such as counters throughout.

Consolidating: Encourage students to predict answers before they check using concrete resources.

Extending: Encourage students to describe their reasoning as they carry out the calculation. Can they link finding unit fractions to dividing?

Stretch zone: *Can you find any other fractions of the 24 cherries?*

Students may find half (12), one quarter (6), three quarters (18), one third (8), two thirds (16). Some students may find eighths or twenty-fourths.

 **Reflection time**

Ask students to share what they found about sharing objects and lengths equally into different-sized groups to make different fractions in the Student Book activities. Ask questions to encourage them to compare fractions of amounts (e.g. *Which is more, $\frac{1}{2}$ of 24 strawberries or $\frac{1}{4}$? How much more? Which is more, $\frac{3}{4}$ of the ribbon or $\frac{1}{4}$? How do you know?*)

Practice Book: Students complete page 93 of the Practice Book. They can do this directly after the main activity, as homework, or as the focus of a separate mathematics session to help students consolidate their learning and build fluency. Students can circle fractions of each amount in the photos or model the amounts using cubes or counters (or similar objects).

Differentiated outcomes	
All students	should find halves and quarters using counters or cubes.
Most students	will carry out all the calculations using concrete resources.
Some students	may find answers by drawing on their knowledge of division facts.

Answers

Student Book page 95

1   12, 6, 18, 12, 8

2   8, 4, 12

3 **a** 6, 6   **b** 3, 9

Practice Book page 93

1	6	**3**	9	**5**	8	**7**	2
2	3	**4**	4	**6**	3	**8**	4

Stretch zone: That $\frac{1}{3}$ of 6 is half of $\frac{2}{3}$ or $\frac{2}{3}$ of 6 is double $\frac{1}{3}$ of 6.

6B Fractions of numbers and amounts

Explore Student Book page 96 • Practice Book page 94

Specific learning foci

- Understand and use fraction notation, recognising that fractions are parts of a whole.
- Begin to relate a fraction of a number to division.

Global skills

- **Creative skills:** exploring/investigating
- **Real-world skills:** presenting information/ interpreting information
- **Interpersonal skills:** communication/teamwork

Key vocabulary

- fraction, equal parts, split, array, row, column

Resources

- cubes or counters

Language support

As you work with students, model the phrase 'to find fractions of ___, put ___ counters in an array'. Ask them to repeat the phrase back to you, filling the gaps with the required number. Focus particularly on the key words 'array', 'row' and 'column'. Ask questions such as: *How many in that column? How many in that row?* Create a poster for class display containing a series of arrays with rows and columns labelled.

 **Introductory activity**

Draw an array on the board with 3 rows of 6. Give each pair of students 18 counters or cubes and ask them to copy the array using their counters. Ask students to split (divide) their counters to show two equal halves. Each half should have 3 rows of 3 counters, showing that 9 is half of 18.

Now ask them to show thirds by splitting the array into three equal parts (this could be one row of 6 for each part if split horizontally or three rows of 2 for each part if split vertically). *What does this show?* (That one third of 18 is 6.)

 **Main activity**

Ask students to look at page 96 in the Student Book. Display the page on the IWB, if possible. Discuss the worked example. *How many counters?* (8) *How did this student find fractions of 8?* Point out how the fractions of 8 are recorded in the table. Ask students to do the same for 12 using their counters/cubes and using what they know from the introductory activity.

Ask pairs to count out 16 counters/cubes and to use them to make an array. Ask students to share what arrays they made (e.g. 2×8, 8×2, 4×4). Ask them to split their array to show two identical halves. *How many counters in each half?* (8) *Now split those halves again to show quarters. How many counters in each quarter?* (4). From this, ask students to work out how many counters in three quarters. They should see that $\frac{3}{4}$ of 16 is 12, or 3 lots of 4. Tell them to record these fractions of 16 in the table in their Student Book.

Pairs now continue to find fractions of amounts to complete the rest of the table on page 96 in the Student Book.

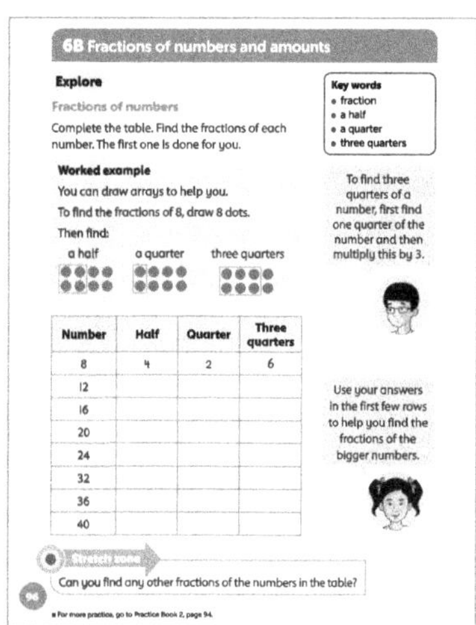

Differentiation

Supporting: Model calculation steps for students, for example, *To find half I split my array of 24 into 2 equal parts. I have 6 rows so I need to find half of 6. That is 3. Now I count how many counters in 3 rows. 12. Half of 24 is 12.*

Consolidating: Encourage students to describe their reasoning as they carry out the calculation.

Extending: Encourage students to predict answers for $\frac{1}{2}$ and $\frac{1}{4}$ before they check using concrete resources.

Stretch zone: *Can you find any other fractions of the numbers in the table?*

Students might use arrays to find thirds of 12, eighths of 16, or fifths of 10 or 20. Ensure that they understand the idea of equal parts or shares (e.g. to find thirds of 12, split an array of 12 counters into 3 equal parts).

 **Reflection time**

Ask students to share how they found three quarters of 24. *What array did they make with 24 counters? How many did they find in each half and then in each quarter?* Ask students why they think there was more than one way to make the arrays. Repeat with 32, 36 and 40.

Practice Book: Students complete page 94 of the Practice Book. They can do this directly after the main activity, as homework, or as the focus of a separate mathematics session to help students consolidate their learning and build fluency. Students can circle fractions of each amount in the drawings or model the amounts using cubes or counters.

Differentiated outcomes	
All students	should find halves and quarters using concrete resources.
Most students	will carry out all the calculations using concrete resources.
Some students	may find answers by drawing on their knowledge of division facts.

Answers

Student Book page 96

Number	Half	Quarter	Three quarters
8	4	2	6
12	6	3	9
16	8	4	12
20	10	5	15
24	12	6	18
32	16	8	24
36	18	9	27
40	20	10	30

Practice Book page 94

1 2   **4** 8   **7** 9

2 4   **5** 12   **8** 6

3 6   **6** 4   **9** 12

Stretch zone: Students should see that if they know how many $\frac{1}{4}$ is, then they need three of those to make $\frac{3}{4}$. For example, if $\frac{1}{4}$ of 12 is 3, then $\frac{3}{4}$ will be three 3s, or 9.

6C Equivalent fractions

Specific learning focus

- Find fractions that are equivalent.

Global skills

- **Creative skills:** exploring/investigating
- **Real-world skills:** interpreting information
- **Interpersonal skills:** communication/teamwork

Key vocabulary

- one half $\left(\frac{1}{2}\right)$, one quarter $\left(\frac{1}{4}\right)$, three quarters $\left(\frac{3}{4}\right)$, one eighth $\left(\frac{1}{8}\right)$, eighths, equivalent

Resources

- large fraction wall showing halves, quarters and eighths (as on Student Book page 98)
- paper circles
- number rods (optional)
- numbered cards, with fractions and their equivalents

Language support

Use a fraction wall and make numbered cards showing fractions and their equivalents, with their names in words. For example, $\frac{2}{4}$ = two quarters = one half = $\frac{1}{2}$. Emphasise the numerator and the denominator when counting fractions, for example 'One half $\left(\frac{1}{4}\right)$ is equivalent to two quarters $\left(\frac{2}{4}\right)$'.

Introductory activity

Ask students to look at the Student Book, page 97. Give out paper circles to each student to follow through the steps in the Student Book as a class.

Ask students to fold their circle in half and then unfold. Ask, *What fractions have you made? How do you know?* They might say they have two equal halves. Ask them to label the halves on their circle. Now ask them to fold the circle in half and then half again. Unfold and ask, *How many equal parts are there now?* (4) *What fraction is each piece?* $\left(\frac{1}{4}\right)$. Draw out the point that two quarters is the same as one half of the circle. Say that one half and two quarters are **equivalent**.

Finally, fold the circle again as before then fold it in half once more. Ask, *How many equal parts are there now?* (8) *What do you think one of these fractions is called?* (one eighth) Show students how to write one eighth as a fraction, $\frac{1}{8}$. *Can you see how many **eighths** are the same as*

a half? (4) *How many eighths are the same as a quarter?* (2) Ask students to answer question 4 in their Student Book.

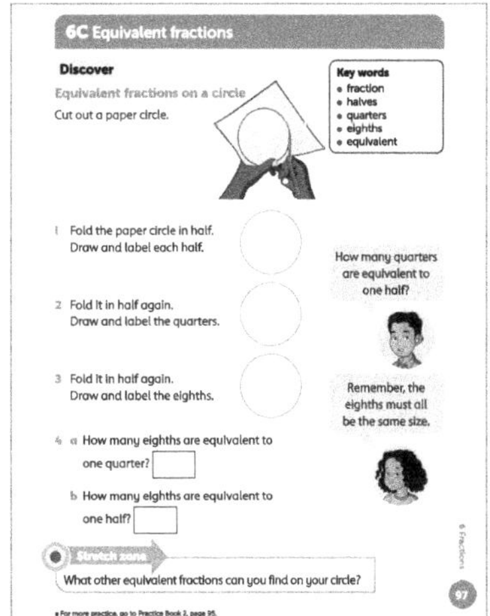

Main activity

Draw or display a large fraction wall on the board, showing a whole, halves, quarters and eighths.

Ask students in pairs to write down four things that they notice about the fraction wall. Take feedback and write some of the students' comments on the board.

Write on the board: _______ is **equivalent** to _______. Explain that 'equivalent to' means having the same value, and they saw some equivalent fractions when they folded paper circles. Ask students to work in pairs to complete the sentence five times using different fractions from the fraction wall. For example, $\frac{2}{4}$ is equivalent to $\frac{1}{2}$.

Differentiation

Supporting: You can help to write the equivalent fractions as students find them on the fraction wall.

Consolidating: Ask students to use concrete resources (number rods, for example) to explore equivalent fractions with eighths.

Extending: Ask students to explore more equivalent fractions using a fraction wall or number rods, for example thirds and sixths, fifths and tenths.

Stretch zone: *What other equivalent fractions can you see on your circle?*

Students might find on the circle that $\frac{2}{4}$ is the same as $\frac{1}{2}$ and that $\frac{3}{4}$ is the same as $\frac{6}{8}$.

 ## Reflection time

Ask students to share some of the equivalences they found and how they know they are equal. Can they use concrete resources? Can they draw a diagram? Can any students find three fractions that are all equal to each other?

Ask students questions to compare the relative size of fractions to reinforce that they are not equivalent and discuss which is bigger or smaller. For example, *Which* *fraction is equivalent to $\frac{1}{3}$ on the fraction wall?* (There is not one.) *Is there one that is bigger?* $\left(\frac{1}{2}\right)$ *Can you tell me one that is smaller?* $\left(\frac{1}{4}\right)$

Practice Book: Students complete page 95 of the Practice Book. They can do this directly after the main activity, as homework, or as the focus of a separate mathematics session to help students consolidate their learning and build fluency.

Differentiated outcomes	
All students	should say that a half and two quarters are the same on a circle and on a fraction wall.
Most students	will find fractions equivalent to halves, quarters and eighths on a circle and on a fraction wall.
Some students	may explore equivalences between other fractions and compare relative sizes of fractions.

Answers

Student Book page 97

4 a 2 **b** 4

Practice Book page 95

1 $\frac{2}{4}$, $\frac{4}{8}$, $\frac{5}{10}$

2 $\frac{2}{8}$

3 $\frac{1}{2}$, $\frac{2}{4}$, $\frac{3}{8}$

4 $\frac{1}{3}$, $\frac{1}{4}$, $\frac{3}{8}$

Stretch zone: Answers will vary depending on fractions chosen in the main part of the activity. Check that any equivalences given are correct.

6C Equivalent fractions

Explore   Student Book page 98 • Practice Book page 96

Specific learning focus

- Find equivalent fractions and use them to add simple fractions.

Global skills

- **Creative skills:** exploring/investigating
- **Real-world skills:** interpreting information
- **Interpersonal skills:** communication/teamwork

Key vocabulary

- half $\left(\frac{1}{2}\right)$, one quarter $\left(\frac{1}{4}\right)$, three quarters $\left(\frac{3}{4}\right)$, one eighth $\left(\frac{1}{8}\right)$

Resources

- fraction wall

Language support

Students may think that $\frac{4}{8}$ is bigger than $\frac{1}{2}$ because of the numbers in the fractions. Use lots of examples to show students that two fractions can be equal even when the numbers in the fractions are different. The language associated with equivalent fractions will support students' understanding of fractions.

Introductory activity

Remind students about the equivalent fractions that they found in the previous lesson. Ask whether any students can recall some of them, for example, can they name a fraction equal to two quarters? *What fraction on the wall was equivalent to $\frac{3}{4}$?*

Main activity

Ask students to look at the fraction wall on page 98 of the Student Book. Display on the IWB, if possible. *What does the top bar on the wall show us?* (1 whole) *How many parts in the second row?* (2) *What does each represent?* (a half) *How many quarters are the same as two halves?* (4) *How many quarters are the same as a half?* (2) *How many eighths are the same as a half?* (4)

Students then answer the questions on page 98 of the Student Book. They can discuss their answers in pairs and make up similar questions and take turns to ask and answer them with their partner using the fraction wall.

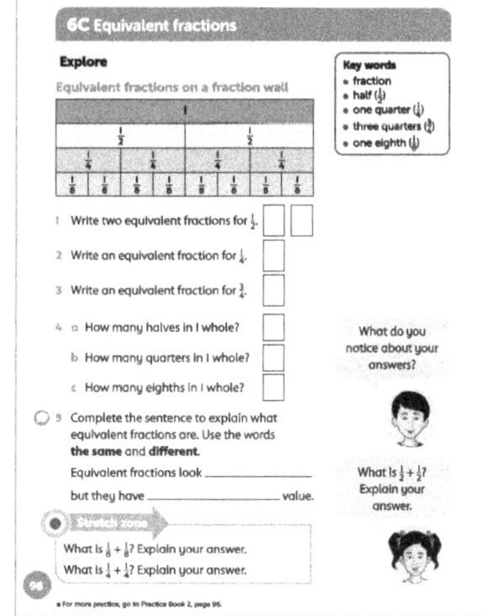

Differentiation

Supporting: You can help students see why two fractions are equivalent by pointing them out on the fraction wall.

Consolidating: Ask students to use concrete resources (e.g. number rods) to model more equivalent fractions.

Extending: Ask students to explore further equivalent fractions using thirds and sixths, or fifths and tenths, extending their fraction wall using number rods.

Stretch zone: *What is $\frac{1}{8}+\frac{1}{8}$? Explain your answer. What is $\frac{1}{4}+\frac{1}{4}$? Explain your answer.*

Check that students can show this using the fraction wall or a suitable drawing and understand that they are adding one eighth and another one eighth to make $\frac{2}{8}$ and that this is the same as one quarter.

 ## Reflection time

Tell students that a younger student has told you that $\frac{1}{8}$ is bigger than $\frac{1}{2}$ because 8 is bigger than 2. *Do you agree? Can you explain your thinking using the fraction wall?*

Give students some time to discuss in pairs, using examples from the fraction wall to support their answer.

Ask a pair to explain their reasoning to the class and ask other pairs to expand on their thinking as necessary.

Practice Book: Students complete page 96 of the Practice Book. They can do this directly after the main activity, as homework, or as the focus of a separate mathematics session to help students consolidate their learning and build fluency. Encourage students to label the missing parts in the fraction wall without looking at the other fraction walls in their Student and Practice Book. They should use these only to check their answers.

Differentiated outcomes	
All students	should say that a half is the same as two quarters.
Most students	will find equivalent fractions between halves, quarters and eighths.
Some students	may explore equivalences between other fractions.

Student Book page 98

1 $\quad \frac{2}{4}, \frac{4}{8}$

2 $\quad \frac{2}{8}$

3 $\quad \frac{6}{8}$

4 **a** 2   **b** 4   **c** 8

5 Equivalent fractions look different but they have the same value.

Practice Book page 96

1 1

$\quad \frac{1}{2}\ \frac{1}{2}$

$\quad \frac{1}{4}\ \frac{1}{4}\ \frac{1}{4}\ \frac{1}{4}$

$\quad \frac{1}{8}\ \frac{1}{8}\ \frac{1}{8}\ \frac{1}{8}\ \frac{1}{8}\ \frac{1}{8}\ \frac{1}{8}\ \frac{1}{8}$

2 $\quad \frac{1}{2}=\frac{2}{4}=\frac{4}{8}$

3 $\quad 1=\frac{2}{2}=\frac{4}{4}=\frac{8}{8}$

4 $\quad \frac{6}{8}$

5 $\quad \frac{2}{8}$

Stretch zone: Students should notice that when the number of parts on the top of the fraction is the same as the total number of parts on the bottom, for example two halves or four quarters, the fraction is equal to 1 whole.

6 Fractions

Big idea

- We can find fractions of shapes, fractions of numbers, fractions of lengths and fractions of amounts.

Global skills

- **Creative skills:** problem solving/exploring/ investigating
- **Real-world skills:** interpreting information
- **Interpersonal skills:** communication/teamwork

Key vocabulary

- fraction, half, halves, quarter, three quarters, numerator, denominator, equal parts

Resources

- counters, cubes, paper circles, paper plates (optional)

Language support

Encourage students to talk within their groups about what they are doing and why. Students may take on different roles in the group such as recorder, cutter, counter. Try to ensure that students do not avoid roles they find difficult and take full part in the task.

 ### Introductory activity

Discuss with students what they know about fractions from what they have been learning. Ask questions to elicit their knowledge and understanding about fractions, such as: *What is a fraction? How do we find a fraction of something? What sort of fractions can we find? When might we use fractions? Can you tell me something you've learned about fractions?*

 ### Main activity

Ask students to look at the problems on page 99 of the Student Book. Display on the IWB if possible. Arrange students in mixed-attainment groups to work on the problems so that less confident language learners hear the appropriate mathematical vocabulary modelled by their peers.

Before students begin, look at each food item in turn together and point out the descriptions for the items in the speech bubble. Together, match each description to the correct item. Ask questions to ensure that students understand what they need to do. For example, *In question 1 how many people are sharing the picnic equally?* Students

may want to use paper shapes, counters or cubes to help them represent the different food items. They should then work on the problems in small groups. You could then ask students from different groups to form pairs to explain to each other how they worked out their answers.

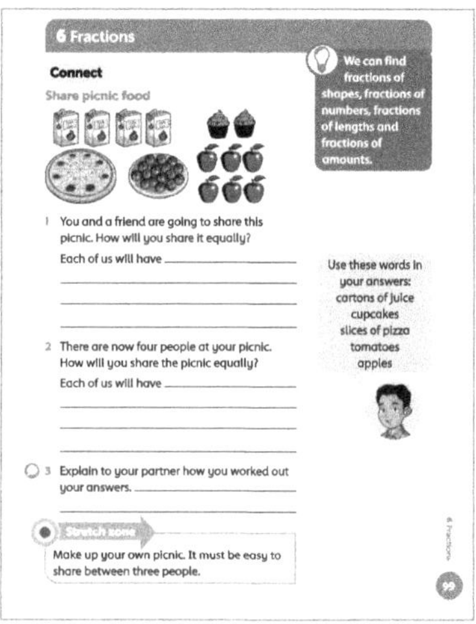

Differentiation

Supporting: Make sure that students take part and use concrete resources to come to an understanding of the process.

Consolidating: Encourage students to share their thinking so that they say their strategies clearly.

Extending: Ask students to explore larger quantities and look for patterns, for example, *What if there were 16 apples and cakes?*

Stretch zone: *Make up your own picnic. It must be easy to share between three people.*

Discuss with students before they begin what 'easy to share between 3 people' means. Agree that they will need to include numbers of food items that are divisible by 3 so that there is no food left over and to avoid any parts being unequal.

 ### Reflection time

Ask students to explain what they did in their groups, how they tackled the problems and how they found their solutions. You could ask a further question: *40 people are coming to a party. They will each have half a sandwich and a quarter of a cake. How many sandwiches and cakes are needed?* Ask a question involving fractions of length within a picnic context (e.g. *The children used paper napkins that were 30 cm wide. They folded them in half width-wise and gave each child half. How wide was the napkin they got? How do you know?*)

Differentiated outcomes	
All students	should take part in the activity, suggesting solutions.
Most students	will use fractions to share whole objects or a number of objects.
Some students	may confidently explain how they find fractions using division to find fractions of an amount.

6 Fractions

Global skills

- **Creative skills:** problem solving/exploring
- **Interpersonal skills:** communication
- **Self-development skills:** reflecting on learning

Student Book

With young children, assessment activities are most effective when carried out as an everyday classroom activity. Students could have counters available to support them. Encourage students to sketch the problems to help them solve them.

Watch as students complete the sentences that represent fraction situations. Check that they understand the language of equal sharing, equal parts and fractions of numbers and shapes, in particular half, quarter, three quarters.

It may help to create flash cards with the fractions and words that are in the review. You can use these as support for some students.

Answers

Student Book page 100

1 10   5   15

2 12   6   18

3 $\frac{1}{2}$ $\frac{3}{4}$ $\frac{1}{4}$

4 Check that the correct fractions have been coloured on each circle.

5 $\frac{1}{3}$ 5 cars

Practice Book

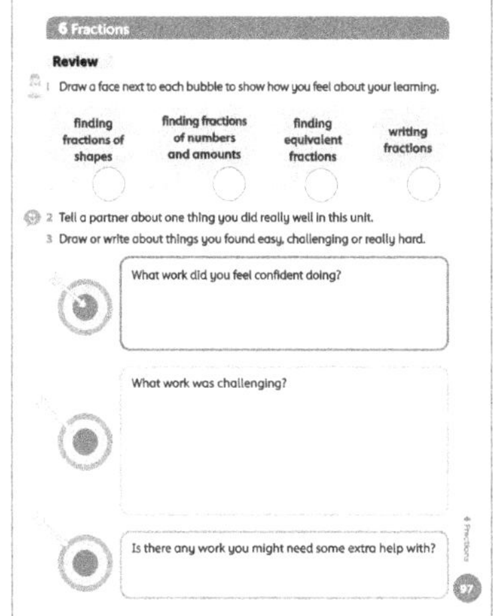

It is appropriate to complete this Practice Book review as a whole-class discussion. You may choose to keep a record of the class discussion or a copy of the review page for your own records. The review provides an opportunity for students to reflect on their learning from the unit, to discuss any areas of mathematics that they feel went particularly well, and any areas that they feel less confident about. Ensure all students have a copy of the Student Book as a reminder of the areas of mathematics that they have worked on in this unit.

Allow students plenty of time for discussion before asking them to complete the Practice Book page individually, and then, if appropriate, to share their responses with the rest of the class. If students complete this self-assessment at home, encourage them to discuss this with adults. Make a note of areas that students still feel unsure about.

Additional material

There are additional end-of-unit assessments available on the *Oxford Owl* website.

7 Length, mass and capacity

Overview

Big idea

One of the Big ideas in this unit is that there are standard units for measuring length, capacity and mass (weight). These are the centimetre and metre, litre and millilitre, kilogram and gram, respectively. These metric units of measure use our base-10 number system and can therefore be added and subtracted in the same way as we calculate with numbers. The focus in this unit, however, is on comparing and weighing to the nearest whole unit of measure. As part of this, students need to be given opportunities to experience measures in different contexts, so they can have a sense of, for example, how long one metre is, or how heavy 20 grams feels. Students need lots of experience of water play as well as exploring capacity through pouring sand, water, rice or something else into various containers before they carry out these activities.

Measuring is not only about units of measurement. Students need to be able to read scales and measuring instruments. Previous experience with non-standard units is essential. Students then need to explore and compare non-standard and standard units before they move on to consider which unit is appropriate in a given situation. A focus on the units rather than the act of measuring will make sure that students recognise and use the appropriate unit after they have had the opportunity to explore the unit and have a physical sense of it. Estimating lengths (and weights and capacities) before measuring them will help to develop a sense of measure.

Students also need to develop a sense of conservation of length. A piece of string may look long when it is laid out straight but is the same length when arranged in a spiral, even though it may look shorter. Moving an object to a different position does not change its length. Students need experience of estimating and measuring objects in different positions.

Look out for

- **Students who lack confidence in reading scales.** They need practice at reading the numbered divisions on a variety of scales, including a ruler. Using a ruler as a number line will make sure that students are already familiar with its layout when they begin to use it for measuring. Reading a scale accurately needs an understanding of the number system and what the space between each mark represents. Frequent practice in drawing lines to a specific length, as well as measuring given lines, is essential.

- **Students who make a range of errors when reading scales.** For example, with liquids, they may bring the container up to their eye rather than putting it down on a flat surface. Further opportunities for practical measuring will help to address this.

- **Students who do not always remember that you need to place the ends of two objects together in order to compare their length.** This includes the object you are measuring and the measuring instrument you are using. Provide many opportunities for practical measuring in engaging and meaningful contexts. For example, an elephant's footprint is often between 40 cm and 50 cm. I wonder how much longer that is than my footprint?

Possible misconceptions

- **Students may think that a taller container will hold more than a shorter one, even when the shorter container is much wider.** Students need to develop a feel for the size of standard units such as 1 kilogram and 1 litre to help them use these as a standard that they compare to.

- **Students may make errors based on misunderstanding of the vocabulary associated with measures.** Big does not have to mean heavy, nor does small have to mean light, so it is important to model the language precisely.

Key vocabulary

- **length:** ruler, tape measure, metre stick, trundle wheel, scale, centimetre, metre, measure, long, longer, longest, short, shorter, shortest, wide, wider, widest, thick, thicker, thickest, thin, thinner, thinnest, tall, taller, tallest, width, narrow, length, accurate

- to the nearest centimetre, actual, rigid, flexible, estimate

- **capacity:** litre, millilitre, measuring jug, container, liquids, hold, scale

- **weight:** weigh, mass, balance scales, weighing scales, heavy, heavier, heaviest, light, lighter, lightest, gram, kilogram

- **temperature:** degrees Celsius (°C)

Learning focus	Learning outcomes (the ENC objectives)
Measuring in centimetres	Choose and use appropriate standard units to estimate and measure length/height in any direction (cm) to the nearest appropriate unit, using rulers.
String measures	Choose and use appropriate standard units to estimate and measure length/height in any direction (cm) to the nearest appropriate unit, using rulers.
Measuring in metres	Choose and use appropriate standard units to estimate and measure length/height in any direction (m) to the nearest appropriate unit, using rulers.
Liquid measures	Choose and use appropriate standard units to estimate and measure capacity (litres/ml) to the nearest appropriate unit, using measuring vessels.
Measuring mass	Choose and use appropriate standard units to estimate and measure mass (kg/g) and temperature (°C) to the nearest appropriate unit, using scales and thermometers.

7 Length, mass and capacity

Big question

- How can I measure things?

Global skills

- **Creative skills:** exploring/investigating
- **Real-world skills:** research
- **Interpersonal skills:** communication/teamwork

Key vocabulary

- measure, metre, centimetre (cm), tall, tallest

Resources

- roll of wallpaper or backing paper
- strips of paper (10 cm long)

Language support

- Begin a measuring display for the classroom. Display a ruler with sentences such as: 'We use a ruler to measure how long something is.' Include one of the paper strips, clearly labelled 10 cm.

Introductory activity

Ask students, *What do we **measure**?* Students may suggest that we measure our height, the temperature, the distance between places, the weight of food when cooking. Discuss a range of examples as a class, then ask students to talk in pairs about the question *How can we measure things?* This will get them thinking about measuring instruments such as rulers, measuring jugs and scales.

Share the results with the whole class.

Main activity

Ask the class, *Who is the **tallest** in the class? How could we find out?* Take students' suggestions. They are likely to suggest standing back-to-back with other students to compare heights. Give students time to arrange themselves in order to find out who is tallest.

Ask students to look at page 101 of the Student Book. Display on the IWB, if possible. Ask students to describe the scene. Read the speech bubble on the far right: *Who is the tallest student in your class? How **tall** are they?* Agree that they know the answer to the first question, but not the second. Ask, *What do we need to do to find out?* Agree that they need to measure how tall they are.

Put students into groups and ask them to first decide who is the tallest in the group and then find out how tall they are. Remind them of work with non-standard units of measure from the previous year and ask them to measure that student in handspans. Each member of the group should measure the height using their own hands. The student to be measured could lie down on a large sheet of paper and a student marks the position of the bottom of their heel and the top of their head. The other students can then place their hands on the paper, starting at one mark and moving their hand carefully in a line to the other mark, and count how many hands tall the student is.

Now ask them to share the height in handspans of the tallest person in their group so that they can compare the tallest person from other groups. Can they compare the heights? Some groups are likely to have come up with more than one height for their tallest student. Can students explain why? If necessary, point out that students have different-sized hands, so they may get a different number of handspans when they measure the same height. This makes it difficult to compare measures.

Share with students that when people measure how tall a horse is they also measure using 'hands' but in this case hands are a fixed length, **standard unit of measure**. Explain that a 'hand' is equal to slightly more than 10 centimetres. Give students 10-cm strips of paper to use to measure the height of the same student in 'hands'.

Differentiation

Supporting: Ask students to repeat the measurements to check their calculation.

Consolidating: Ask students to estimate how many 10 cm strips will be needed to measure an individual's height.

Extending: Ask students to estimate heights in multiples of 10 centimetres, knowing that one hand is approximately 10 cm.

Reflection time

Compare the heights in 'hands' of the tallest students from all the groups and order them. *Can you describe how tall they are in hands?* Remind students that 1 hand = 10 cm and ask them to now work out the heights of these students in centimetres. (They should count up in tens.) Show them how to record the measurements using **'cm'** for **centimetres** rather than writing the full word. Agree as a class who is the tallest student in the class and how tall they are.

Return to the Student Book and read the middle question, *Do you know any units of measure?* Record 'hand' and 'centimetres' and any more that students suggest (e.g. **metre**). Say that you will add to this list throughout the unit.

7A Measuring in centimetres

Discover   Student Book page 102 • Practice Book page 98

Specific learning focus

- Estimate, measure and compare centimetre lengths.

Global Skills

- **Creative skills:** investigating
- **Interpersonal skills:** communication/teamwork

Key vocabulary

- metre, centimetre, ruler, tape measure, metre stick, scale, trundle wheel, long, longer, longest, short, shorter, shortest, tall, tallest, wide, width, narrow, thick, thin, thickest, thinnest, estimate, length, to the nearest centimetre, accurate

Resources

- rulers, tape measures, metre sticks, trundle wheels (one of each item per group if possible)
- pencil, glue stick, crayon, paper clip (one set for each pair of students)

Language support

Take some photographs of students measuring length and add these to the measuring display. Add pictures of other people measuring, for example, a carpenter measuring a length of wood or displays of lengths of wood at a store. Include a labelled picture for each of the words 'wide', 'narrow', 'thick' and 'thin' in the measuring display.

Introductory activity

Show students a **ruler, tape measure, metre stick** and **trundle wheel**. Ask, *Do you know what we use these for?* Agree that they are all used to measure how **long** things are. If you have enough, give each group one of each of the measuring items. Ask students, *What is the*

same and what is different about them? Share ideas such as they all have numbers and ticks on them but some have more than others. Some are hard (i.e. the ruler and metre stick) and other can bend (i.e. the tape measure and trundle wheel). Ask pairs to discuss what they would use each tool to measure and then share ideas (e.g. you may use a trundle wheel to measure how wide a road is).

Main activity

Tell students to think about their paper strips from the previous lesson and hold their hands 10 cm apart. Explain that because they don't know exactly how far apart 10 cm is they need to **estimate**, that is, make an intelligent guess. Students take it in turns in pairs to hold their hands 10 cm apart. Then their partner measures how far apart the hands actually are with a ruler to see how **accurate** they were. Repeat for 20 cm and 30 cm. Ask students whether they are getting better at estimating lengths.

Ask students to look at the objects on page 102 of the Student Book. *What is the best tool to measure how long these objects are?* As none of the items is very long, agree that a ruler is the most suitable.

Students should complete the activity in the Student Book in pairs. Suggest that they think about how long 10 cm is to help them make a good estimate for the **length** of each of the objects. Talk through with them how to use a ruler to measure in centimetres. Point out the numbering on the ruler and how to line up the zero mark with one end of the object before reading off the length against the other end. Make sure that they know that measuring does not start from the end of the ruler, but from the zero mark. By looking at the marks on the ruler, help students to see how long a centimetre is. This will help them begin to estimate lengths in centimetres.

It may be more suitable for students to measure objects using metre sticks or tape measures even when measuring in centimetres, depending on which objects they choose to measure to complete question 2. Help them to read the **scale** and explain that although there are centimetres marked on these tools, they also measure a metre (or more).

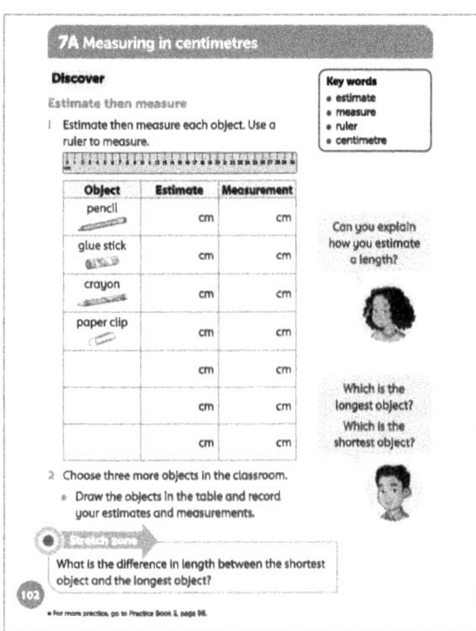

Practice Book: Students can complete page 98 of the Practice Book. They can do this directly after the main activity, as homework, or as the focus of a separate mathematics session to help students consolidate their learning and build fluency. Check that students can tell you what the < and > symbols represent before they begin the activity.

Differentiated outcomes	
All students	should measure with some accuracy using a ruler, for example lining up objects correctly with 0.
Most students	will estimate and measure with some accuracy using a ruler.
Some students	may refine estimates and measure more accurately using a range of measuring equipment.

Answers

Student Book page 102

Check that students' estimates and measures are reasonably accurate. If they are not, then revisit the task and support them with further modelling and questioning.

Practice Book page 98

Check that students' estimates and measures are reasonably accurate.

Differentiation

Supporting: Help students to estimate and measure to the nearest centimetre.

Consolidating: Ask students how they have estimated.

Extending: Ask students about their estimates, for example, *What is the difference between each of your estimates and the actual length? Did your estimates improve? If so, why?*

Stretch zone: *What is the difference in length between the **shortest** object and the **longest** object?*

Check that students have correctly found the difference in the lengths of the two objects.

 Reflection time

Ask students to list the things that were shorter than 10 cm and those that were 10 cm or more. Explain that we often focus on measuring in one direction, that is, how long something is, but that there are other ways to measure length as well. Talk through measuring in more than one direction, using words such as **wide** and **width**, **narrow**, **thick** and **thin**. Ask students to measure the width of each object. Follow this up with questions such as: *Is your longest object also the **widest**? Is your shortest object also the most narrow?*

Explore
Student Book page 103 • Practice Book page 99

Specific learning focus
- Estimate, measure and compare centimetre lengths.

Global skills
- **Creative skills:** investigating
- **Interpersonal skills:** communication/teamwork

Key vocabulary
- centimetre, ruler, scale, long, longest, short, shortest, tall, tallest, estimate, accurate, nearest, wide, wider, widest, to the nearest centimetre

Resources
- three long sticks (or similar) of different lengths, three shorter objects of different lengths (e.g. pencil, glue stick, ruler), three tall objects of different heights (e.g. tin cans, empty cartons or boxes)
- 30 cm rulers, pencil, glue stick, crayon, paper clip

Language support
Draw up a grid of measuring words and talk through how the word changes from when we are talking about a single object, to when we are comparing two objects and when we are comparing three or more.

Add the grid to the classroom display.

	Comparing 2	Comparing 3 or more
long	longer	longest
short	shorter	shortest
tall	taller	tallest
thin	thinner	thinnest
thick	thicker	thickest
narrow	narrower	narrowest
wide	wider	widest

 **Introductory activity**

Write 'long' on the board. Show students one stick and say, *This stick is long.* Hold up a second, longer stick and say, while lifting it higher, *This stick is **longer**.* Hold up a third stick, and compare it with each of the other two sticks and say, *This stick is the **longest**.* Repeat each statement and write 'long', 'longer' and 'longest' on the board. Repeat with other objects for **short/shorter/shortest** and **tall/taller/tallest**.

Ask students to take it in turns to say a sentence about themselves – or an object around the room – using one of these words. For example, 'I am taller than my sister' or 'My arm is shorter than my leg'.

Draw a line on the board (or use an IWB tool if possible) that is close to, but less than, 30 cm (make sure it is not a whole number of centimetres). Ask, *Can you estimate how long the line is?* Take estimates and record them on the board. Now ask a student to come to the front of the class to measure the line using a ruler. As they do, remind them how to align the ruler with the object. *How many centimetres is it? Does the length of line match up exactly with a 1-cm mark on the ruler?* Agree that it does not. Ask, *How can we measure a line that is not exactly on a centimetre mark?* If no students suggest it, explain that you can measure to the closest or nearest centimetre. Ask students in pairs to come to the front of the class to see how the line is measured and ask them to say which 1-cm mark it is closest to. Record the length on the board and write that the line is *x* cm long **to the nearest centimetre**.

Ask students to look at the lines on page 103 in their Student Books. Display it on the IWB, if possible. Ask, *Which line is longer, A or B? Which is the shortest line? Which is the longest?* Explain that they are going to measure each line and record the lengths to the nearest centimetre. Students should complete the activities in pairs so that they can check each other's measurements.

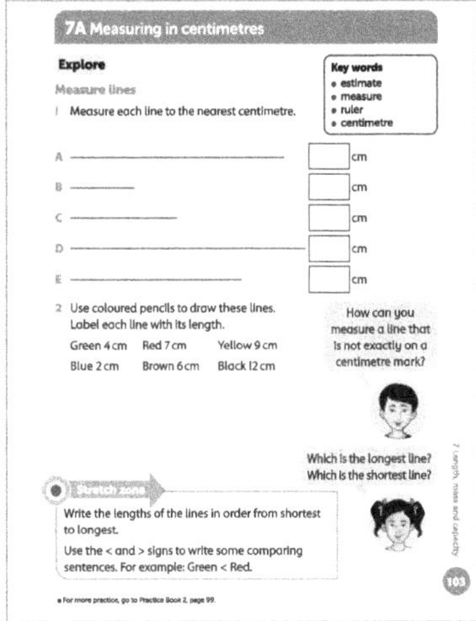

Differentiation
Supporting: Help students to measure the lines to the nearest centimetre by ensuring the correct positioning of their ruler.

Consolidating: Ask students how they would explain to a younger student how to measure lines in centimetres.

Extending: Ask students the question in the speech bubble, *How can you measure a line that is not exactly on a centimetre mark?* Students may provide a variety of responses (e.g. fractions of a centimetre); discuss mm only if students suggest them themselves.

Stretch zone: *Write the lengths of the lines in order from shortest to longest.*
Use the < and > signs to write some comparing sentences. For example: Green < Red.

Check that students can recall what each of the < and > symbols represents before they begin.

Reflection time

Ask questions about the lines students drew for question 2 on page 103 of the Student Book: *What is the difference in length between the blue line and the green line? How long would a yellow and a black line be altogether?* Display a ruler on the IWB, if possible, and show students how they can use rulers as they do a number line to help them with these calculations. For example, start at the length of the blue line (2 cm) and count on 2 cm to the length of the green line (4 cm).

Practice Book: Students now complete the activity on page of 99 of the Practice Book. They can do this directly after the main activity, as homework, or as the focus of a separate maths session to help students consolidate their learning and build fluency. Before they begin, ask students to tell you what they know about the word 'wide'. Introduce **wider** and **widest**, if necessary.

Differentiated outcomes	
All students	should measure with some accuracy using a ruler.
Most students	will estimate and measure with reasonable accuracy using a ruler.
Some students	may estimate and measure with accuracy using a ruler.

Answers

Student Book page 103

Check that students have measured the drawn lines correctly for question 1 and drawn lines of the required lengths for question 2.

Practice Book page 99

1 1 cm wide — Finger

2 10 cm long — Mouse

3 90 cm long — Desk and chair

4 50 cm long — Stride

5 20 cm wide — Book

6 3 cm wide — Cricket

Stretch zone: Six sharpeners will fit along the notebook. Students may have found this by seeing how many times they can jump 3 cm along their ruler to get to 18 cm.

7B String measures

Discover Student Book page 104 • Practice Book page 100

Specific learning foci

- Estimate the distance around objects in centimetres.
- Measure the distance around objects in centimetres using a piece of string and a ruler.

Global skills

- **Creative skills:** exploring/investigating
- **Real-world skills:** presenting information
- **Interpersonal skills:** communication/teamwork

Key vocabulary

- centimetre, ruler, tape measure, scale, estimate, long, longest, short, shortest, tall, tallest, actual, rigid, flexible

Resources

- pieces of string (about 30 cm long), rulers, tape measures
- at least one ball, a selection of tins of different widths

Language support

Explain the words 'rigid' and 'flexible'. Ask students to name some things that are rigid and some things that are flexible. Include these words in the measuring display. You could label a ruler and a tape measure to reinforce understanding of the words.

Introductory activity

Explain that a friend is making you a bracelet and needs to know how big your wrist is. Model trying to measure it with a ruler. Ask students how you could do this as a ruler does not work. *What other measuring tools do you know?* If necessary, suggest using a tape measure. Show students how to measure your wrist using a tape measure, taking care to show them how to read the tape measure where it meets, or overlaps, the start of the tape measure. Explain that the measuring scale starts from the very end of the tape measure. There is no small gap before the measuring scale starts, so there is nowhere to label zero. Walk around the class asking students to take turns to measure the circumference of objects. Select a student to read the length and another to check.

Ask, *If I didn't have a tape measure but I did have a ruler and a piece of string do you think I could measure my wrist?* Give each pair of students a piece of string, a ruler and access to a measuring tape. Give pairs some time to explore trying to measure with the string and ruler. Help them by asking questions such as, *How long is your piece of string? How about half of your piece of string? Can you use your ruler to find out? If you can find the length of your string, could that help you find the length of something else?* After a while, show, or ask a pair to demonstrate, how to measure your wrist using string, then measuring the given length of string against a ruler. Explain that rulers are **rigid**, they do not bend. This means they are not suitable for measuring everything. A tape measure is **flexible**, so you can wrap it around the thing you want to measure to be sure that you are measuring accurately.

Ask students to estimate the distance around their own wrist. Will it be bigger or smaller than yours and by how much? Ask each pair to use their string and ruler (or tape measure) to measure each other's wrists. They should check each other's measurements. Students should then go on, in their pairs, to estimate, then use string and rulers to measure, around the remaining items listed on page 104 in the Student Book, as well as three objects of their choice. Remind them to estimate first and explain that they only need to measure to the nearest centimetre.

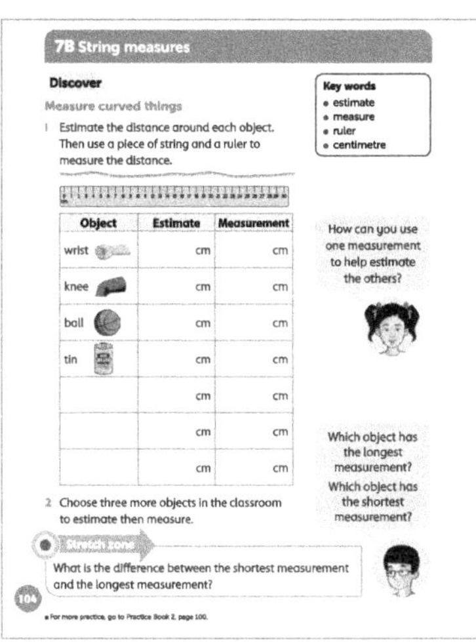

Differentiation

Supporting: Help students measure curved objects by modelling how to use string laid straight against a ruler.

Consolidating: Ask students, *How can you use one measurement to help estimate the others?*

Extending: Ask students to compare the lengths and describe them in terms of being the longest or shortest, for example.

Stretch zone: *What is the difference in length between the shortest measurement and the longest measurement?*

Check that students have correctly found the difference between two measurements.

Ask students to look at page 104 of the Student Book. Display on the IWB, if possible. Ask *Which object has the longest measurement? Which has the shortest? Did anyone choose an object that was longer? How about shorter?*

Ask, *What did you notice about your estimates and your **actual** measurements?* (Explain that actual means 'real'.) If no one suggests it, explain that when you measure around something, as they have just done, the measurement is usually longer than you expect. Show students a coiled piece of string. Ask them to talk to their partner and agree an estimate for how long the string is. They should then unwind the string and measure it. *How close were your estimates?*

Practice Book: Students now complete the activity on page of 100 in the Practice Book. They can do this directly after the main activity, as homework, or as the focus of a separate maths session to help students consolidate their learning and build fluency. Explain that they will need to measure round a number of household items; if they cannot find these particular items, they can be substituted with something else.

Differentiated outcomes	
All students	should use string and a ruler to measure curved objects with support.
Most students	will use string and a ruler to measure curved objects independently.
Some students	may use string and a ruler to measure curved objects with increasing accuracy.

Answers

Student Book page 104

Check that students' estimates and measures are reasonably accurate to the nearest centimetre.

Practice Book page 100

Check that students' estimates and measures are reasonably accurate to the nearest centimetre.

7B String measures

Explore Student Book page 105 • Practice Book page 101

Specific learning foci

- Estimate the length of ribbons in centimetres.
- Measure their lengths in centimetres using a piece of string and a ruler.
- Compare lengths of ribbons and describe the difference in centimetres.

Global skills

- **Creative skills:** exploring/investigating
- **Real-world skills:** presenting information
- **Interpersonal skills:** communication/teamwork

Key vocabulary

- centimetre, ruler, scale, long, longer, longest, short, shorter, shortest

Resources

- a coiled piece of string stuck to card, scissors
- string, rulers, tape measures
- ribbons of different lengths

Language support

Add some of the ribbons to the classroom display. Label each ribbon with its length in centimetres. Some ribbons could be labelled with a sentence including the full word 'centimetre' and others could be labelled using the abbreviation 'cm'.

 Introductory activity

Remind students that they have used string and tape measures to measure things. *Why were they useful?* Both were very useful when the measuring tool needed to be flexible enough to bend. Show students a coiled piece of string stuck to card. Instead of estimating how long the string is, ask pairs of students to work together to try to make their own copy of the coiled string by making a similar shape from the string and cutting it to size. When they think they have replicated the coiled string, ask students to uncoil their string and compare it with the string produced by another pair. *Who had the longer piece? Who had the shorter piece? How much longer? How much shorter?* Ask the two pairs of students to measure the lengths of their two pieces to find the difference in length. Remind students of their earlier work with difference, modelling how they could use a ruler like a number line to find the difference, for example, counting on from the shorter length to the longer length. You could then remove the original string from the card and measure it. Check whether its length is close to those made by the students.

 Main activity

Ask students to look at the five ribbons on page 105 of the Student Book (or display on the IWB, if possible). Ask, *How long do you think each ribbon is? Which looks the longest? Which looks the shortest?* Explain that they should estimate the length of each of the ribbons and then measure them to find the actual length. After they measure each ribbon, they can adjust any of the estimates for ribbons they haven't yet measured so that they can make their estimates more accurate. Students should use string to measure the coiled ribbons. Tell them to record both their estimate and the actual length in the table on page 105. Ask students to look back at the table on page 104 of the Student Book. *What is different about the two tables?* Students should notice that 'cm' is written in each cell in one table but not in the other. Explain that it is very important to include the unit of measure when we record lengths. Students may measure independently but encourage them to work in pairs to find the difference in lengths, discussing strategies.

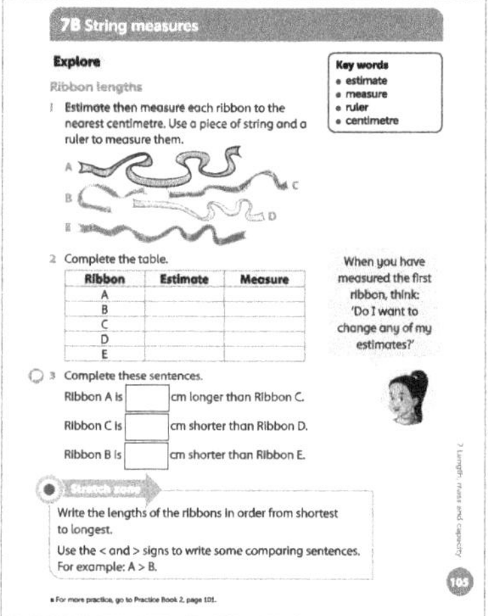

Differentiation

Supporting: Help students with matching the different lengths of string to see which is longer, and with finding the difference between lengths.

Consolidating: Ask students to describe the steps of measuring the length of a ribbon and to explain how they estimated the length of one of the ribbons.

Extending: Ask students to tell you the possible length of a ribbon between two given estimates.

Stretch zone: *Write the lengths of the ribbons in order from shortest to longest.*
Use the < and > signs to write some comparing sentences. For example: A > B.

Check that students are using the < and > symbols correctly.

 Reflection time

Discuss what was easy or hard about measuring the ribbons. What did students do that made it easier? What did they need to remember when they were measuring the length of string (e.g. line up one end with 0)? Ask students whether the length would be different if they started measuring from the 'other' end? Give them some time to discuss in pairs and explore with strings and rulers.

Practice Book: Students can now complete the activity on page of 101 in the Practice Book. They can do this directly after the main activity, as homework, or as the focus of a separate maths session to help students consolidate their learning and build fluency. Explain that, as in the main activity, they will need to find the actual length of an object to the nearest centimetre, but that in this case they already have an estimate but no object. They need to find objects with lengths close to the estimate in the table. Show the first couple of lengths on a ruler and then ask students to brainstorm some possible objects, for example, a rubber (5 cm), a crayon (10 cm).

Differentiated outcomes	
All students	should line up the string with a ruler and read the length from the ruler with support.
Most students	will estimate and measure to the nearest centimetre with reasonable accuracy using a ruler.
Some students	may adjust their estimates to make them more accurate based on previous estimates and measures.

7C Measuring in metres

Discover Student Book page 106 • Practice Book page 102

Specific learning focus

- Estimate, measure and compare lengths, choosing and using centimetres and metres and appropriate measuring instruments.

Global skills

- **Creative skills:** exploring/investigating
- **Real-world skills:** presenting information
- **Interpersonal skills:** communication/teamwork

Key vocabulary

- metre, centimetre, ruler, metre stick, long, longest, short, shortest, tall, tallest

Resources

- metre sticks (one per pair)
- string
- vocabulary cards for the key vocabulary

Language support

Make a set of vocabulary cards for the key measuring words listed in the vocabulary above. Demonstrate one of the words in some way. Show students two of the cards and ask them to choose the correct word.

Student Book page 105

Check that students' estimates and measures for the ribbons, and their comparisons in question 3. Accept some level of inaccuracy, given the challenges of measuring the ribbons. The most important aspect of the activity is that students record a length and an estimate for ribbon A that is greater than their estimate and length for ribbon C, and that B has the smallest length.

Practice Book page 101

Check that students have picked reasonable objects based on the estimates given and that the actual measures for these objects are reasonable.

Stretch zone: Students might, for example, make sure that the ends of the string line up with the length being measured, place the string straight against the ruler, starting from 0, and not stretch the string.

 **Introductory activity**

Tell students that many things are measured in metres. Today, they are going to be measuring things that are more or less than 1 metre. Show students a metre stick and explain that 1 centimetre is not very long, so for bigger objects we measure using the metre, which is the same as 100 cm. Show students the marking on the metre stick so they can see that 50 cm is $\frac{1}{2}\left(\frac{2}{4}\right)$ of a metre, 25 cm is $\frac{1}{4}$ of a metre, 75 cm is $\frac{3}{4}$ of a metre. Ask them to also notice how many 30-cm rulers line up with a metre stick (3 with 10 cm left over).

Ask them whether they can name anything that they think is longer than 1 metre (e.g. door height, height of teacher, length of the classroom and so on). Then ask for things that are shorter than 1 metre (e.g. book, pencil, length of their arm and so on).

 **Main activity**

Give each pair a metre stick. Ask them to look at the scale on the stick and find different markings. For example, *Show me 10 cm on your metre stick. Where is 75 cm? What is 5 cm more? Show me on your metre stick. What is 10 cm less than 1 metre?*

Explain that you would like students to explore in pairs, finding things to measure with their metre stick. There is space in the Student Book page 106 for them to record things that are less than a metre long and things that are more than a metre long by writing or drawing. As students work, ask pairs to describe to you how they chose objects that they thought were shorter that I metre.

Pairs should then discuss question 3, using their metre sticks and what they know about fractions from earlier work to answer the questions. Once they have completed the questions, they can look for an object that measures exactly 2 metres.

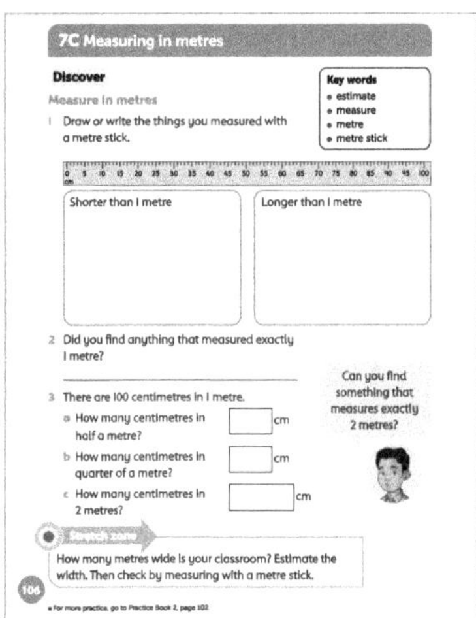

Differentiation

Supporting: Ask students to point out objects that they think are either longer or shorter than a metre.

Consolidating: Ask students to show you how they know they have answered each part of question 3 correctly. Can they tell you how many centimetres in 3 metres?

Extending: Ask students to measure some objects in metres and centimetres and record these measurements in the Student Book.

Stretch zone: *How many metres wide is your classroom? Estimate the width. Then check by measuring with a metre stick.*

Check that students have estimated and measured reasonably accurately.

 Reflection time

Ask students whether they found anything exactly 1 metre long, other than the metre sticks themselves. Explain that although a centimetre and a metre are standard units of measurement, most things that we measure are not exactly those lengths. We usually need both metres and centimetres to begin to measure length accurately. Put two metre sticks together to measure the height of a student to show that they are 1 metre and some centimetres tall. Students could go on to measure each other's height.

Practice Book: Students can complete page 102 of the Practice Book. They can do this directly after the main activity, as homework, or as the focus of a separate mathematics session to help students consolidate their learning and build fluency. Explain that they need to find six objects that are longer, wider or taller than 1 metre but these need to be objects that they can realistically measure with a tape measure. If the object is more than 1 metre, they record a tick. If it is shorter than a metre, they record an x.

Differentiated outcomes	
All students	should compare objects to a metre stick to find out whether an object is longer or shorter than 1 metre.
Most students	will estimate with greater accuracy and have an improving sense of how long 1 metre is.
Some students	may measure in metres and centimetres.

Answers

Student Book page 106

Check that the students have recorded the items correctly in questions 1 and 2.

3 **a** 50 cm **b** 25 cm **c** 200 cm

Practice Book page 102

Check that students have chosen objects that were likely to be more than 1 metre and that they have marked an 'x' or a tick next to items that were less or more than a metre once measured.

Stretch zone: Check that students have ordered their lengths from longest to shortest, including the unit of measure, and have used the < and > symbols correctly.

7C Measuring in metres

Explore Student Book page 107 · Practice Book page 103

Specific learning foci

- Choose appropriate units of measure (centimetres or metres).
- Estimate and measure to the nearest whole metre.

Global skills

- **Creative skills:** exploring/investigating
- **Interpersonal skills:** communication/teamwork

Key vocabulary

- metre, centimetre, ruler, metre stick, trundle wheel, long, longest, short, shortest, tall, tallest

Resources

- metre sticks, tape measures, rulers, trundle wheel

Language support

Highlight the language 'length', 'height' and 'width'. Check understanding and which word would be used if we were comparing two, or more than two, lengths, widths or heights. Take some photographs of students measuring to include in the classroom display on measuring.

 **Introductory activity**

Ask students to hold their hands apart to show you 1 cm and then 1 m. Give one student a metre stick to quickly check some of the 1-m estimates and another a ruler to check the 1-cm estimates. Repeat to see whether students can get closer to estimating 1 centimetre and 1 metre. Explain that now they have a good idea of how long 1 metre and 1 centimetre are, they can start thinking about when it is useful to use each of them.

 **Main activity**

How long do you think the classroom is? Ask students to discuss in pairs and estimate the length of the classroom, allowing them to walk around to get 'a feel' for how long the classroom is. Ask, *Will it make a difference where you measure from?* The answer will vary depending on the shape of your classroom. *What unit of measure will you use? Can you explain why?* Explain that when we are deciding whether something is measured in centimetres or metres, it is useful to think what we would use to do the measuring with. If we would use a metre stick or trundle wheel, we are likely to be measuring metres. If we only need a ruler or tape measure, then we are likely to be measuring centimetres.

Show students how to use the trundle wheel to measure something that is several metres long. Ask one student to use a piece of string 1 m long to confirm that the outside of the trundle wheel, the part that touches the floor, is 1 m. Measure the length of the classroom, carefully counting the clicks together as each metre length is measured. Be sure to include the word 'metre' in your count to reinforce the idea of including the unit of measure when describing lengths. Compare the measurement to their estimates. *Was your estimate reasonable?* Repeat for the width of the class. *Was your estimate better this time? Why do you think that was?*

Students should complete the activity on page 107 of the Student Book in pairs to allow for discussion. They should decide between them how they might measure each item and which unit would be most appropriate. As they work, listen to their discussions and ask, for example, *If you think you would use metres, can you estimate how many metres it might be?*

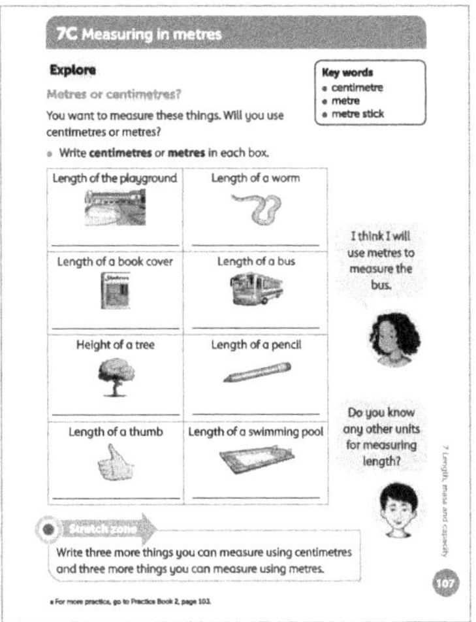

Differentiation

Supporting: Help students to understand which unit is best suited to each object and to explain why.

Consolidating: Ask students to explain why an object is best measured in metres or centimetres.

Extending: Ask students to estimate the lengths of each item in metres and centimetres and explain how they reached their estimate.

Stretch zone: *Write three more things you can measure using centimetres and three more things you can measure using metres.*

Check that students have chosen suitable objects. *Can you estimate to the nearest whole metre or centimetre how long, wide or tall they may be?* If possible, have them check their estimates by measuring or doing research (e.g. to find out how many metres tall a house might be).

 **Reflection time**

Choose one of the items from the Student Book to measure (e.g. the length of the playground). Take rulers, tape measures, metre sticks and trundle wheels. Ask one group to measure with tape measures, another with metre sticks and another with a trundle wheel.

Do you think we will all get the same answer? Why or why not? Students should recognise that having to use a metre stick or tape measure several times is more likely to lead to errors than using the trundle wheel continuously. However, the actual length of the playground is the same, regardless of the tool used.

Practice Book: Students can now complete the activity on page of 103 in the Practice Book. They can do this directly after the main activity, as homework, or as the focus of a separate maths session to help students consolidate their learning and build fluency. For this activity, students could work with an adult to help them improve their accuracy of measuring, still measuring only to the nearest centimetre or metre, with no mixed units.

Differentiated outcomes	
All students	should select some appropriate units to measure length.
Most students	will select the most appropriate units to estimate lengths with increasing accuracy.
Some students	may estimate accurately in metres or centimetres.

Student Book page 107

Length of playground: metres

Length of a book cover: centimetres

Height of a tree: metres

Length of a thumb: centimetres

Length of a worm: centimetres

Length of a bus: metres

Length of a pencil: centimetres

Length of swimming pool: metres

Practice Book page 103

Check students' objects and measures for the estimates given.

Stretch zone: Zahid could measure his desk in metres but it will probably be no more than 1 metre wide, so to find its actual length centimetres would be better.

7D Liquid measures

Discover Student Book page 108 • Practice Book page 104

Specific learning focus

- Estimate, measure and compare capacities, choosing and using suitable uniform, non-standard and standard units and appropriate measuring instruments.

Deeper learning

- **Creative skills:** exploring/investigating
- **Interpersonal skills:** communication/teamwork

Key vocabulary

- capacity, litre, millilitre, measuring jug, liquids, container, hold

Resources

- litre jugs; empty and washed 1-litre milk or drinks containers (each marked with the fill line for 1 litre)
- set of cups of the same size
- range of containers of different shapes and sizes
- food colouring, water (or sand, rice, lentils)

Language support

Display a picture of a jug with a clear 1-litre scale. Add statements such as: 'We measure liquid in litres and millilitres', '1 litre = 1000 millilitres', 'We shorten litres to l and millilitres to ml' and so on.

 **Introductory activity**

Show students a jug that you have filled with 1 litre of water. Have some empty cups all of the same size and then fill one cup from the jug. *Do you think we could fill another cup? And another? How many cups do you think we can fill from the jug?* Pour the water into the cups in turn and count to see how many it will fill. Fill the jug again and repeat with a different-sized container.

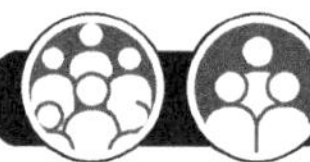 **Main activity**

Explain that it is useful to know how much of something we can put into a **container**. Choose two containers with the same capacity, one tall and thin, the other one short and fat. Fill one container with water and ask students whether they think there is enough room in the other container for all the water. Discuss their opinions without confirming any answer as right or wrong.

Pour the water into the other container and explain that different shapes can be confusing. We can say that both containers have the same **capacity**, that is, they hold the same amount. Explain that we can use sand, water, rice and lots of other things to compare capacities, but when we measure **liquids**, we measure in **litres**. Show students a litre **measuring jug** and point out the scale on the side, focusing on where '1 litre' is marked. Fill the jug to the 1-litre mark. Add a drop of food colouring to make it easier for students to see the water. Explain that when reading a liquid measuring scale, you must have the container on a flat surface so that the liquid surface is level. Show students how the level can be misread when the container is tipped.

Organise students in mixed-attainment groups. Give each group a selection of containers of different shapes and sizes, and a litre measuring jug (or a litre carton with a fill mark). *Do you think that all your containers will **hold** 1 litre of water? Do they have a capacity of at least 1 litre?* (Some will but not all.)

Tell students that you would like them to decide whether each of their containers will hold more or less than 1 litre of water and record it in their Student Books on page 108. Some students may measure out a litre of water in a measuring jug and then pour it into the other containers to check. Others may fill each container with water, then pour it into a 1-litre jug. If all the water goes in, then the container holds less than 1 litre (or exactly a litre). If it doesn't all pour in, then the container holds more than 1 litre.

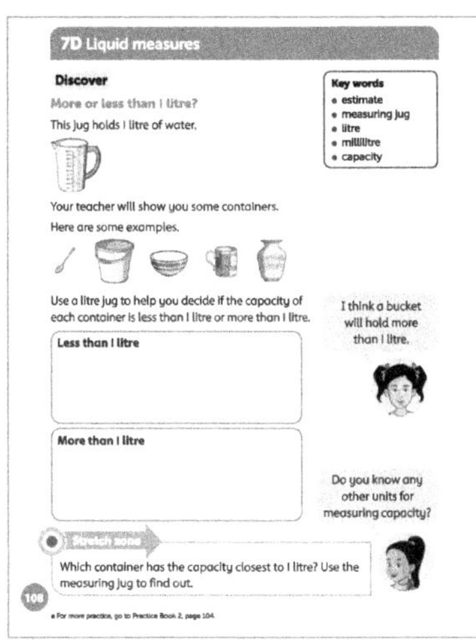

Differentiation

Supporting: Help students to measure 1 litre accurately and to record the name of each container correctly.

Consolidating: Ask students to predict whether a capacity will be more or less than 1 litre before measuring.

Extending: Ask students to consider the following problem: *You have two containers that each hold almost 1 litre of water. How could you use your measuring jug to work out which holds more water?*

Stretch zone: *Which container has the capacity closest to 1 litre? Use the measuring jug to find out.*

Check that students have measured accurately. Encourage students to think about comparing how much is left after filling each container to compare.

 Reflection time

Ask students to tidy up the containers by sorting them onto three different tables, one for containers that hold less than a litre, one for containers that hold more than a litre and the third one for containers that hold one litre, if there are any. Ask whether anyone knows any other units of measure for capacity. If no one says millilitres, tell students that for amounts less than 1 litre we can measure them in **millilitres**.

Ask students to look at the pictures of containers on page 108 of the Student Book. *Which of the containers do you think will hold more than a litre? Which containers may hold less than a litre?*

Practice Book: Students now complete the activity on page of 104 in the Practice Book. They can do this directly after the main activity, as homework, or as the focus of a separate maths session to help students consolidate their learning and build fluency. Set aside time after students have completed this activity for discussion. *Was it easy or hard to find containers with a capacity more than 1 litre? What can make it easy or hard to find out the capacity? Which containers in your house do you think have a capacity of about 1 litre?*

Differentiated outcomes	
All students	should be able to measure 1 litre accurately using a measuring jug.
Most students	should be able to fill a measuring jug with water to the 1-litre mark accurately.
Some students	may compare capacities by comparing how much is left in the 1-litre jug.

Answers

Student Book page 108

Check that the students have recorded their containers correctly.

Practice Book page 104

Check that students have correctly identified containers with capacities of more than 1 litre and any that do not.

7D Liquid measures

Specific learning focus

- Estimate, measure and compare capacities, choosing and using millilitres and litres and appropriate measuring instruments.

Deeper learning

- **Creative skills:** exploring/investigating
- **Interpersonal skills:** communication/teamwork

Key vocabulary

- capacity, litre, millilitre, measuring jug, scale

Resources

- litre measuring jugs, ten plastic cups of the same size, three plastic cups pre-labelled with a mark to show 100 ml, 200 ml, 250 ml
- range of containers of different shapes and sizes
- food colouring, water (or sand, rice, lentils)

Language support

Display some simple litre problems and their solutions. These could be illustrated by students. Be careful to use words such as 'capacity' and 'litre' correctly. Capacity is the maximum amount a container can hold. There is no need to talk about volume (the amount of space an object takes up) and how we measure volume. The focus is on the litre as a unit of measure of liquids.

 ## Introductory activity

Explain that there are 1000 **millilitres** in a **litre**. Count from 0 to 1000 millilitres in steps of 100 millilitres and back to zero. To get a sense of a millilitre, show students 5-ml measuring spoons and ask them to think about how small a millilitre is. *Do you think you will measure 1 ml very often?* Agree that they will not.

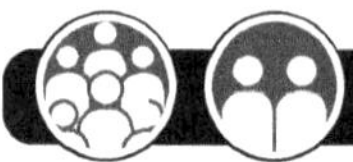 ## Main activity

Explain that students are going to do some more thinking about millilitres and litres. Ask students if they have ever tried lemonade, and explain, if necessary, that it is a drink made with water, sugar and lemons. *If I poured you a glass of lemonade, would it be more or less than 1 litre?* (less) *More or less than 10 ml?* (more) Show students a 1-litre jug of water (with a little food colouring) and ask them to imagine that it is lemonade. Show them 10 empty plastic cups. *If I poured the same amount of lemonade into each cup, using all the lemonade in the jug, how much lemonade would be in each cup?*

Explain that you are going to draw a diagram to show the problem. Draw a bar model with the top bar labelled 1 litre and the bottom bar showing 10 equal parts to represent 10 cups. *How can we divide 1 litre into 10 equal parts?* Can anyone remember how many millilitres are in 1 litre? (1000) Amend your bar to include the label '= 1000 ml'. *We can divide 1000 by 10.*

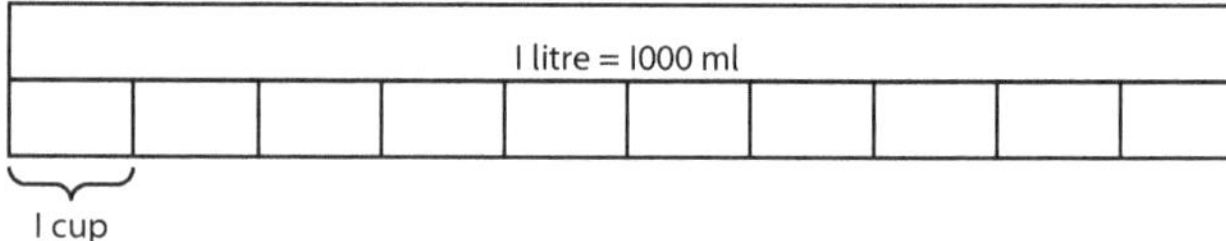

Draw or display a 0–1000 number line the same width as your bar model. Explain that we can use the number line to help us.

How is this number line labelled? (in hundreds) *How many sections are on this number line?* (10) *How many parts are we sharing our 1000 ml of lemonade into?* (10) *How many millilitres in each equal part?* (100) Complete the bar model, recording 100 ml in each box in the bottom bar.

Give each pair of students a litre measuring jug. Ask them to look at the **scale** and then back at the number line on the board. *What do you notice?* (The jug has a similar number line but it goes up and down, not left to right.) Read through the questions on page 109 of the Student Book. Encourage students to draw each problem as a bar model. This is quite a challenging activity so you may prefer to work through questions as a class.

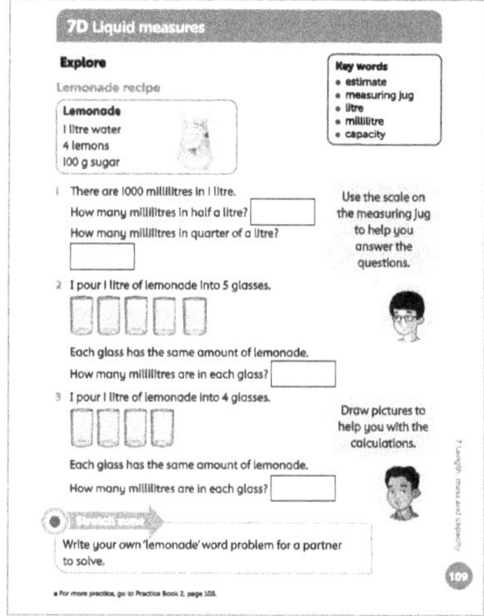

Differentiation

Supporting: Work with students to model solving the questions practically.

Consolidating: Ask students to calculate answers and then check using practical materials.

Extending: Ask students to create additional word problems for each other.

Stretch zone: *Write your own lemonade word problem for a partner to solve.*

Encourage students to use litres and millilitres in their problems.

Check that students have devised suitable problems.

Reflection time

Show students three plastic cups, marked with level 100 ml, 200 ml and 250 ml, respectively. Invite selected students to come to the front of the class to demonstrate how many times they can fill each capacity of cup using 1 litre of water (10 times, 5 times and 4 times, respectively). Remember to add some food colouring to the water so that students can see the levels more easily.

Practice Book: Students can now complete the activity on page of 105 in the Practice Book. They can do this directly after the main activity, as homework, or as the focus of a separate maths session to help students consolidate their learning and build fluency. This is a hands-on measuring task where students will need access to a measuring jug and a variety of containers with different capacities, so it is best done with the support of an adult.

Differentiated outcomes	
All students	should understand that there are 1000 millilitres in 1 litre and be able to count up a number line in 100 ml steps to 1 litre.
Most students	will, with support, use a bar model to represent how many millilitres in a glass and work out the answer.
Some students	may draw a bar model to represent problems and work out how many millilitres in a glass.

Answers

Student Book page 109

1 500 ml, 250 ml

2 200 ml

3 250 ml

Practice Book page 105

Check students' objects and measures for the estimates given.

Stretch zone: 1 l = 1000 ml

7E Measuring mass

Discover Student Book page 110 • Practice Book page 106

Specific learning focus

- Estimate, measure and compare masses, using appropriate measuring instruments.

Global skills

- **Creative skills:** exploring/investigating
- **Interpersonal skills:** communication/teamwork

Key vocabulary

- scales, balance scales, weighing scales, weigh, grams (g), kilograms (kg), heaviest, heavier, lightest, lighter, weight, mass

Resources

- balance scales and kitchen weighing scales (1 per group, if possible)
- ball of modelling dough with a mass of 100 g to represent an egg (one per group)
- small classroom objects to weigh, including interlocking cubes, books, counters, paperclips, pencils

Language support

Display some simple kilogram problems and their solutions. These could be illustrated by students.

Remind students of the words used when measuring length – 'long', 'longer', 'longest'; 'short', 'shorter', 'shortest'. Draw up a similar chart for 'heavy', 'heavier', 'heaviest' and 'light', 'lighter', 'lightest' to add to the classroom display.

Introductory activity

Choose two classroom items to compare (e.g. a book and pencil). Hold one in each hand and try to balance them. Raise the hand with the pencil higher than the book and say, *The book is **heavier** than the pencil. The pencil is **lighter** than the book.* Choose a third item (e.g. a large water bottle). Repeat, comparing the bottle to the book and then the pencil. Line up the three items and say, *The water bottle is the **heaviest**. The pencil is the **lightest**.* Choose three more classroom items of similar size. Pick two and ask *Which one is heavier?* Invite a student to come to the front and balance the two items to decide which is heavier. Repeat with other students, asking which items are heaviest/lighter/lightest. Repeat with another set of items. Throughout, ask students to compare using complete sentences (e.g. The pencil pot is heavier than the paper clip.)

Students need lots of experience of making weight comparisons with balance scales before they **weigh** with a kitchen scale as part of the activity on page 110 of the Student Book. Show students some **balance scales**. Check that they understand that when something is in each pan, the heavier item presses down more so the lower side of the scales contains the heavier item. Ask students to choose two items and estimate which is heavier. Ask them to place the item thought to be the heavier in one scale pan. *What will happen if we put one of the other items in the other pan?* Place the item in the pan to test which of the two items is heavier. Repeat with the third item. Were students correct?

If you have enough balance scales, ask students to work in groups to practise weighing items, estimating which item will be heavier or lighter first. Otherwise, continue as a class, inviting students to weigh the items.

Next, show students a **weighing scale** and give students guidance and practice in weighing items. Explain that **g** is an abbreviation of **gram** and that a gram is a unit of measure we use to describe how much an object weighs. Say that we also use **kilograms** to describe how much an object weighs and the abbreviation is **kg**. Heavier objects are measured in kilograms.

Ask students to imagine that they are holding an egg. Say that a large egg weighs about 100 grams. If possible, give each group a ball of modelling dough with a mass of 100 g to act as a substitute for an egg. Students are going to work in groups to compare the **masses** of different items to the mass of an egg and then weigh them on a weighing scale. Ask students to look at page 110 in the Student Book. Ensure that they understand where they need to record their estimates and actual masses.

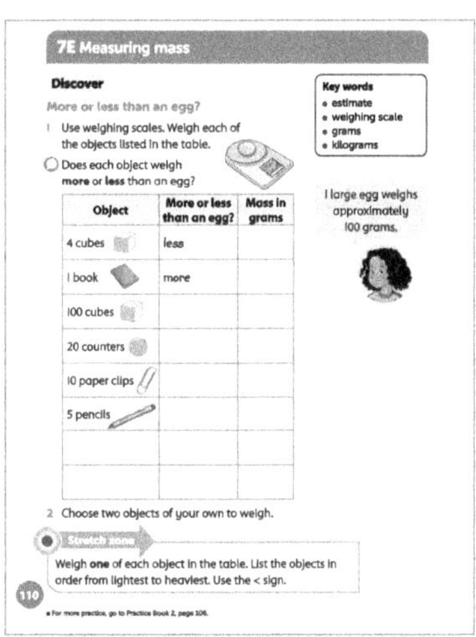

Differentiation

Supporting: Work with students to model the use of balance scales and weighing scales.

Consolidating: Ask students to predict outcomes before using the balance scales and weighing scales. *How did they predict the mass?*

Extending: Ask students to measure masses accurately using weighing scales and compare with the mass measured using a balance scale.

Stretch zone: *Weigh one of each object in the table. List the objects in order from lightest to heaviest. Use the < sign.*

Check that students have ordered the items and used the < sign correctly. Ask them to explain how they know they put the items in the correct order.

 **Reflection time**

Discuss with students their results from the Student Book activity. *Were you able to predict which objects weighed more or less than the egg? How could you tell?* For example, they may have used their earlier results of things weighing less or more than an egg to help them estimate later objects.

Practice Book: Students can now complete the activity on page of 106 in the Practice Book. They can do this directly after the main activity, as homework, or as the focus of a separate maths session to help students consolidate their learning and build fluency. This activity requires weighing scales so you will need to take this into consideration when deciding whether to give this as homework or not.

Differentiated outcomes	
All students	should use balance scales and weighing scales accurately.
Most students	will predict weights and use balance scales and weighing scales accurately.
Some students	may estimate in grams with some accuracy.

Answers

Student Book page 110

While students are working, walk around the classroom and observe their measuring. Check that students record whether each object weighs more or less than the egg.

Practice Book page 106

Check that the objects chosen are likely to weigh less than 1 kg, and that students have completed the sentences correctly.

7E Measuring mass

Explore Student Book page 111 · Practice Book page 107

Specific learning focus

- Estimate, measure and compare masses, choosing and using appropriate measuring instruments.

Deeper learning

- **Creative skills**: exploring/investigating
- **Real-world skills**: interpreting information
- **Interpersonal skills**: communication/teamwork

Key vocabulary

- scales, grams (gm), kilograms (km), heaviest, lightest, weight, degrees, Celsius (°C)

Resources

- Resource sheet 3 (temperatures)
- place-value counters
- recipe books, food packaging (cartons, boxes)

Language support

Provide recipe books for students to look through and share. You could also display food and other packaging with the weight highlighted.

Introductory activity

Ask students to share what some of their favourite food dishes are. Do they know how they are made? Does the person who makes them follow a recipe? Explain that a recipe is a list of ingredients and a set of instructions for how to make a food dish. Ask students to look at Grandma's sweet cake recipe on page 111 of the Student Book. Display on the IWB, if possible. Read the recipe together. Ask, *What numbers can you see? What does the 'g' stand for?* Remind students if necessary that it is the abbreviation for grams and that the abbreviation for kilograms is kg. Focus on the baking instructions and read the sentence aloud: *Bake for 20 minutes at 180 °C (degrees Celsius). What do you think this sentence tells us?* Explain that it tells us how hot the oven needs to be to cook the cake and that we can measure temperature in **degrees Celsius**. The short way of writing this is: **°C**. Look at the instructions about the butter and read the speech bubble. Explain that room temperature means how warm or cold the air is in a room. The instruction for butter to be at room temperature means that it will be the perfect softness for baking a cake.

Main activity

If you have real thermometers, pass these around or bring students to the front of the class to look at them. Discuss what they notice and ask what temperature students think they show. Focus only on degrees from 0 upwards.

Show students an example of an oven thermometer or use an image from Resource sheet 3 (displayed on the IWB, if possible). Count up together in multiples of 50 degrees and then count up in multiples of 10 degrees and back.

Give students a copy of Resource sheet 3 to complete individually and check in pairs.

Next, ask students to read page 111 in the Student Book in pairs. They should discuss what they think the task involves. After 5 minutes, take feedback and make sure that all students understand the task: they are going to complete the recipes for different quantities of ingredients and numbers of cakes made. Explain that you will work on this together as a class with teacher direction and support. Compare the recipes as a whole class. Ask students to look at Grandma's recipe and then look at recipe in question 1. Ask *What is the same and what is different?* Students may notice that the amounts of butter are 100g and 200g, so, for question 1, the amounts need to be doubled. The weights of butter, flour and sugar are double what they are in the original recipe. Therefore, they will need double the number of eggs and then they will make double the number of cakes.

For the recipes in questions 2 and 3, help students to work out the additional amounts needed. They could use place-value counters in hundreds to help work out the amounts. Explain or remind students that 1000 g = 1 kg.

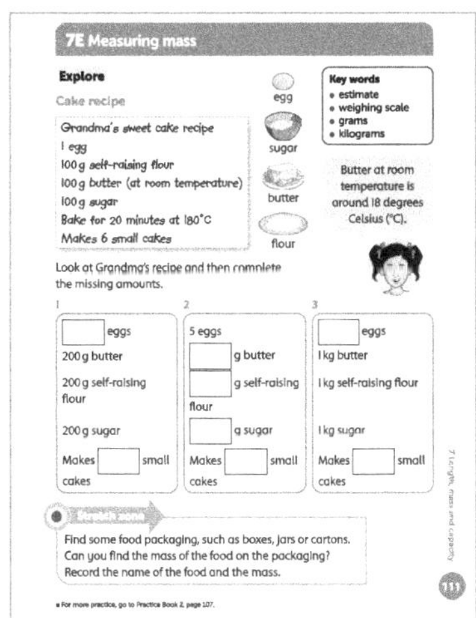

Differentiation

Supporting: Work with students to support them in carrying out the calculations.

Consolidating: Ask students to explain their strategies.

Extending: Ask students to develop their own word problems based on the recipes that you have given them.

Stretch zone: *Find some food packaging, such as boxes, jars or cartons. Can you find the mass of the food on the packaging? Record the name of the food and the mass.*

Check that students have the name and mass recorded correctly. *Was it more or less than you expected? Were the heaviest packages you chose always the biggest?*

 **Reflection time**

Give students some simple calculations to write a weighing or temperature story involving doubling and halving. For example, 'I carried a sack of flour weighing 5 kg in each hand. I carried 10 kg altogether', 'One bucket of sand weights 5 kg so 2 buckets weigh 10 kg' or 'I had 200 g of flour and used half to make a loaf of bread, so I had 100 g left.

Practice Book: Students can now complete the activity on page of 107 in the Practice Book. They can do this directly after the main activity, as homework, or as the focus of a separate maths session to help students consolidate their learning and build fluency. This activity requires weighing scales so you will need to take this into consideration when deciding whether to set this as homework or not.

Differentiated outcomes	
All students	should complete the recipes with teacher support.
Most students	will complete the recipes using teacher support as necessary.
Some students	may complete some of the recipes independently.

Answers

Student Book page 111

1 2 eggs makes 12 small cakes.

2 500 g butter, 500 g self-raising flour, 500 g sugar, makes 30 small cakes.

3 10 eggs makes 60 small cakes.

Practice Book page 107

Check students' objects and measures for the estimates given.

Stretch zone: 1500 g

7 Length, mass and capacity

Connect Student Book page 112

Big idea

- We can measure length, mass and capacity using standard units of measure.

Global skills

- **Creative skills:** exploring/investigating
- **Real-world skills:** research
- **Interpersonal skills:** communication/teamwork
- **Self-development skills:** reflecting on learning

Key vocabulary

- longer, shorter, more, less, weigh, hold

Resources

- lunchboxes, drink bottles, school bags, scales, measuring jugs, rulers/tape measures

Language support

Revisit all the key vocabulary for this unit to check understanding and pronunciation.

 **Introductory activity**

Write 'cm', 'm', 'g', 'kg', 'l', 'ml' on the board. Challenge children to work in pairs to think of one item at school and one item at home that they would measure with each of these units.

Record their examples in lists on the board.

 **Main activity**

Ask students to look at page 112 of the Student Book or display it on the IWB, if possible. Talk through the problem with students. Ask a student to read the text in the speech bubble: 'Think about what equipment you will need to measure.' Take suggestions, asking students to explain their choices.

Put students into small mixed-attainment groups to work on the problem so that less-confident language learners hear the appropriate mathematical vocabulary modelled by their peers. In their groups, they should decide whose lunchbox, bag and drink bottle to use and work together to answer the questions. However, they should each record the answers in their own book. For question 2, remind students that they will be completing three sentences – one comparing lengths, one comparing capacity and one comparing mass.

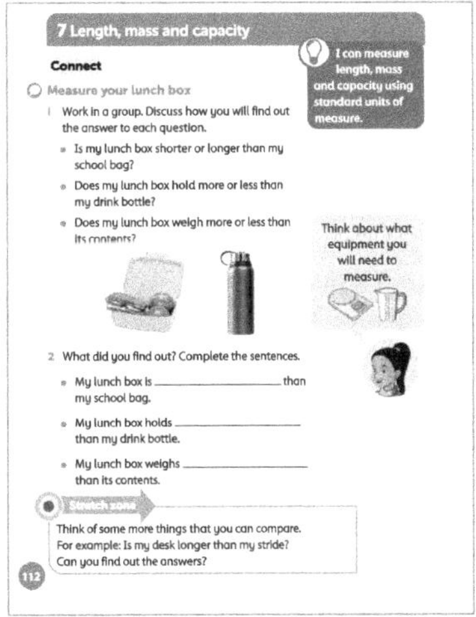

Differentiation

Supporting: Make sure that students take part and use measuring devices such as rulers, scales and jugs to help them answer the questions.

Consolidating: Encourage students to share their thinking so that they say their strategies clearly.

Extending: Ask students to measure three or four lunchboxes 3 and arrange them in order of size using the different measures.

Stretch zone: *Think of some more things that you can compare. For example: Is my desk longer than my stride? Can you find out the answers?*

Check that students have made some appropriate comparisons.

 **Reflection time**

Share solutions. Did groups reach similar answers, but in different ways? Were their answers very different in some cases? Can they explain why?

Differentiated outcomes	
All students	should take part in the measuring and suggest solutions.
Most students	will understand the concept of comparing measures.
Some students	may be able to confidently explain the process to other students.

7 Length, mass and capacity

Review Student Book page 113 · Practice Book page 108

Global skills

- **Creative skills:** problem solving/exploring
- **Interpersonal skills:** communication/teamwork
- **Self-development skills:** reflecting on learning

Student Book

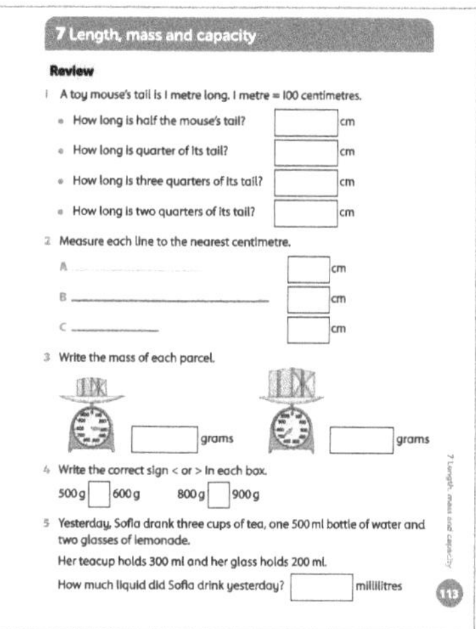

With young children, assessment activities are most effective when carried out as an everyday classroom activity. Students should have rulers, tape measures available to them.

Watch as students complete the sentences that represent situations involving length, mass and capacity. Check that they understand the language of measuring, including estimation and correct units.

As students work, ask questions to assess their understanding: *What would I use to measure how much water a sink holds? What does to the nearest centimetre mean? I want to find out how wide something is. What measuring tools could I use?*

Answers

Student Book page 113

1 50 cm   25 cm   75 cm   50 cm

2 Check measures have been completed accurately.

3 300 g   700 g

4 500 g < 600 g, 800 g < 900 g

5 1800 ml

Practice Book

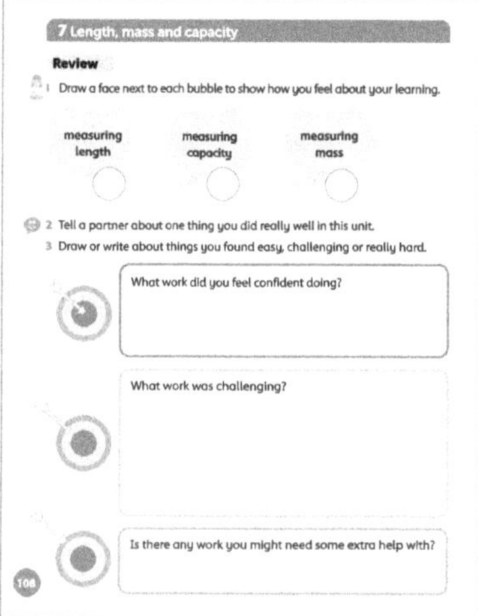

It is appropriate to complete this Practice Book review as a whole-class discussion. You may choose to keep a record of the class discussion or a copy of the review page for your own records. The review provides an opportunity for students to reflect on their learning from the unit, to discuss any areas of mathematics that they feel went particularly well, and any areas that they feel less confident about. Ensure all students have a copy of the Student Book as a reminder of the areas of mathematics that they have worked on in this unit.

Allow students plenty of time for discussion before asking them to complete the Practice Book page individually, and then, if appropriate, to share their responses with the rest of the class. If students complete this self-assessment at home, encourage them to discuss this with adults. Make a note of areas that students still feel unsure about.

Additional material

There are additional end-of-unit assessments available on the *Oxford Owl* website.

8 Money

Big idea

To students, money often has no connection to measuring. It is used only for buying: they pay with notes and coins for what they want, giving the number of notes or coins they are asked for – £5.50, £3.25, 85p for example. The idea that money measures the exchange value of goods is a very advanced idea. Students only know that a price is attached to goods and what notes or coins to offer to pay for them. This means that we must treat money differently from the properties of length, mass and capacity that have been covered earlier.

Instead of choosing a unit that gives us a measure by repeating along a line, we must accept that a number of different notes and coins are already familiar to students and that they need to learn the relationship between the different denominations.

Increased use of bank cards is reducing students' opportunities to observe money being handled, so it is crucially important that teachers introduce students to 'real' money in the classroom so that students can learn the currency.

The three key ideas for students to learn about and understand money are: coin recognition, equivalence and practical situations.

Look out for

- **Students who lack experience of handling currency** We assume the values of the coins as given, and learn the relationships between the different notes and coins, but students have less and less experience of handling notes and coins, and therefore less awareness of the values and the exchange process when using money. They may have a sense of counting coins, but not knowing the relative values. They may not understand the fact that two coins have different values. Provide lots of opportunities for handling real coins, discussing their value and their relation to other coins (*A dime has a value of 10 cents. Show me 10 cents using other coins*).

Possible misconceptions

- **Students may think that the bigger the coin, the greater its value.** Work with coins that challenge this assumption (e.g. 5p coins and 2p coins).
- **Student may think that the greater the number of coins, the greater the value.**
- **Students may think that once money has been handed over to pay for something, the transaction is complete.** They may not understand that change needs to be given and might wonder why a shopkeeper is giving money back.

Key vocabulary

- money, notes, coins, cost, buy, shop, pay, price
- cents, dollars, $, ¢, value
- how much, cheapest, most/least expensive, spend, change
- counting on, counting back
- grams, kilograms, litres, millilitres

Coverage in lessons

Learning focus	Learning outcomes (the ENC objectives)
Amounts of money	Recognise and use symbols for pounds (£) and pence (p); combine amounts to make a particular value.
	Find different combinations of coins that equal the same amounts of money.
Giving change	Solve simple problems in a practical context involving addition and subtraction of money of the same unit, including giving change.

8 Money

Engage Student Book page 114

Big question
- What notes and coins are used in my country?

Global skills
- **Real-world skills:** financial literacy
- **Interpersonal skills:** communication/teamwork

Key vocabulary
- money, notes, coins, buy, shop, pay, price

Resources
- examples of local coins and notes

Language support

Begin a money display for the classroom. Display images of notes and coins with labels such as: 'We use notes and coins to buy things in shops.' Include labels for the notes and coins showing their values, including local variation in names. For example, in the UK, include the vocabulary 'pounds' and 'pence', while in the USA include 'dollars', 'cents', 'nickels ', 'dimes', 'quarters' and 'buck'.

 ## Main activity

Ask students to look at page 114 of the Student Book. Display on the IWB, if possible. DIscuss each of the photos in turn. *What can you see? Why would you visit these places? If you wanted some tomatoes what would you need to do before you could take them home?* Write down any key words for reference, particularly those relating to shopping or money (e.g. **price**, **shop**, **pay**).

Choose students to read each of the questions in the speech bubbles in turn and discuss. Give them time to discuss quickly with a partner before discussing as a class. Ask them to think about things they **buy** (e.g. food, drink or toys). You can use known prices and then see how they would be paid for using the local currency and discuss whether the students think things are cheap or expensive.

 ### Reflection time

Discuss why there are different notes and coins. For example, how difficult would it be to pay for something like a bicycle only using 1p or 1 cent coins?

Look at the two sides of some coins and see what they show. *Whose pictures are on the coins? Where does it say what each is worth?* Ask students to design their own coin. What would they put on it?

Introductory activity

*What can you tell me about **money**? Do you know what our local **notes** and **coins** are called?* Record students' ideas on a flipchart or similar. Write down the names of the different denominations and have examples of each to pass round. If possible, show screen images of the coins and notes on your IWB.

"/>

Discover
Student Book page 115 • Practice Book page 109

Specific learning foci
- Recognise coins and notes.
- Use coins to make totals.

Global skills
- **Creative skills:** exploring
- **Real-world skills:** financial literacy
- **Interpersonal skills:** communication/teamwork

Key vocabulary
- value, coins, notes, cents, dollars, $, ¢ (or other symbols relating to local currency), cost

Resources
- coins and notes in your local currency
- US coins (optional)
- mini whiteboards and markers
- 0–100 number line: large one for front of class and tabletop one for each pair

Language support
Add enlarged images of each local coin and note to the classroom display. Label each with its value.

 **Introductory activity**

Pass around a range of coins and notes in the local currency for students to look at. Ask students to talk to their partner about what they notice, for example, *How do you know the **value** of each coin? Which coin is the biggest? What is its value? Which coin has the highest value? What other numbers can you see on some coins? What do they tell you?* (the year the coin was minted)

Taking each local coin or note in turn, ask students to write the value of that coin or note on their mini whiteboards. Check that students write **$** in front of the amount for **dollars** and **¢** after the amount for **cents** (or introduce whichever symbols and convention of positioning them is used for your local currency).

 **Main activity**

Ask students to look at page 115 of the Student Book. Display on the IWB, if possible. Tell students that they are going to pretend they are doing some shopping for toys and they need to decide which coins they can use to pay the exact price for each toy. Explain that you will be using local currency rather than US currency, if necessary. Discuss the selection of coins being used, and their

values. Supply each pair with coins as students benefit from physically making the amount to pay for each toy. Use real coins rather than plastic copies, if possible.

Work through the first example on page 115 together. *How much does the horse **cost**?* Remind students to include the denomination (i.e. 43 *cents*). *Which coins are used? Let's check whether they are correct using a number line.* Display a 0–100 number line and show each coin as a jump.

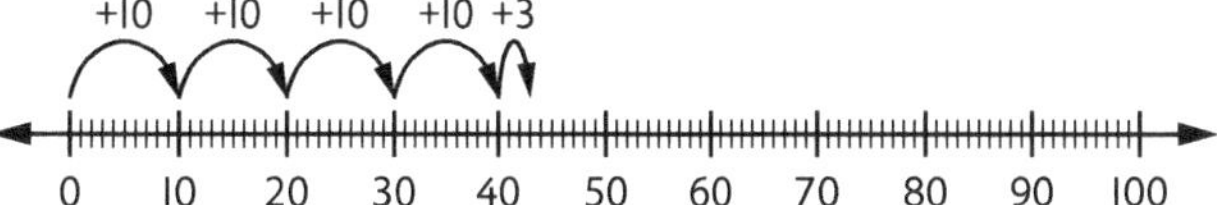

Count up saying, *ten, twenty, thirty, forty, forty-one, forty-two, forty-three.*

Is there another way we could make 43¢? What's the biggest coin we could start with to make 43¢? (25¢) Mark this as a jump on the number line. *What coin could we use to get us closer to 43¢?* Whatever students choose, mark on the number line and continue until you make 43¢.

Students then work in pairs on the remaining activities, working out how to 'pay' for the other four toys. Give them access to tabletop 0–100 number lines to help them make or check amounts.

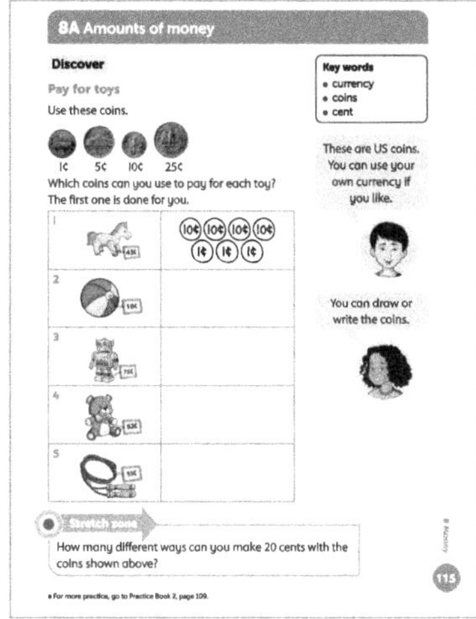

Differentiation
Supporting: Help students to pick a starting coin and count up from that amount to the required total using the number line for additional support.

Consolidating: Ask students to describe which coins they used and how many of each.

Extending: Ask students to find the way of making the total that uses the least number of coins.

Stretch zone: *How many different ways can you make 20 cents with the coins shown?*

Challenge students to find all nine ways by working systematically: 20 1-cent coins; four 5-cent coins; two 10-cent coins; one 5-cent coin, 15 1-cent coins; one 10-cent coin, ten 1-cent coins; one 10-cent coin, one 5-cent coins, five 1-cent coins; one 10-cent coin, two 5-cent coins; two 5-cent coins, ten 1-cent coins; three 5-cent coins, five 1-cent coins.

 ## Reflection time

Call out amounts for students to write on their mini whiteboards and then make the amounts using coins. Ask them to share different ways that they made the amounts. Can they explain how they know they have made the correct amount?

Practice Book: Students can now complete the activity on page of 109 in the Practice Book. They can do this directly after the main activity, as homework, or as the focus of a separate maths session to help students consolidate their learning and build fluency. Prices can be amended to reflect local currency, if appropriate.

Differentiated outcomes	
All students	should make a total when given a starting number and using a number line for support.
Most students	will correctly make the total for each toy.
Some students	may find different ways of making totals and select the simplest.

8A Amounts of money

Explore Student Book page 116 • Practice Book page 110

Specific learning focus

- Solve real-life money problems in the context of holiday spending.

Global skills

- **Creative skills:** problem solving/investigating
- **Real-world skills:** financial literacy
- **Interpersonal skills:** communication/teamwork

Key vocabulary

- how much? cost, cheapest, most/least expensive, spend

Resources

- price lists of foods and holiday items
- concrete resources (e.g. interlocking cubes, base-10 equipment), 0–20 number lines
- large 0–20 number line for display

Language support

As students to discuss costs of items, and reinforce the use of 'cheapest' and 'most expensive'. You can also include comparatives such as 'cheaper', 'more expensive', 'less expensive' to help students compare the value of two different amounts.

Student Book page 115

Students' answers will vary. For example:

Horse, 43¢: four 10¢ coins and three 1¢ coins

Ball, 18¢: one 10¢ coin, one 5¢ coin, and three 1¢ coins

Robot, 75¢: two 25¢ coins, two 10¢ coins, and one 5¢ coin

Teddy, 82¢: two 25¢ coins, three 10¢ coins, and two 1¢ coins

Skipping rope, 55¢: two 25¢ coins and one 5¢ coin

Practice Book page 109

Check that the prices given are made up using coins correctly.

 ## Introductory activity

Show students a list of items with their prices (in whole dollars or cents but no mixed amounts), perhaps food items from a shop. Show them in random order. Ask students to say how much each item costs. Then ask them to say which item has the lowest price and how they know. Explain that when we compare the prices of three or more items and we are talking about the lowest price, we can say the item is the **least expensive** or **cheapest**. Say that the item with the highest price is the **most expensive**. *Which item is the most expensive? How do you know?* Ask students to work with a partner to decide the order of the items from cheapest to most expensive. Take feedback from pairs and agree the order.

 ## Main activity

Ask students to look at page 116 of the Student Book. Display it on the IWB, if possible. Read the names and prices of each of the holiday items. Remind students of what the $ sign means, if necessary. Ask questions about the items to get students used to answering questions about prices, and say that students should answer in full sentences.

How much do the flipflops cost? (The flipflops cost twelve dollars.)

Which is more expensive, the suitcase or the sunglasses?
I spend less than 8 dollars. What did I buy? Is there more than one possible answer?

I buy the beach ball and the towel. How much did I spend altogether? How could I use a number line to help me find the answer?

Choose a student to show how to mark the higher of the two prices ($9) on a large number line and count on the second price ($5) to find the total.

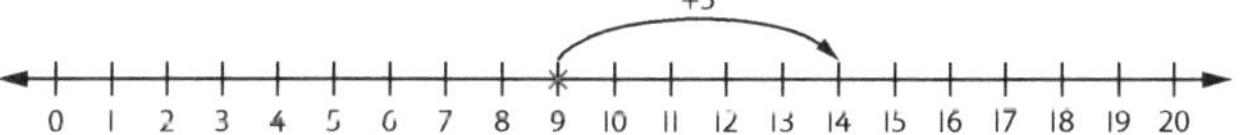

Explain that students should work together in pairs to answer more questions like this about the holiday items on page 116 of the Student Book, discussing each of the questions in turn before recording their answers. Students may choose to use ones-cubes and tens-rods (or similar concrete resources) to support their calculations, as well as or instead of number lines. Encourage them to think about their number bonds to 20 to see whether they can calculate the answers mentally.

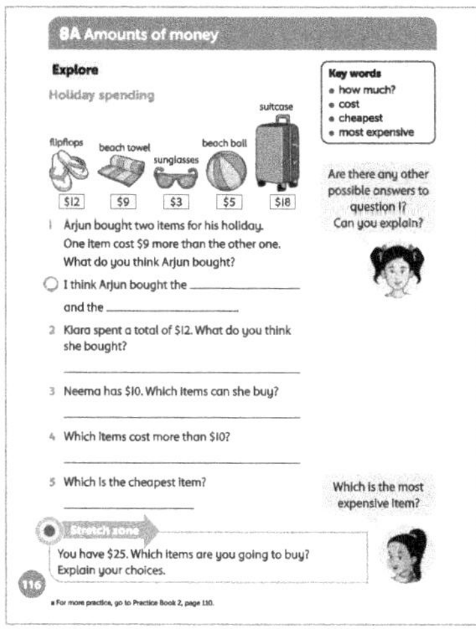

Differentiation

Supporting: Students continue to use concrete resources to support their calculations.

Consolidating: Ask students to check whether there is more than one possible answer for each question. Can they say how they know they have found them all?

Extending: Encourage students to use their knowledge of number bonds to 20 to find the answers and use inverse operations to check.

Stretch zone: *You have $25. Which items are you going to buy? Explain your choices.*

Check that students have calculated accurately and not spent more than $25.

 **Reflection time**

Which is the most expensive item? (the suitcase) *What combination of two other items could you buy that would be more expensive?* (the flipflops and the beach towel) *How much more?* ($3)

Invite students to come up with their own problems about the holiday items for the class to solve.

Practice Book: Students now complete the activity on page of 110 in the Practice Book. They can do this directly after the main activity, as homework, or as the focus of a separate mathematics session to help students consolidate their learning and build fluency. This provides further practice solving money problems with amounts not exceeding $20. Tell students that they may buy more than one of each item. For example, they can buy two stickers and two boxes of crayons, spending exactly $6.

Differentiated outcomes	
All students	should compare money amounts and answer questions using concrete resources.
Most students	will find at least one possible answer, and sometimes more for each question.
Some students	may work out most answers mentally and find all possible answers.

Answers

Student Book page 116

1 suitcase and beach towel **or** flipflops and sunglasses

2 flipflops **or** beach towel and sunglasses

3 beach towel, sunglasses or beach ball

4 flipflops and suitcase

5 sunglasses

Practice Book page 110

1 Various answers are possible, for example: football and book, **or** doll and crayons.

2 Various answers are possible, for example: football and 3 books, or doll, football, crayons.

3 book, crayons or stickers

4 Various answers are possible, for example, book and crayons.

Stretch zone: Answers will vary but the total should be $10.

8B Giving change

Discover
Student Book page 117 • Practice Book page 111

Specific learning focus

- Work out the amount of change from spending.

Global skills

- **Creative skills:** exploring/investigating
- **Real-world skills:** financial literacy
- **Interpersonal skills:** communication/teamwork

Key vocabulary

- how much?, coin, note, change

Resources

- coins and notes of local currency, real or otherwise
- 0–100 number lines: large one for front of class and tabletop ones for each pair
- base-10 equipment

Language support

Add to the classroom display a selection of items with a price label and the change received if the item was paid for with $1 or $5, as appropriate.

Introductory activity

Adapt these activities to use local currency if you prefer. Say to students that you are going to buy an item for 30¢. Explain that you only have a 50-cent coin. *Can I still buy the item?* Agree that if the amount of money you have is more than the price of the item, you can buy it. *What happens after I give the shopkeeper my 50-cent coin?* If no students suggest it, explain that you will get the difference between the money paid and the cost of the item back, and this is called **change.** Ask questions for students to decide whether they can buy an item and, if so, whether they will get change. *My item costs 15¢ and I want to pay with a 25-cent coin. Can I buy the item?* (yes) *Will I get change?* (yes) Include an example where they will not have enough money, for example, *I have a dime and I want to buy an item that cost 18¢. Can I buy the item?* (no)

Main activity

Return to the first example (buying an item for 30¢ with a 50-cent coin). *How much change should I get?* Ask students to work out how much change and say how they did it. They may subtract 30¢ from 50¢ to leave 20¢, using known number bonds to help them to calculate or they may count on from 30¢ to get to 50¢. Model on a 0–50 number line, counting on from the cost to the amount paid, or with two rows of ones-cubes or two rows or tens-rods to show the difference.

Agree that the change would be 20¢. *Which coins might you get back as 20¢ change?* Encourage students to find all possibilities. There are nine.

Ask students to look at page 117 of the Student Book. Display on the IWB, if possible. Explain that they will continue to find change and that they can record the change by writing an amount (e.g. 5¢) or drawing a picture of the coins they would receive as change. Focus on question 5. *What is being used to pay?* (1 dollar) *How can you find change from a dollar?* Explain or remind students that $1 = 100¢. *This means that you can find the difference between 100 and 85 to find out how much change you get.* Provide students with a 0–100 number line and coins and notes to support their calculating. Students complete the activities in pairs.

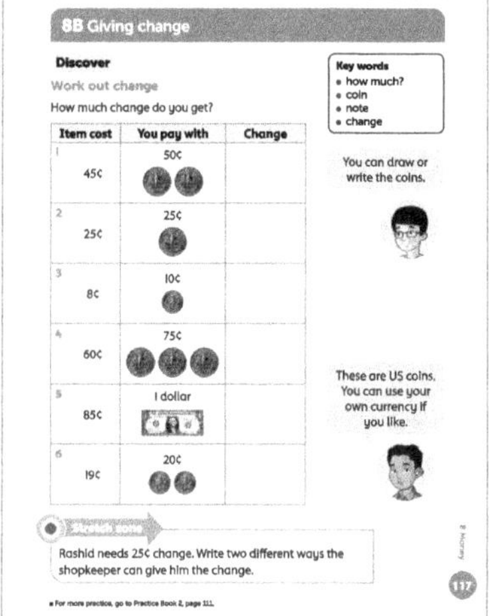

Differentiation

Supporting: Encourage students to use a number line to work out the change. Model again how to count on from the item cost to the amount paid.

Consolidating: Ask students to say, in complete sentences, how much an item cost, how much they used to pay and the change they received. Can they show their method?

Extending: Ask students to find different ways of making the right amount of change.

Stretch zone: *Rashid needs 25¢ change. Write two different ways the shopkeeper can give him the change.*

Check that students can correctly show different ways to make the change using the available coins.

 Reflection time

Ask students to describe how they worked out the change for each question in the Student Book. *Which way used the fewest coins?* Ask why shopkeepers might not always give change using the fewest coins. (They may want to give smaller coins as change because they take up lots of space.) Ask students who did the stretch activity how they worked it out.

Practice Book: Students can now complete the activity on page of 111 in the Practice Book. They can do this directly after the main activity, as homework, or as the focus of a separate mathematics session to help students consolidate their learning and build fluency. Ask students to look at the number line at the top of the page. *What do they notice?* If no students suggest it, point out that the line goes up in 10-cent steps and then after 90¢ it changes to $1. Do they know why? Remind them that this is because 100¢ = $1, so you could also write 100¢ there.

Differentiated outcomes	
All students	should identify when change is needed.
Most students	will work out how much change and what coins or notes can be given.
Some students	may find more than one way to make the change amount.

Student Book page 117

Which coins students choose to make each amount will vary.

1 5¢   **3** 2¢   **5** 15¢

2 0¢   **4** 15¢   **6** 1¢

Practice Book page 111

1 40¢   **4** 10¢

2 20¢   **5** 5¢

3 55¢

Stretch zone: Check that students have written a word problem that results in change of 15¢.

8B Giving change

Explore   Student Book page 118 • Practice Book page 112

Specific learning foci

- Recognise all coins and notes; use money notation.
- Work out the amount of change from spending.

Global skills

- **Creative skills:** exploring
- **Real-world skills:** financial literacy

Key vocabulary

- coins, notes, cents, dollars, change, counting on, counting back

Resources

- blank number lines marked with ten intervals: large one for front of class and small one for each student
- coins and notes in the local currency
- mini whiteboards and markers

Language support

Display some items with a price label (e.g. 70¢ and 26¢) and the change if the item was paid for with $1 (e.g. 30¢ and 74¢). Add a number line with corresponding jumps to show how to reach the answer.

Introductory activity

Give each student some coins in their own currency. Ask them to make different amounts (e.g. 45¢, 60¢, 82¢). Can they make the amounts in more than one way? (e.g. to make 45¢, they might use 20¢, 20¢ and 5¢, or 20¢, 10¢, 10¢ and 5¢.) *Which way used fewest coins?*

Main activity

How many cents are equal to one dollar? Remind students that $1 = 100¢, if necessary. Display a blank number line marked with ten intervals, but unlabelled. Give each student a similar blank number line as well. Explain that you would like to work together to label the number line. Label the left end 0¢ and the right end 100¢. Ask students to do the same on their own number lines. **Count on** together in steps of 10¢ from 0 to 100¢. When you reach 100¢, ask students if there is another way that you could label this end mark. Agree $1. Label the remaining marks on the number line 10¢, 20¢, …, 90¢ and ask students to do the same on their own lines.

If you paid for something costing 20¢ with $1, how much change would you get? Ask students to use the number line to help them calculate 100 – 20. Explain that this is the same as 100¢ – 20¢, or $1 – 20¢. Repeat with a different calculation that does not involve only multiples of ten (e.g. $1 – 73¢). Ask a student to explain how they worked out the amount of change by using jumps on the number line. For example, 'First I jumped 7 from 73 to get to 80. Then I did two jumps of 10 to get to $1.' Ask any other students who used a different method (e.g. **counting back** from 100) to share it with the class and show it on the number line. Discuss and compare how efficient the methods were. Model counting on and back to find change from a $1 if you bought an item for 15¢.

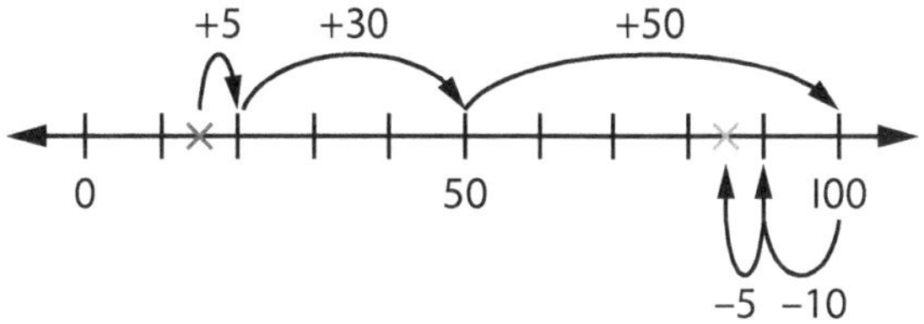

Ask students to work independently to complete the questions on page 118 of the Student Book. After calculating change for each item, ask them to make up a story to go with each calculation.

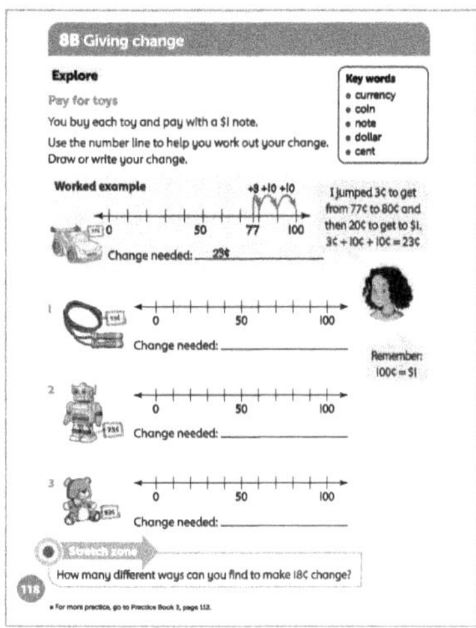

Practice Book: Students now complete the activity on page of 112 in the Practice Book. They can do this directly after the main activity, as homework, or as the focus of a separate mathematics session to help students consolidate their learning and build fluency. Encourage students to look at the price of each item and consider which would be more efficient, counting on or back.

Differentiated outcomes	
All students	should find change in units of 10 by counting on.
Most students	will find change for $1 by counting on using a number line.
Some students	may make up a change number story that can be solved by subtraction.

Answers

Student Book page 118

1 45¢   **2** 25¢   **3** 18¢

Practice Book page 112

Item	Cost	Change from $1
Fruit	25 cents	75 cents
Ball	64 cents	36 cents
Balloon	12 cents	88 cents
Pencil	75 cents	25 cents
Juice	50 cents	50 cents
Hair clip	93 cents	7 cents

Stretch zone: Students may offer different responses here based on their personal knowledge.

Differentiation

Supporting: Change the prices in the Student Book activity so that they are multiples of 10 only.

Consolidating: Ask students to explain their strategies for counting on.

Extending: Ask students to write their own change problem, coming up with their own toy and price, and sharing with the whole class to solve.

Stretch zone: *How many different ways can you find to make 18¢ change?*

Check that students have found different ways of making 18¢. (There are six possible ways.)

 **Reflection time**

What change did you get when you bought the robot? Which coins could the shopkeeper use for this? Record the different possibilities students suggest. Ask those students who made up word problems to share these with the rest of the class to solve. Ask a student to model their calculation method on the number line.

8 Money

Connect   Student Book page 119

Big idea

Each country uses different notes and coins. When we buy things, we need to make sure that we get the correct change.

Global skills

- **Creative skills:** problem solving/exploring/investigating
- **Real-world skills:** interpreting information/financial literacy
- **Interpersonal skills:** communication/teamwork
- **Self-development skills:** reflecting on learning

Key vocabulary

- grams, kilograms, litres, millilitres, dollars, cents

Resources

- ball or bean bag
- balance scales, real coins
- tens-rods, 100-flats or interlocking cubes

Language support

Support students with the language associated with a real-life context of purchasing snacks and drinks, with a focus on the ingredients used in each of the recipes in the Student Book.

Gather students in a circle. Tell them that they will be counting in hundreds forward and back. Throw a ball to a student to catch and say, *one hundred*. They catch the ball and then throw it to another student saying the next number in the count (200) and so on. Continue until you reach 1000 and then count back. Next, choose a student to start with any multiple of 100 less than 1000 and start counting forward or back. Shout 'change' occasionally so students change the direction of their count.

Students may benefit from continuing to use scales, weighing amounts and working with coins. Provide scales and coins if possible.

Ask students to look at page 119 of the Student Book. Display it on the IWB, if possible. Read together the information at the top of the page. Discuss the context (a class party) and recipes. You may want to show students photos of the different ingredients to support their understanding.

Ask questions about the units of measure for the ingredients. For example, *How much butter is in the flapjack recipe? Is that more or less the amount of rolled oats? How much sugar is in the lemonade? Is that more or less than the amount of sugar in the flapjacks?*

Next look at the list of food items and their prices. Ask questions about the money amounts and measures. *What does the $ symbol tell us? What units of measure can you see? Which is more, 500 g or 1 kg? How do you know? Which food item is the same price as the golden syrup? How much would 2 kg of sugar cost?*

Go back to the first section of text on page 119 of the Student Book and ask students to identify the key pieces of information. Record them on the board.

Read and work on question 1 together as a class with students collaborating in small groups. *How can we find out if we have enough ingredients? What do we need to do first? What information will we need to use?* Agree that they need to work out how many flapjacks and glasses of lemonade they will need for the number of students in the class. *How can we work out how many glasses of lemonade we will need?* (Double the number of people in the class.) Give groups time to work out the answer. Compare this number to the number each recipe makes and decide, if necessary, by how much you need to increase the recipe. Repeat for the flapjack recipe.

Students then continue to work in their groups to answer questions 2 and 3. Question 3 is quite open ended. Encourage students to think about all they know about food and drink prices and quantities to help them decide on a price.

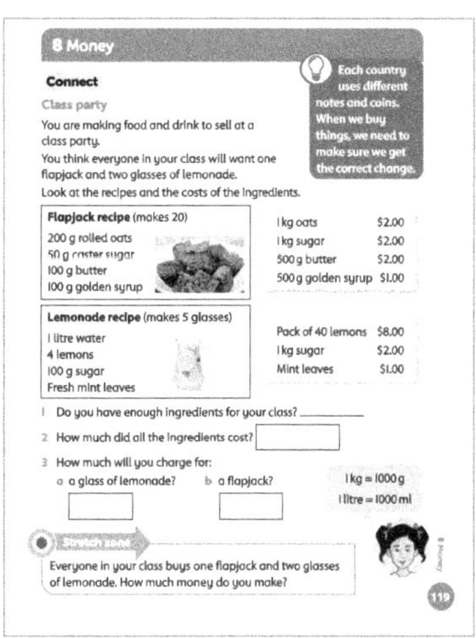

Differentiation

Supporting: Ask students to explain strategies for some of the simpler totals.

Consolidating: Ask students to explain their strategies for increasing the recipes and calculating whether they had enough ingredients.

Extending: If the class has enough ingredients, ask them to work out how much will be leftover.

Stretch zone: *Everyone in your class buys one flapjack and two glasses of lemonade. How much money do you make?*

Check that students have calculated correctly for the number in the class.

 Reflection time

Bring together the results from the various groups to find out how much the ingredients cost. Discuss the prices they decided on for each item and how they decided on their prices.

Ideally, do some cooking with students and have a small party or tasting session at the end of this unit.

Differentiated outcomes	
All students	should contribute to the group discussion.
Most students	will listen to each other's calculations, correcting if necessary.
Some students	may explain their calculations clearly for others to understand.

Answers

Student Book page 119

Check that students have answered questions 1 and 2 correctly and that they can explain their reasoning for their choice of prices for each item in question 3.

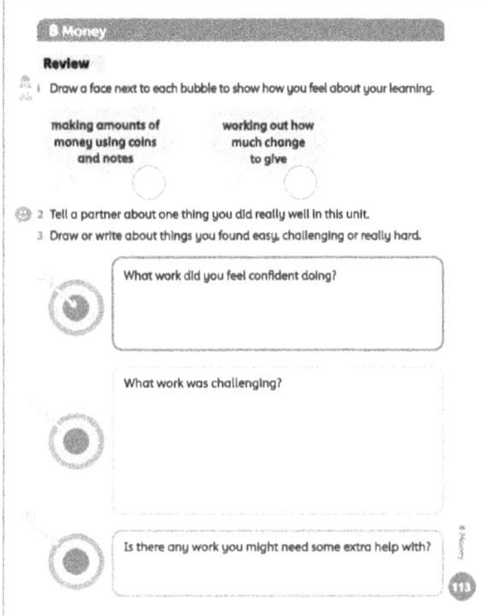

8 Money

Global skills

- **Creative skills:** problem solving
- **Real-world skills:** financial literacy
- **Interpersonal skills:** communication
- **Self-development skills:** reflecting on learning

Student Book

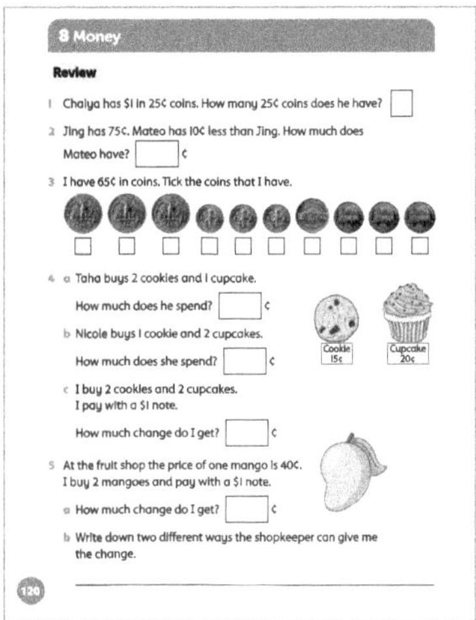

With young children, assessment activities are most effective when carried out as an everyday classroom activity. Students could have number lines and coins available with both the values and the numbers in words so they can refer to this to support them, if necessary.

Watch as students decide on the best way to make up different amounts using the coins available. Ask students how they decided to make up the amounts. What was their strategy? Did they use larger value coins first or did they choose to repeat smaller value coins? Can they show their method using jumps on a number line?

Encourage students to use the coins. Extend by asking a range of similar questions or asking students to make their own money problems, for example, *Which amount needs fewest coins, 23¢ or 41¢?* or *How much is the total cost of 2 drinks at 25¢ and a snack at 35¢?*

Answers

Student Book page 120

1 4

2 65¢

3 2 × 25¢, 1 × 10¢, 1 × 5¢

4 a 50¢   **b** 55¢   **c** 30¢

5 a 20¢   **b** 10¢ + 10¢, 10¢ + 5¢ + 5¢, for example.

It is appropriate to complete this Practice Book review as a whole-class discussion. You may choose to keep a record of the class discussion or a copy of the review page for your own records. The review provides an opportunity for students to reflect on their learning from the unit, to discuss any areas of mathematics that they feel went particularly well, and any areas that they feel less confident about. Ensure all students have a copy of the Student Book as a reminder of the areas of mathematics that they have worked on in this unit.

Allow students plenty of time for discussion before asking them to complete the Practice Book page individually, and then, if appropriate, to share their responses with the rest of the class. If students complete this self-assessment at home, encourage them to discuss this with adults. Make a note of areas that students still feel unsure about.

Additional material

There are additional end-of-unit assessments available on the *Oxford Owl* website.

9 Time

Overview

Big Idea

The Big idea for this unit is that time can be measured in agreed units that are related to each other, for example, seconds, minutes, hours, days, weeks and months. Students will have heard many of these units in everyday contexts but will still be learning to understand how long each unit 'feels' with experience.

Students are usually very familiar with the days of the week, particularly in the ways that they relate to themselves and their activities. The length of a day is easily understood because each time we wake up after a night's sleep, it is a new day. Referring to the date and relating the months to the student's experiences will help them to learn the order.

Look out for

- **Students who confuse the hands on a clock or watch face.** They may experience confusion between the hour hand and the minute hand, perhaps because hours are longer than minutes but the minute hand is longer than the hour hand. For clocks or watches with a second hand, this confusion can be even worse.

- **Students who have difficulty reading scales, including those on clocks.** They are likely to be more familiar with a digital time display, rather than an analogue display on a clockface, because they see it on computers, televisions, microwaves and mobile phones. Using both types of clock together will help students to understand the words we use when talking about and telling the time. A student who can read 10:30 on a digital clock does not necessarily understand that this means the time is halfway through the hour. Students usually have a good understanding of half of a circle, so we can use this to help explain what we mean by half past. The minute hand on an analogue clock points straight down for half past because it has travelled halfway around the circle.

- **Students who may not interpret the time they can read in real contexts.** Even though students may learn to 'tell the time' on a clock, they may not be able to relate that to the actual time of day, for example, seeing a clock that shows 'half past 3' when the time of could be 10 o'clock, and so on.

- **Students who find months of the year difficult because of the much bigger timescale than everyday times.** You can support students in this by referring to calendars and noticing that the month in the date only changes rarely in their experience.

Possible misconceptions

- **Students may think the numbers on a clockface tell them 'how many minutes',** for example, they read the time 3:20 as '3 oh four'. Focus on reading in intervals of 5 minutes using a geared clock so students can see more clearly that when the minute hand is pointing at, say, 5 that this means 25 minutes, not five minutes. The same number on the face is used for different purposes, for example, the 4 is to show 4 o'clock, but also represents 20 minutes past the hour.

- **Students may think that the hour hand still points exactly to the hour when, for example, it is half-past four.** Use a geared clock to model how the hands move as an hour passes.

- **Students may think that the hour is always the number the hour hand is closest to,** for example, they read 9:55 as 10:55. Starting on the hour, move the hand forward slowly on a geared clock, stopping at key times (e.g. quarter and half past, quarter to) and talk about what time it is and describe where the hour hand is pointing.

Key vocabulary

- second, minute, hour, o'clock, half past, day, week
- year, time, today, yesterday, tomorrow
- how long?, stopwatch, sand timer, digital, clockface, on time
- a.m., p.m., minute hand, hour hand, middday, noon
- the nearest five minutes, quarter past, quarter to
- days of the week: Monday, Tuesday, Wednesday, Thursday, Friday, Saturday, Sunday
- month, months of the year: January, February, March, April, May, June, July, August, September, October, November, December, calendar

Learning focus	Learning outcomes (the ENC objectives)
Telling the time	Tell and write the time to five minutes, including quarter past/to the hour, and draw the hands on a clockface to show these times.
Days of the week	Compare and sequence intervals of time.
Calendars	Compare and sequence intervals of time.
How long?	Compare and sequence intervals of time.
Comparing units of time	Know the number of minutes in an hour and the number of hours in a day.

9 Time

Engage
Student Book page 121

Big question

- How do I tell the time and how long do things take?

Global skills

- **Real-world skills:** presenting information
- **Interpersonal skills:** communication/teamwork

Key vocabulary

- time, second, minute, on time, stopwatch, sand timer, months of the year, calendar, days of the week

Resources

- digital timer, stop watches, other timers (mobile phone, computer and so on)

Language support

Begin a time display for the classroom. Take photos of students at important times of the day, for example washing their hands for lunch, and display these with a sentence label such as 'Lunch time is at 12:30' and 'We start school at 9 o'clock.'

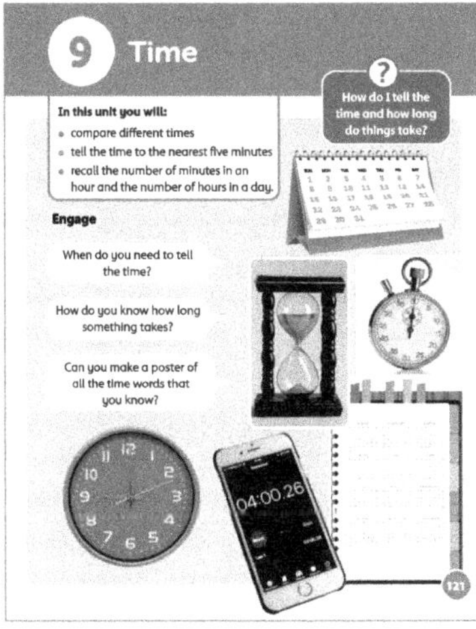

 **Main activity**

Ask students to work in mixed-attainment groups. Choose the most confident writer to help write the words for the group. Ask the group to think of as many words related to time as they can. After ten minutes' discussion, collect all the words and write them on a flipchart. Ask each group to take responsibility for a collection of words – such as **days of the week, months of the year,** words for duration of time – and create a poster that can be displayed throughout the unit.

Ask students to look at page 121 of the Student Book. Display it on the IWB, if possible. *What can you see in the photos?* Ask students to talk to a partner to discuss for a few minutes. As a class, discuss what the photos show, sharing any less-familiar vocabulary, e.g. **sand timer** and **stopwatch**. Ask questions about each of the photos, e.g. *How many hands are on the pink clock? What do you notice about the numbers on the stopwatch?* (they go up in fives) *What do the numbers on the calendar mean?* (days of the month) *What do you think the first two digits on the mobile phone display mean?* (the number of minutes)

Differentiation

Supporting: Model the correct pronunciation of key vocabulary.

Consolidating: Ask for linked words. For example, if a student suggests a day of the week, ask for the day before or the day after.

Extending: Ask students to explain the meanings of words to others in their group.

 Reflection time

The small groups should present their posters to the whole class. As they do, ask questions to make sure that they understand the meanings of the words. Ask more questions, and encourage other students as well, to find out what else they can tell you about the information on their posters, *for example, What is the second month of the year? Which day comes before Wednesday? Which is longer, a minute or an hour?*

Introductory activity

Explain that when we arrive somewhere at – or when something starts at – a specific, agreed **time** we say, **on time.** Share an example, for example, *I planned to leave the house at 7 o'clock and I left at 7 o'clock. I left on time.* Ask students to discuss in pairs when it is important to be on time. Share ideas, for example, when you have an appointment at the doctor or dentist, for the start of the school day, when catching a bus, train or flight. *What is the opposite of on time?* (late)

9A Telling the time

Specific learning focus

- Read and record the time on analogue and digital clocks, to the nearest quarter hour for both the 12-hour and 24-hour clock.

Global skills

- **Real-world skills:** interpreting information
- **Interpersonal skills:** communication

Key vocabulary

- a.m., p.m., minute hand, hour hand, digital, clockface, o'clock, half past

Resources

- geared analogue clock for front of class and smaller analogue clocks for each pair

Language support

Support students by helping them describe what they do at different times and then model for them how to say this in a full sentence. For example, say 'wake up' and the full sentence is 'I wake up at six fifteen a.m.'

 ### Introductory activity

Ask students *What time is it?* and focus on showing **digital** times only. *Do you know what a.m. means when we say the time is 11* **a.m.***? What about 7* **p.m.***?* Students may say that a.m. means morning and p.m. means afternoon. Explain that a.m. is before **midday**, or before 12 **noon**, and p.m. is after midday, or 'after noon'. Ask students questions about times at different parts of the day – morning, afternoon, evening and night – they can suggest activities they do in each part of the day and put a time to each, for example, 'I eat breakfast at 7 in the morning' and so on.

 ### Main activity

Display an analogue **clockface** with the time showing 12 **o'clock**. Ask students whether they know what time this is. How do they know? Can they tell you that the big hand is on 12, showing that it's an 'o'clock' time, and the little hand is on 12, showing that it's 12 'o'clock? Remind students that the big or long hand is called the **hour hand** and that the little or short hand is called the **minute hand**. Read time to the hour, quarter hour and half hour for different hours, but just saying, for example, four fifteen at this point. Show how to draw times to the hour, quarter hour and half hour.

Set the analogue clock to half past 7. Ask students whether they know how to write this as a digital time. Does this clock tell us whether it is a.m. or p.m.? (No) How do we write 7.30 a.m. in digital? (07:30 a.m.) Repeat with a different digital time, asking each student to show this on an analogue clock. They may need support with this – more confident students can help. Take this opportunity to remind students that each number on the clock is separated by five minutes.

Students should complete the activity in the Student Book page 122 in pairs. Help them, if necessary, to think about what they might be doing at the different times by asking questions such as, *Is 12.45 p.m. the morning, afternoon or evening?* (afternoon) *We eat lunch at 12 p.m. What do we do immediately after lunch?*

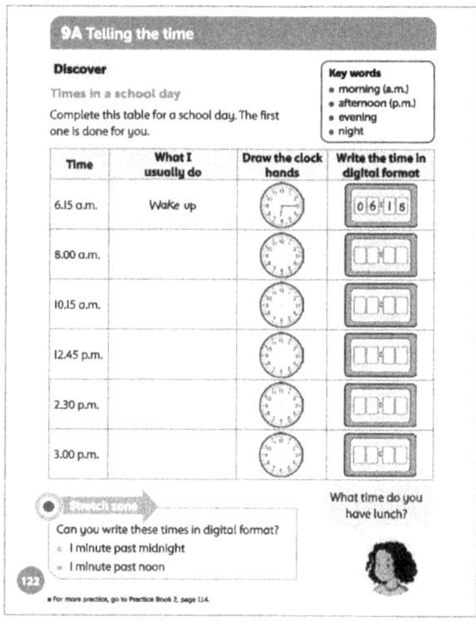

Differentiation

Supporting: Help students to make the times on an analogue clock, talking the process through; they copy it in their Student Book.

Consolidating: Ask students to tell you the process they use to tell the time.

Extending: Ask students to tell you digital times for each analogue time.

Stretch zone: *Can you write these times in digital format?*

- *1 minute past midnight*
- *1 minute past noon*

Follow up with questions such as, *What's the same? What's different? What would these two times look like on a clockface? How will they be different?* (They won't!)

 ### Reflection time

Ask students to describe how they completed the times in the Student Book activity. Ask how they knew where to draw the hands on the analogue clock and how they knew which digits to fill in for the digital clock. Ask questions such as *How did you know where to draw the hour hand and minute hand? What is the same and what is different about 2.30 p.m. on the analogue clock and on the digital clock?*

Practice Book: Students now complete the activity on page of 114 in the Practice Book. They can do this directly after the main activity, as homework, or as the focus of a separate mathematics session to help students consolidate their learning and build fluency. Before they begin, discuss what time 'to the nearest 15 minutes' means, including examples to show the possibilities within an hour (e.g. 5.00, 5.15, 5.30, 5.45).

Differentiated outcomes	
All students	should read and write analogue and digital times in quarter hours with support.
Most students	will read and write analogue and digital times in quarter hours.
Some students	will read and write digital times for an analogue time in a.m. or p.m.

Answers

Student Book page 122

Answers will vary according to what events students have matched to the different times. However, check that the times have been shown correctly on the analogue and digital clocks.

Practice Book page 114

Answers will vary according to when students do the activities listed and according to the activities they add to the table. Check that the digital times have been written correctly and that the times drawn on the analogue clocks match the digital times.

9A Telling the time

Explore Student Book page 123 • Practice Book page 115

Specific learning foci

- Read the time on analogue and digital clocks to the nearest five minutes.
- Read times as quarter to, quarter past, half past and o'clock.

Global skills

- **Real-world skills:** presenting information/ interpreting information
- **Interpersonal skills:** communication

Key vocabulary

- a.m., p.m., the nearest 5 minutes, quarter past, quarter to, o'clock

Resources

- geared analogue clock to model times for students
- mini whiteboards and markers
- worksheet for students (Resource sheet 4)

Language support

Display a clockface showing a time with labels showing all possibilities, for example. What time is it? 4.15 a.m., 4.15 p.m., quarter past 4, 04:15.

 **Introductory activity**

Display an analogue clock showing 8 o'clock. Ask, *Is this before or after noon? Would I write the time with a.m. or p.m.?* Students should say that they cannot know.

Now show them a digital time such as 07:45 and ask how we say this time. (7.45) Can they say where the hands would be on an analogue clock? Ask a student to come to the front of class and show the time.

Explain that there is another way to say it. Introduce **quarter to** and say that 7.45 is a quarter of an hour before 8 o'clock, so we can say 'quarter to 8'.

Repeat with, say, 07:15, leading to **quarter past,** using similar questions. Repeat with more examples and ask students to say the time.

 **Main activity**

Show students an analogue clockface. Move the hands to different times saying all the quarter and half times. Use a geared clock so they can see that the hour hand is not pointing directly at the hour except at o'clock times; it moves gradually from one number to the next.

Then move on to counting in fives around the clockface again using the geared clock and stop at some times to **the nearest 5 minutes** and ask, *What time is it?* Students should be able to say, for example, 'eight thirty-five'. Show on a clock that this can also be said as '25 minutes to 9'.

Give each student a copy of the worksheet (Resource sheet 4) to fill in the missing information. Discuss the answers and tell students to keep this worksheet to refer to as they work through the questions on Student Book page 123.

Students should work through these questions in pairs, explaining to each other why they think matches are correct.

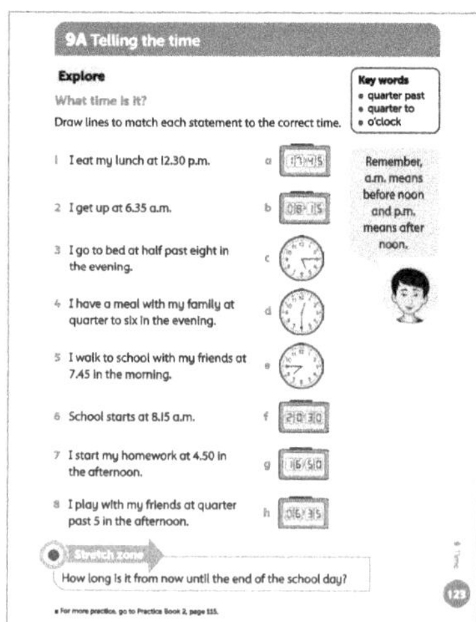

Differentiation

Supporting: Model how to say each time for quarter to and quarter past, for students to copy and explain.

Consolidating: Ask, *How do you know that you matched the time to the right clockface? Can you explain your thinking?*

Extending: Ask students for different ways of writing or saying some of the times in the Student Book. For example, for question 6, 'school starts at a quarter past 8'.

Stretch zone: *How long is it from now until the end of the school day?*

Ask students to think about their work on subtraction and think about what type of problem this is (difference). *How did we solve these?* (We counted along a number line.) They find the current time to the nearest 5 minutes and count up to the end of the school day. Help them draw a number line to help with this if necessary and total the hours first and then the minutes.

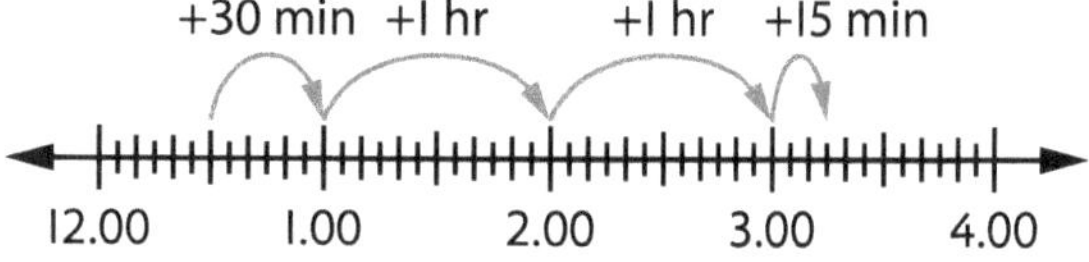

 **Reflection time**

Show students a clockface. Ask them to discuss in pairs what they can see, then come together to share as a class. Ask for example, *What time is the clock showing? How do you know? What does each hand tell you? Can you say the time in digital time?*

Practice Book: Students now complete the activity on page of 115 in the Practice Book. They can do this directly after the main activity, as homework, or as the focus of a separate mathematics session to help students consolidate their learning and build fluency. Encourage students to use Resource sheet 4 for support.

Differentiated outcomes	
All students	should match written times with digital and analogue times with support.
Most students	will match the written time with the correct digital or analogue time.
Some students	may be able to say and write the time in several ways and read digital times.

Answers

Student Book page 123

1 d	**3** f	**5** e	**7** g
2 h	**4** a	**6** b	**8** c

Practice Book page 115

1 Quarter to 11

2 Ten past 3

3 Twenty-five to 5

4 Twenty to 9

5 Five past 9

6 Twenty-five past 9

7 Ten to 8

8 Five to 6

Stretch zone: Students will offer different answers depending on their home routine. A typical response could be 4 hours.

9B Days of the week

Discover Student Book page 124 · Practice Book page 116

Specific learning foci

- Know and order the days of the week and the months of the year.
- Know the relationships between consecutive units of time.

Global skills

- **Creative skills:** exploring
- **Real-world skills:** presenting information/ interpreting information
- **Interpersonal skills:** communication/teamwork

Key vocabulary

- day, week, Monday, Tuesday, Wednesday, Thursday, Friday, Saturday, Sunday, today, yesterday, tomorrow, calendar

Resources

- calendars, planners
- days-of-the-week cards, cut out from Resource sheet 5 copied onto card

Language support

Regularly chant or sing the days of the week. Support students to use the correct pronunciation, particularly for Tuesday and Thursday. List words that have the same starting sounds: for Tuesday – tube, tulip, tuna, tutor, tune; for Thursday – third, thirsty, thermal.

 **Introductory activity**

Ask, *What **day** is it **today**? What day was it **yesterday**? What day will it be **tomorrow**? Can you think of any other days of the **week**?* Give the corresponding day card to each student who answers a question and invite them to the front of the class. Ask questions until all seven days-of-the-week cards have been distributed.

Ask the rest of the students to move the seven 'days-of-the-week' students so that they are in the correct order for the days of the week. Find a 'days of the week' song or poem in a book or from the internet to teach to students. The students holding the days-of-the-week cards can raise the card when their day is said or sung.

 **Main activity**

Display a **calendar** for the current month. Ask students to describe in pairs what they notice about the calendar. Ask, for example, *What does this calendar show? Can you find the days of the week on the calendar? What is the first day of this month? What day is the last day of this month? How many Tuesdays are there this month?*

Draw a chart on the board with seven columns labelled with the days of the week. Ask students to name something that happens on each day. After you have recorded several ideas for each day of the week, ask questions such as: *What happens on [Monday]? What do you do on [Tuesday]?* Provide framework sentences to help students answer in complete sentences, for example, 'On [day of the week] we ______.' and 'I ________ on [day of the week].'

Ask students to complete the activity in the Student Book page 124 individually and then join up in pairs. One student reads out the question and the other gives their answer. They then swap roles. Encourage them to ask questions out of order to help develop their listening skills. Once they have finished, ask them to look at the speech bubble in the Student Book and discuss which is their favourite day of the week with their partner and explain why.

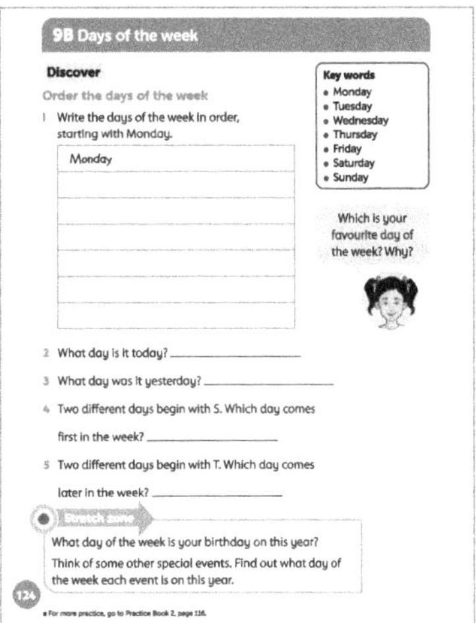

Differentiation

Supporting: Ask students to name the days of the week in order. Use different starting points.

Consolidating: Ask students to name days before and after given days.

Extending: Ask students to name days that are two or three days before or after a given day. Ask them to tell you, for example, the fifth day of the week.

Stretch zone: *What day of the week is your birthday on this year? Think of some other special events. Find out what day of the week each event is on this year.*

Students will need access to a 12-month calendar to complete this question. Check that they are confident using one.

Reflection time

Have a set of days-of-the-week cards for students to order by attaching to a 'washing line' or placing on a table or desk. You could start with Wednesday or Thursday and ask questions such as: *Which day comes before Wednesday? After Wednesday?* Continue until the week is complete. Repeat using a different start day. Use this as an opportunity to also revisit ordinal numbers, for example, *What day is second in the week?*

Practice Book: Students complete page 116 of the Practice Book. They can do this directly after the main activity, as homework, or as the focus of a separate mathematics session to help students consolidate their learning and build fluency. Students need a copy of the current year's calendar (full year to a page). Check that they are confident using one. Ask students to refer to the key word list on page 124 of the Student Book to check that they have spelt each day of the week correctly.

Differentiated outcomes	
All students	should name the days of the week in order.
Most students	will solve and answer simple questions about the days of the week.
Some students	may ask and answer more complex questions about days of the week.

9B Days of the week

Explore
Student Book page 125 • Practice Book page 117

Specific learning foci

- Know and order the days of the week and the months of the year.
- Know the relationships between consecutive units of time.

Global skills

- **Interpersonal skills:** communication/teamwork

Key vocabulary

- day, week, Monday, Tuesday, Wednesday, Thursday, Friday, Saturday, Sunday

Resources

- mini whiteboard and marker
- 2 sets of days-of-the-week cards (cut out from Resource sheet 5, copied onto thin card), an envelope or feely bag (per pair of students)

Answers

Student Book page 124

1 Check that the chart has been filled correctly:

Monday	Friday
Tuesday	Saturday
Wednesday	Sunday
Thursday	

2 Check that students have written the correct day name for today.

3 Check that students have written the correct day name for yesterday.

4 Saturday

5 Thursday

Practice Book page 116

1 Saturday   5 Monday

2 Thursday   6 Wednesday

3 Tuesday   7 Sunday

4 Friday

Stretch zone: Students' answers will vary. Check that they have written the day name correctly.

Language support

Continue to focus on saying 'Tuesday' and 'Thursday' correctly. Look at the word 'Wednesday' together. Cross out the first 'd' and explain that the word is said as if the 'd' is not there, more like 'wensday'. Focus on saying each day individually to establish the correct sounds.

Introductory activity

Play 'guess which day'. Write a day of the week on a mini whiteboard, concealing it from students. Students ask questions to which you can only answer yes or no. For example, 'Is it a school day?', 'Is it the day we have music/PE/something else?' After playing this a few times as a whole class, you could ask students to play in pairs.

Main activity

Tell students that they are going to play a game in pairs and that they will record it on page 125 of their Student Books. Give each pair of students an envelope or feely bag and two sets of (foldable) days-of-the-week cards. Ask them to fold the cards (so that the day cannot be read) and put them into the envelope or feely bag.

Take them through the rules for playing the game.

Model the steps, choosing a student to be your partner, and play a couple of rounds of the game. Ask students to look at page 123 of the Student Book (display on the IWB, if possible), so that they can see how you would record a day.

- Players take it in turns to take a piece of paper out of the envelope. Each player should write the first day they pick out in the first space in their column in the Student Book.
- They continue picking days alternately and write each day of the week in the correct space so that all the days are written in order; It doesn't matter which day comes first as long as they are in the correct order.
- If they pick out a day that they have already written, they must return it to the envelope and miss a turn.
- The winner is the first person to write the whole week!

Pairs should play the game twice.

As students play the game, ask questions to get them talking about the order of the days, for example, *Which days have you recorded but your partner has not? Can you tell me the second day on your list? What day will come before this day? If Wednesday is in the second row of your list, what day will be in the fourth row?*

Differentiation

Supporting: Ask students to use the list of the days of the week in order, on page 124 of the Student Book, as support.

Consolidating: Ask students to name days before and after given days. Can they say which days they are still looking for to complete the table?

Extending: Ask students to name days that are two or three days before or after a given day.

Stretch zone: *Can you make up some questions that have a day of the week as the answer?*

Give students key words to incorporate into their questions, if appropriate (e.g. before, after, 6th , between).

Check that students can make up suitable questions.

 **Reflection time**

Ask students to think about the days of the week and the order they come in. Arrange the students in a circle and choose one student to start. They choose a day of the week and you then go round the circle, each student saying the next day of the week. Ask questions that make students think about the sequence, for example, *If yesterday was Thursday, what day is it tomorrow? If we skip these two students' turns, what day is it? How do you know?*

Practice Book: Students now complete the activity on page of 117 in the Practice Book. They can do this directly after the main activity, as homework, or as the focus of a separate mathematics session to help students consolidate their learning and build fluency. Before they begin, remind students that in English we always capitalise the first letter of the days of the week.

Differentiated outcomes	
All students	should name the days of the week in order starting from Sunday or Monday.
Most students	will answer simple questions about the days of the week and say them in order starting from any day.
Some students	may answer more complex questions about days of the week.

Answers

Student Book page 125

Check that the days of the week have been written in the correct order in each grid. For this activity, it doesn't matter which day is the first day in the grid.

Practice Book page 117

1 Saturday

2 Sunday

3 Tuesday

4 Friday

5 Wednesday

6 Check that each day has a suitable activity filled in.

Stretch zone: Answers will vary.

Discover
Student Book page 126 • Practice Book page 118

Specific learning foci
- Know and order the days of the week and the months of the year.
- Use a calendar to find how many days there are in each month.
- Know the relationships between consecutive units of time (e.g. the month that comes before April).

Global skills
- **Creative skills:** exploring/investigating
- **Real-world skills:** research/interpreting information
- **Interpersonal skills:** communication/teamwork

Key vocabulary
- day, week, month, year, January, February, March, April, May, June, July, August, September, October, November, December, calendar

Resources
- calendars, full year on one page (one for each pair)

Language support
Always have a calendar on display and talk about today's date with the students every day. This could be at the beginning of each day or at the start of a session, or both.

 ### Introductory activity

Show students a calendar and give out enough copies for one per pair. In pairs, students should identify dates that are special for them. They should share these dates with the rest of the class. Discuss how to say them properly, for example, December the 8th, the 8th of December.

 ### Main activity

Students work in pairs. Each pair has a 12-month calendar. Give them several statements and ask them to discuss and decide whether they are always true, sometimes true or never true. For example:

Each month starts on a Monday.
A month has 31 days.
There is a month with 27 days in it.
There are four Sundays in every month.

Ask students to investigate each statement in their pairs and then discuss their answers as a class.

Students should work in pairs to complete the activities in the Student Book page 126, using their calendars to help them. Encourage them to use the key words list to spell each month of the **year** correctly.

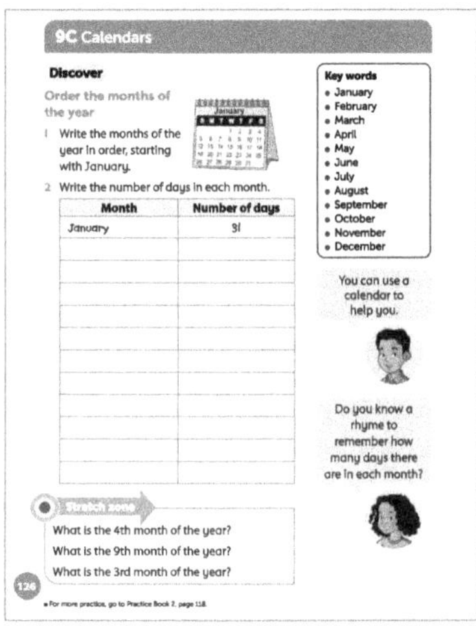

Differentiation
Supporting: Ask students to name the months in order. Use different starting points.

Consolidating: Ask students to name the months before and after given months.

Extending: Ask students to name months that are two or three months before or after a given month.

Stretch zone: *What is the 4th month of the year?*
What is the 9th month of the year?
What is the 3rd month of the year?

Check that students have identified April, September and March. Can they write two more things about each of these months?

 ### Reflection time
Tell students that there is an old rhyme in English to help them recall how many days each month has. Write it on the board with information missing.

[30] days hath September,
April, June and November,
All the rest have [31],
Except February alone,
And this has [28] days clear,
But 29 in a leap year.

Can they use the information they gathered in their Student Books to complete it?

Once they have completed it, read the rhyme together several times.

Practice Book: Students complete page 118 of the Practice Book. They can do this directly after the main activity, as homework, or as the focus of a separate mathematics session to help students consolidate their learning and build fluency.

Differentiated outcomes	
All students	should know the month names using the calendar for support.
Most students	will know the month names in order with different starting points.
Some students	will be able to ask and answer questions about the months of the year on a calendar.

9C Calendars

Explore Student Book page 127 • Practice Book page 119

Specific learning foci

- Know and order the months of the year.
- Use a calendar to answer questions about months of the year.

Global skills

- **Creative skills:** exploring
- **Real-world skills:** interpreting/ presenting information
- **Interpersonal skills:** communication/teamwork

Key vocabulary

- day, week, month, year, January, February, March, April, May, June, July, August, September, October, November, December, calendar

Resources

- calendars (full year to a page – one for each pair)

Language support

Make a class timeline focusing on years and months. Include past events, such as when students were born, as well as current events about this year and some future events.

Answers

Student Book page 126

January	31	July	31
February	28 or 29	August	31
March	31	September	30
April	30	October	31
May	31	November	30
June	30	December	31

Practice Book page 118

Check that students have written a suitable activity for each month of the year.

Stretch zone: Check that students have recorded a date and a day of the week correctly.

 ## Introductory activity

Play 'calendar hunt'. Give each pair of students a calendar for the current year (simple one-page calendars can be found on the internet). Call out a day of the week and a month, for example *A Monday in August*. Students have to find a matching date on the calendar and read the full date, for example, Monday the 25th of August 20xx. Ask for other examples (e.g. Monday the 18th of August 20xx and so on). Repeat with another day and month.

Main activity

Ask students to work in small groups to create a poster of the months of the year. They should make a table with the months of the year and either a description or an image which relates to each month. This could relate to a festival or weather, such as the start of the rainy season. Allow time to display the posters and for students to walk around to look at each group's poster and ask questions.

Students then complete the activity on page 127 of the Student Book individually. Students should have access to calendars for support. Continue to encourage students to write the names of the month accurately.

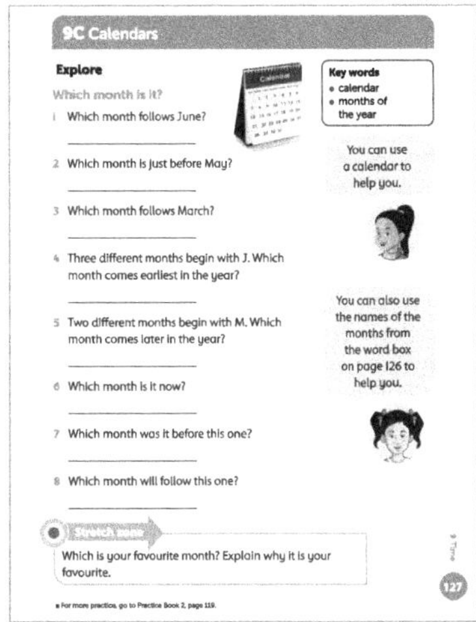

Differentiation

Supporting: Ask students to name the months in order. Use different starting points. They can refer to the key word list of months on page 126 of the Student Book for additional support.

Consolidating: Ask students to name the months before and after given months.

Extending: Ask students to name months that are two or three months before or after a given month.

Stretch zone: *Which is your favourite month? Explain why it is your favourite.*

This is a very accessible stretch zone. Encourage all students to complete this activity. They can discuss with a partner.

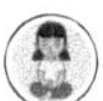 Reflection time

Which day of the week does each month start on this year? Ask students to use a calendar to find which day of the week each month begins on. They may also look at old calendars and planners too. *What do you notice?* Students should recognise that the first day of the month isn't always the same day. *Can you explain why that is?* Because each month does not have a whole number of weeks.

Practice Book: Students now complete the activity on page of 119 in the Practice Book. This can be done directly after the main activity, as homework, or as the focus of a separate mathematics session to help students consolidate their learning and build fluency. Students will need to use a calendar to answer these questions.

Differentiated outcomes	
All students	should know the month names using the calendar for support.
Most students	will know the month names in order with different starting points.
Some students	may recall the numbers of days in each month.

Student Book page 127

1 July

2 April

3 April

4 January

5 May

6 Check that students have written the correct current month.

7 Check that students have written the correct month for the previous month.

8 Check that students have written the correct month for next month.

Practice Book page 119

1 July

2 February

3 December

4 August

5 April

6 September

7 January

8 November

Stretch zone: Answers will vary depending on the student's birthday.

9D How long?

Specific learning focus

- Estimate and measure the duration of different events.

Global skills

- **Creative skills:** exploring /investigating
- **Real-world skills:** research/presenting information
- **Interpersonal skills:** communication/teamwork

Key vocabulary

- minutes, seconds, hours, how long?

Resources

- stopwatch or similar timer for minutes and seconds (one per group of six)
- pieces of music for students to listen to

Language support

Make a classrrom display showing some examples of things that take a number of minutes, for example, 2 minutes for brushing teeth. Display some well-known English idioms about minutes, for example, 'in a minute', 'at the last minute', 'wait a minute'.

Introductory activity

Ask students to listen as you play two pieces of music. One should be bright and quick, the other much slower, but both should be the same length. After students have listened to both pieces, ask them which they think was the longer and which was the shorter, and why.

Discuss with students how time seems to pass more quickly when they are doing things they enjoy.

Ask students how long they think a **minute** is. They might say it is how long it takes for the minute hand to move one place on the clockface, and some may know it is the same as 60 **seconds**. Set a stopwatch for 60 seconds, displayed on the IWB if possible. Count up to 1 minute together in seconds. Count back to 0 in seconds from 1 minute.

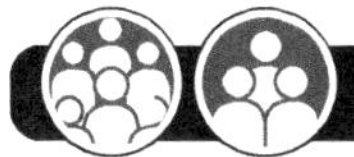

Main activity

Suggest to students that they might take one minute to put on their shoes. Ask students to close their eyes. Tell them they are going to estimate how long a minute is from the time you say 'go'. When they think 60 seconds

has passed, they should raise a hand. Start a stopwatch or timer as you say 'go' and say 'stop' when 60 seconds is passed. Try to keep a rough track of when students put their hands up.

How many students were close to 60 seconds when they raised their hand? Ask students to think of other things they think take about 60 seconds to do, for example, getting their books and pencils ready at the start of a lesson.

Students can now repeat the activity in groups of six and record their timings on page 128 of the Student Book. For the second question, they work in pairs within their group.

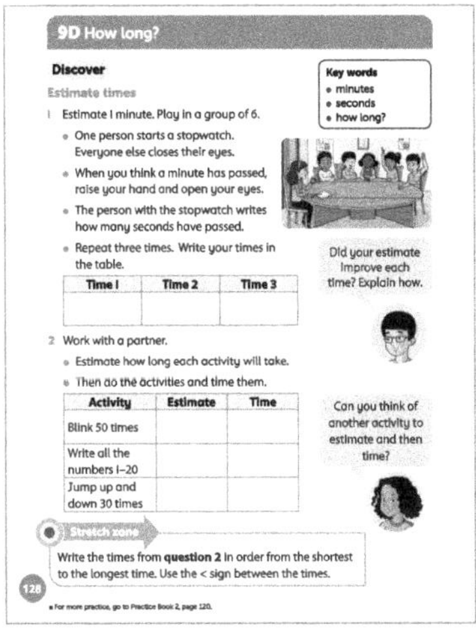

Differentiation

Supporting: Help students count in seconds, using for example '1 elephant, 2 elephants, …'

Consolidating: Ask students to think about other activities they could estimate and then time.

Extending: Ask students to think of pairs of activities that last roughly the same amount of time.

Stretch zone: *Write the times from question 2 in order from the shortest to the longest time. Use the < sign between the times.*

Check that students have correctly ordered their timings.

Reflection time

Ask students to share their results from the timing activities. Ask them to describe how easy or difficult they found each task, and what they could do to make their estimates or timing better.

Practice Book: Students complete the activities on page 120 of the Practice Book. They can do this directly after the main activity, as homework, or as the focus of a separate mathematics session to help students consolidate their learning and build fluency. This activity is quite challenging as these are multi-step word problems. So you may wish to complete this in a separate class session as a teacher-led activity when students have a good grasp of time. Use a geared clock to show time advancing as you total times. A time number line can also be useful to model the problems.

Differentiated outcomes	
All students	should estimate how long 1 minute takes to pass.
Most students	will estimate and measure activities that take 1 minute.
Some students	may estimate with increasing accuracy how long an activity lasts in minutes.

Student Book page 128

The students will be estimating and timing the passing of 1 minute and carrying out three activities. Check that their estimates are reasonable.

Practice Book page 120

1 Yes (55 min) **2** 12.45 **3** Lena < David < Usman

Stretch zone: Students' answers will vary, based on their own breakfast routine.

9D How long?

Explore Student Book page 129 • Practice Book page 121

Specific learning focus

- Use start and end times to work out the duration of different activities.

Global skills

- **Creative skills:** exploring/investigating
- **Interpersonal skills:** communication/teamwork

Key vocabulary

- minutes, seconds, hours, how long?

Resources

- stopwatch or timer for minutes and seconds
- large 0–100 number line for front of class, analogue clocks

Language support

Encourage students to describe the start and end times of activities they do, leading to calculating how long each thing took. When students are thinking about how long an event might last, such as lunchtime or the mathematics lesson, they can say for example, 'lunchtime starts at 12.15 p.m. and finishes at 1 p.m., so lunchtime lasts 45 minutes' and so on.

 Introductory activity

Tell students you are going to time them to see how many times they can write their name neatly in 1 minute. Ask them to estimate it first and write down their estimate. Time them for 1 minute as they write their name, then they should count how many times they have written their name. Ask why some students have written their name more times than others. Ask who was very close with their estimate.

 Main activity

Look at page 129 of the Student Book together. Display it on the IWB, if possible. Ask students to think of a suitable start time for each of the activities in question 1, then think of a stop time. For example, morning playtime may start at 10.30 a.m. and end at 10.50 a.m. Ask students to work out the interval so they can say that morning playtime lasts 20 minutes. Look together at times on an analogue clock, counting up in fives, then also counting on a number line as a class to calculate the interval. Ask students to think of two more activities as a class that they can find start and end times for. Then they should work out each duration independently. Students can also draw the times on blank clockfaces to get more practice with that before doing question 3 independently.

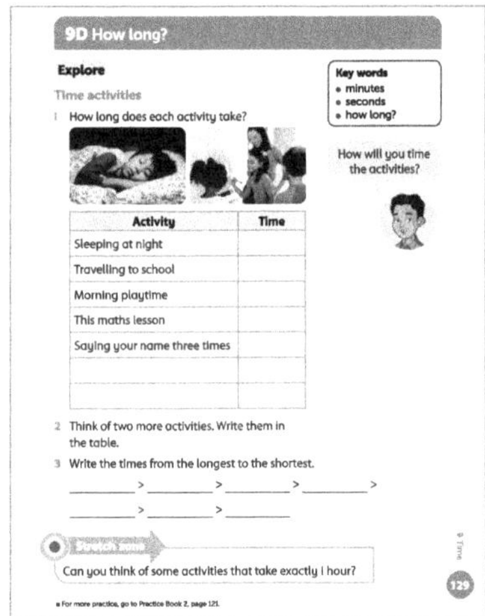

Differentiation

Supporting: Help students identify start and end times using a clock and count with them using a clock or number line.

Consolidating: Ask students to explain to you how they put the times in order. Can they explain how they know they were correct?

Extending: Ask students to tell you how long an activity would last if it lasted, for example, 5 minutes or 2 hours longer.

Stretch zone: *Can you think of some activities that take exactly 1 **hour**?*

Check that students have chosen some reasonable activities that last 1 hour. How did they know they last an hour? Can they think of anything that lasts half an hour?

 **Reflection time**

Ask students to share their responses to questions 2 and 3 in the Student Book. Ask for example, *Can you think of other activities that last as long as the ones on page 129?*

Practice Book: Students now complete the activity on page of 121 in the Practice Book. They can do this directly after the main activity, as homework, or as the focus of a separate mathematics session to help students consolidate their learning and build fluency. Ask students to think about the exact times they start and stop these activities to work out how long the activities usually take. Tell them to ask an adult to share with them approximate start and stop times. They can then draw a number line to help them work out the durations.

Differentiated outcomes	
All students	should find the duration of an activity with support.
Most students	will use start and end times to calculate activity durations using number lines and clockfaces for support.
Some students	may order several events by how long they last and be able to describe approximately the difference in times.

Answers

Student Book page 129

Check that the times recorded are reasonable and that the ordering is correct.

Practice Book page 121

Check that the times recorded are reasonable and that the ordering is correct.

Stretch zone: Students' answers will vary, based on their own home routine.

9E Comparing units of time

Discover Student Book page 130 • Practice Book page 122

Specific learning focus
- Convert between units of time.

Global skills
- **Creative skills:** exploring
- **Interpersonal skills:** communication/teamwork

Key vocabulary
- seconds, minutes, hours, days, weeks, years

Resources
- calendar, clocks or stopwatches, 0–100 number lines
- conversion guide (see language support)

Language support

Display a conversion guide showing the equivalences between seconds and minutes, minute and hours and so on (see below). Refer to these as students are working on the Student Book examples.

1 minute = 60 seconds
1 hour = 60 minutes
1 day = 24 hours
1 week = 7 days
1 year = 365 days
1 year = 12 months
1 year = 52 weeks

 **Introductory activity**

List some ways we use time measures in everyday life, for example, birthdays, school start and end times and school lunchtimes. Ask students to think of other ways in which they use measures of time.

List on the board a number of activities or times. For example,

Next Sunday	End of the school day	Your 9th birthday
This time tomorrow	The start of the next lesson	The next school year

How long until these times? How could you measure how long? Could you use months for each? What about minutes? Why or why not? Give students some time to discuss in pairs and then discuss as a class. Take their suggestions for each and agree that, for example, minutes are better for measuring very short times and years are better for long amounts of time.

 **Main activity**

Work through a few word problems involving time together. For example, say to students that it takes you 18 minutes to walk from home to the station. If you leave home at 8.25 a.m., will you get to the station in time for the 8.45 a.m. train?

Give pairs some time to work on the problem, using clocks and number lines as in the last lesson to support them in their calculations. Agree that you will get to the station in time but with only 2 minutes to spare.

Say that you started reading a book on 1 March and it took 7 weeks to read. What date was it when you finished reading the book? Help students to locate 1 March on the calendar and then count on 7 weeks – so you finished reading on 19 April.

Ask students to continue to work in pairs to answer questions about 'how long' on page 130 of the Student Book. Students may find these questions challenging. Walk around the class supporting students to answer the questions as necessary.

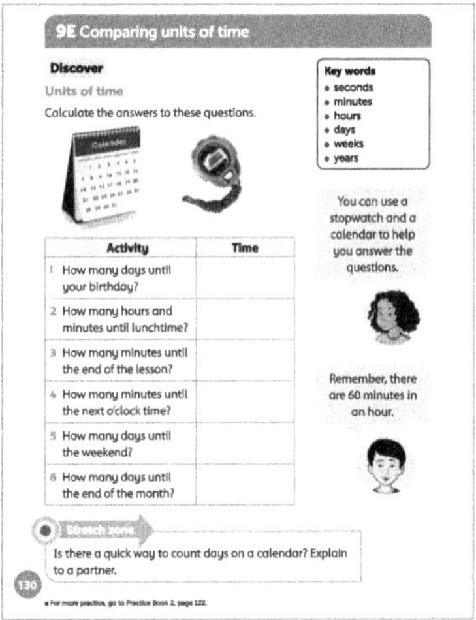

Differentiation

Supporting: Ask students to work on fewer questions, one involving days and another involving minutes.

Consolidating: Ask students to explain how they worked out how long.

Extending: Ask students to record their answer using more than one unit of time: for example, how many months or weeks until their birthday.

Stretch zone: *Is there a quick way to count days on a calendar? Explain to a partner.*

Students should mention that counting in weeks is faster but that they need to count up in steps of 7, which is challenging. Tell them that there are 14 days in 2 weeks and 28 days in 4 weeks. They could pick a date and try counting up in 2- or 4-week steps.

 **Reflection time**

Read the first speech bubble on page 130 of the Student Book. Then ask, *Which of these six activities would it be useful to use a stopwatch for? Which would be very difficult to use a stopwatch for?*

Ask students to consider which of the activities in the table is going to happen soonest. Which is going to happen the farthest in the future? Draw out that minutes are the shortest unit of time and years the longest.

Pairs can share their answers to each of the questions and explain how they reached their answers. Did any pairs use a different strategy? Take time to consider how the strategies are similar and different.

Practice Book: Students now complete the activity on page of 122 in the Practice Book. They can do this directly after the main activity, as homework, or as the focus of a separate mathematics session to help students consolidate their learning and build fluency. However, this is a challenging practice activity so it may be best done as a teacher-led activity when students are confident with units of time. You may like to build on their knowledge of doubling to find out how many minutes in two hours. Double this again to find how minutes in 4 hours. Can they use these two amounts to find out how many minutes in 6 hours? Can they use their minute equivalents to find how many minutes in 10 hours?

Differentiated outcomes	
All students	should know which unit of time is most appropriate for working out how long.
Most students	will begin to apply an appropriate strategy to find out how long until a set time.
Some students	may choose an appropriate strategy to find out how long until a set time and apply it successfully to reach a correct answer.

Answers

Student Book page 130

The answers to these questions will depend on your school and individual students' birthdays. Check that these are correct for your local situation.

Practice Book page 122

1 60

2 120

3 240

4 360

5 600

6 7

7 14

8 28

9 42

10 70

Stretch zone: Students' answers will vary, based on the date and time of completing the task.

9E Comparing units of time

Explore Student Book page 131 • Practice Book page 123

Specific learning focus
- Compare times by using conversions.

Global skills
- **Creative skills:** problem solving/investigating

Key vocabulary
- seconds, minutes, hours, days, weeks, years

Resources
- base-10 equipment
- geared analogue clocks
- calendars
- conversion guide

Language support
Continue to use the displayed conversion guide showing the equivalences between seconds and minutes, minutes and hours and so on. Refer to these as students are working on the Student Book activities.

 **Introductory activity**

Call out a pair of times. *Which is longer? 20 minutes or 20 hours? 300 days or 300 hours?* Call out other pairs (e.g. 10 years or 12 months, 24 hours or half a day). Can students explain how they know?

 **Main activity**

Ask students to recall the number of seconds in a minute, 60 seconds = 1 minute. Now ask *How many seconds in 2 minutes?* Pairs can work it out with base-10 equipment, on a number line or by counting around a clockface in fives.

Write these time equivalents on the board:

24 hours = 1 day

60 minutes = 1 hour

14 days = 2 weeks

Are these correct? How can you show me?

Leave pairs some time to consider how they can explain or show that these are correct. Encourage them to use equipment used across the unit to support their explanation (e.g. a clock or calendar).

Ask students, in pairs, to complete the questions on page 131 of the Student Book – these also involve thinking about time equivalences. They should use the conversion guide display to help them where necessary. As they work, go around the class and ask students to recall the various conversions without looking at the display, for example, ask *How many hours in a day?*

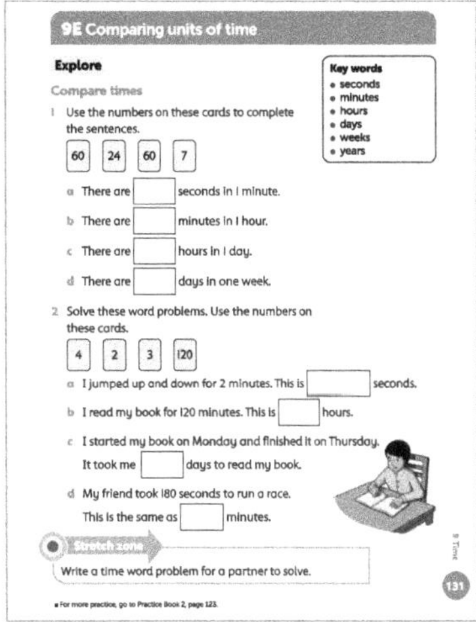

Differentiation
Supporting: Help students to identify the correct conversion between time units.

Consolidating: Ask students to convert from whole units.

Extending: Ask students to make their own questions requiring conversion of time units.

Stretch zone: *Write a time word problem for a partner to solve.*

Students should use realistic contexts for their word problems.

 **Reflection time**

Ask students to calculate a conversion starting from the smaller unit and converting to a larger unit, for example, *How many days is 72 hours?* Give students some time to think about this, then discuss together as a class how it is done. (24 hours in 1 day, so 2 days will be 24 + 24 = 48 hours, and 3 days is 48 + 24 = 72 hours). They can also ask a similar question of a partner.

Practice Book: Students now complete the activity on page of 131 in the Practice Book. They can do this directly after the main activity, as homework, or as the focus of a separate mathematics session to help students consolidate their learning and build fluency. Remind students of the key conversions before they begin (e.g. 1 minute = 60 seconds). Share a new one with them to help them with question 2: 5 hours = 300 minutes.

<table>
<tr><td colspan="2">Differentiated outcomes</td></tr>
<tr><td>All students</td><td>should begin to recall that 60 minutes = 1 hour, 24 hours in a day, half an hour = 30 minutes.</td></tr>
<tr><td>Most students</td><td>will convert between some time units.</td></tr>
<tr><td>Some students</td><td>may be able to use the conversions of time units to answer some word problems mentally.</td></tr>
</table>

Answers

Student Book page 131

1 **a** 60 **b** 60 **c** 24 **d** 7

2 **a** 120 **b** 2 **c** 3 **d** 3

Practice Book page 123

1 120 **2** 600 **3** 120 **4** 60 **5** 30

Stretch zone: Students will make up their own problems based on their experience.

9 Time

Connect Student Book page 132

Big idea

- Seconds, minutes, hours, days and months are units of time. I can tell the time using clocks and calendars.

Global skills

- **Creative skills:** problem solving/investigating
- **Interpersonal skills:** communication/teamwork
- **Self-development skills:** reflecting on learning

Key vocabulary

- second, hour, day, week, month, year, Monday, Tuesday, Wednesday, Thursday, Friday, Saturday, Sunday; January, February, March, April, May, June, July, August, September, October, November, December; calendar

Resources

- calendars, clocks

Language support

Make the link between the days of the week or months of the year and the ordinal numbers. For example, If Monday is the 1st day of the week, support students in naming the 4th day or the 6th day. Display a list of the months next to their ordinal position: 1st is January, 2nd is February, and so on.

 **Introductory activity**

Give each pair of students a copy of the calendar for the current year. A photocopied single sheet downloaded from the internet is sufficient. Call out a date (e.g. Wednesday 25th March). Students need to check the date on their calendar and answer true or false. After a few turns, invite a student to take over the calling out. Encourage them to say the date using ordinal numbers.

 **Main activity**

Explain that students will be working on time problems in small groups (the groups should be mixed-attainment). *What words do you think you will need to use? What do you know about different measures of time that might help you?* (e.g. 1 hour = 60 minutes) Take suggestions to make a list of key words and time conversions and record them on the board.

Ask students to look at the activities on page 132 of the Student Book. Say that while they may use examples from earlier lessons for suggesting activities in questions 2 and 3, you would like them to try to think of new examples if possible. Look at question 2 and ask, *How many is a few?* Agree that it is a number more than 2 but not a lot more. Provide calendars for question 1, and clocks, for support.

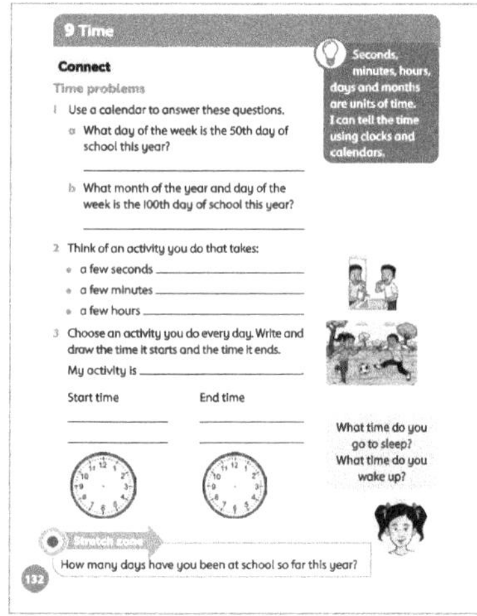

Differentiation

It is important that students work in mixed-attainment groups so that less confident learners can work alongside their more confident peers. This supports all learners as the lower-attaining students hear developed vocabulary and strategies from their higher-attaining peers, and the more confident students consolidate their understanding through explaining their learning and thinking to others.

Stretch zone: *How many days have you been at school so far this year?*

Check that students have calculated correctly.

 ## Reflection time

Discuss with the students how they found the answers to the problems on page 132 of the Student Book. Ask them to explain how they started and how they decided what they had to do. For example, did they count on from the 50th day to get to the 100th day?

Differentiated outcomes	
All students	should answer simple time word problems using a calendar.
Most students	will answer time word problems with increasing accuracy.
Some students	will answer time word problems involving months, weeks and days accurately.

9 Time

Review Student Book page133 • Practice Book page 124

Global skills

- **Creative skills:** problem solving/exploring
- **Interpersonal skills:** communication
- **Self-development skills:** reflecting on learning

Student Book

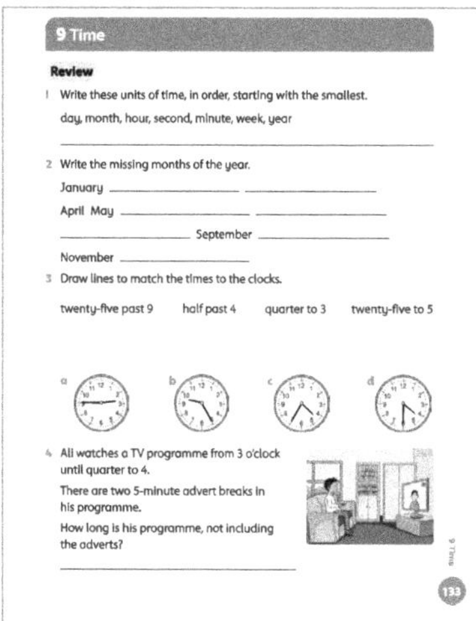

With young children, assessment activities are most effective when carried out as an everyday classroom activity. Some students may need the additional support of clocks and calendars to answer some of these questions.

Watch as students complete the activities that require days or months to be identified. Check that students understand the language of time: seconds, minutes, hours, days and months and other vocabulary associated with telling the time or measuring time.

Encourage students to use the clock or calendar for the questions. Extend by asking a range of similar questions or asking the students to make their own problems. For example, *Which is longer, an hour or a minute? How many weeks in December and January?* and so on.

Answers

Student Book page 133

1 second, minute, hour, day, week, month, year
2 January, February, March, April, May, June, July, August, September, October, November, December.
3 25 past 9 – b, half past 4 – d, quarter to 3 – a, 25 to 5 – c
4 35 minutes

Practice Book

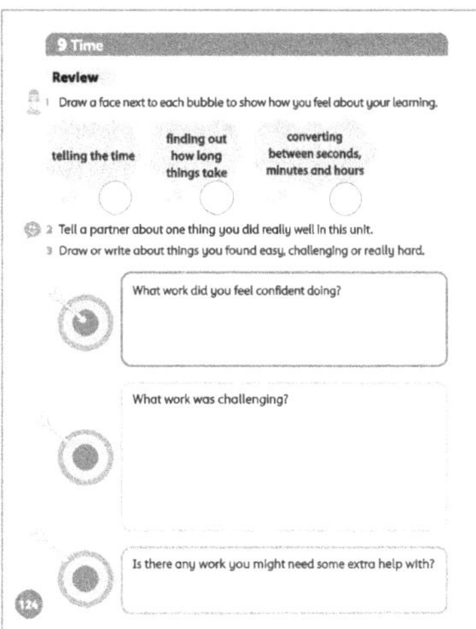

It is appropriate to complete this Practice Book review as a whole-class discussion. You may choose to keep a record of the class discussion or a copy of the review page for your own records. The review provides an opportunity for students to reflect on their learning from the unit, to discuss any areas of mathematics that they feel went particularly well, and any areas that they feel less confident about. Ensure all students have a copy of the Student Book as a reminder of the areas of mathematics that they have worked on in this unit.

Allow students plenty of time for discussion before asking them to complete the Practice Book page individually, and then, if appropriate, to share their responses with the rest of the class. If students complete this self-assessment at home, encourage them to discuss this with adults. Make a note of areas that students still feel unsure about.

Additional material

There are additional end-of-unit assessments available on the *Oxford Owl* website.

10 Geometry – properties of shapes

Overview

Big Idea

The Big idea in this unit is that we can use the properties of 2D and 3D shapes to classify and name them. Shapes can be compared and sorted according to their properties. Exploring what shapes have in common and what is different about them helps students to understand these properties.

Students need lots of experience of handling shapes, talking about them and comparing them. Many of the activities in this unit require students to talk about what they see. It is important that students recognise that 2D shapes have no thickness whereas 3D shapes do. A further Big idea is that many shapes and objects are symmetrical.

Look out for

- **Students who only recognise shapes when they are shown in a standard orientation.** For example, they may recognise squares and rectangles when they have a horizontal base, but not when they are positioned at an angle. Show students a variety of examples of the same shape in different orientations.

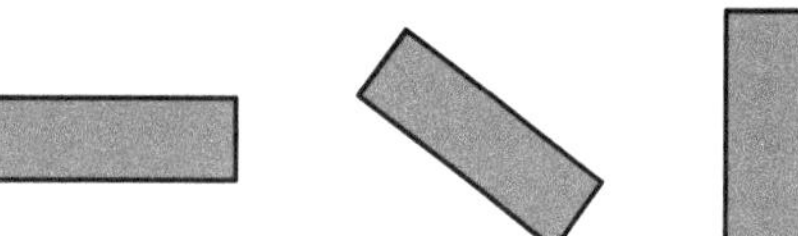

- **Students who only recognise 'regular' shapes that are familiar to them.** For example, they may not recognise scalene triangles as triangles, or only recognise regular polygons. Expose students to a wide variety of types of 2D shapes from the start. Refer to less-common examples by their proper names (e.g. rather than saying this 'shape', say this 'pentagon').

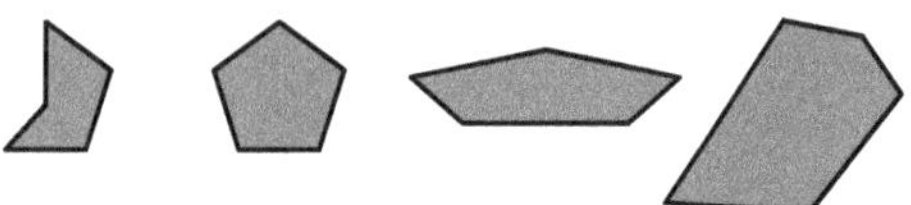

Possible misconceptions

- **Students may think that shape names such as 'circle' and 'square' can be used to describe 3D shapes such as spheres and cubes.** Be sure to use the proper names for 3D shapes. Emphasise that, for example, the faces of a cube may be square but the 3D shape is called a cube.
- **Students may think that squares and rectangles are two distinct shapes.** Discuss their properties, stressing that squares are rectangles that have sides the same length.

Key vocabulary

- shape, property(ies), regular, irregular
- 2D shape, square, circle, triangle, rectangle, hexagon, pentagon, quadrilateral
- curved, side, straight, flat, vertex, vertices, edge, face, surface
- 3D shape, cube, cuboid, cone, square-based pyramid, quadrilateral, sphere
- criteria, sort, intersection, Venn diagram, Carroll diagram
- line of symmetry, symmetrical, line symmetry

Coverage in lessons

Learning focus	Learning outcomes (the ENC objectives)
2D shapes	Identify and describe the properties of 2D shapes, including the number of sides.
3D shapes	Identify and describe the properties of 3D shapes, including the number of edges, vertices and faces.
Sorting shapes on a Venn diagram	Compare and sort common 2D and 3D shapes and everyday objects.
Sorting shapes on a Carroll diagram	Compare and sort common 2D and 3D shapes and everyday objects.
Line symmetry	Identify and describe the properties of 2D shapes, including line symmetry in a vertical line.

10 Geometry – properties of shapes

Engage Student Book page 134

Big question

- How can I describe the 2D and 3D shapes I see around me?

Global skills

- **Creative skills:** exploring/investigating
- **Real-world skills:** presenting information
- **Interpersonal skills:** communication/teamwork

Key vocabulary

- 3D shape, properties, face, edge, flat, curved, vertex, cube, cuboid, cone, cylinder, square-based pyramid, sphere, 2D shape, straight, side, pentagon, hexagon, triangle, quadrilateral, rectangular

Resources

- a range of foods and other household goods in packages approximating common 3D shapes
- common 3D shapes: sphere, cube, cuboid, cylinder, cone and square-based pyramid
- a large piece of paper for each group

Language support

Add the name of the 3D shape and a line drawing of it to each list prepared by students in the introductory activity. Display these lists as posters to help students to learn the names of the shapes.

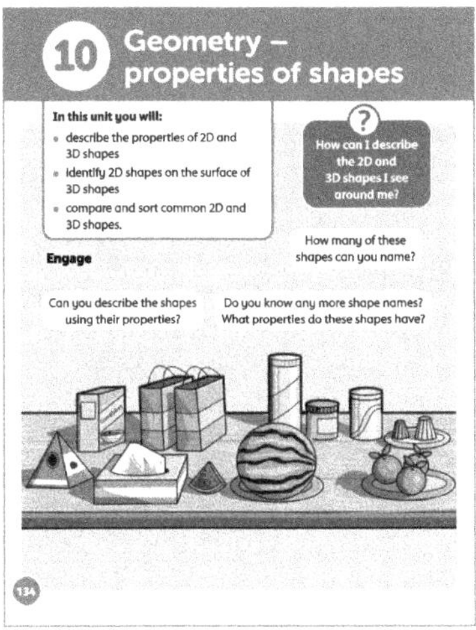

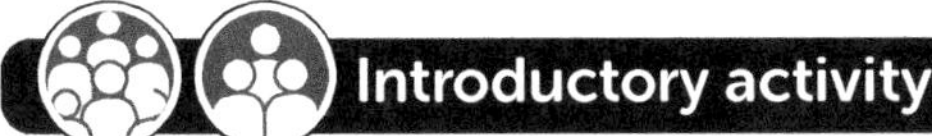 **Introductory activity**

Put the food and household goods on a table with the **3D shapes**. Ask students to take a good look at the household items and the shapes. Ask them to think about how they are they similar and different. After a few minutes, put each 3D shape on a different table and ask students to take a household item from the main table and put it with the shape it most closely matches. Once all shapes have been put on a table, split students into groups and ask them to record in a list the properties that the objects on a table have in common. They can also write down any other items they can think of that are the same shape.

Discuss each list in turn. At this stage, naming all properties is not necessary, nor is the use of exact language; this will be developed across the unit. Students can suggest other points to add to each list if they wish.

 **Main activity**

Ask students to look at page 134 of the Student Book. Display it on the IWB, if possible. *What type of objects are these? Can you name them? Which 3D shapes are they?*

After the names of the objects have been shared, ask questions to encourage students to think about the properties of each of the items. Include the questions in the speech bubble on page 134 and ask others, for example,

What do the oranges and the watermelon have in common? (They are curved.)

How many **cylinders** *can you see?* (3)

How many **rectangular faces** *do you think the tissue box has?* (6)

Differentiation

Supporting: Ask students to name as many shapes in the classroom as they can.

Consolidating: Ask students to describe properties of the shapes that they name.

Extending: Encourage students to name a wide range of shapes and describe their properties.

Reflection time

Ask two members of different groups to come to the front of the classroom with their 3D shape (e.g. cylinder, cuboid). Ask all students to discuss in small groups: *What is the same and what is different about these two shapes?* Share ideas as a class. Repeat for different pairs of shapes.

Discover
Student Book page 135 • Practice Book page 125

Specific learning focus

- Sort, name, describe, visualise and draw 2D shapes referring to their properties; recognise common 2D shapes in different positions and orientations.

Global skills

- **Creative skills:** exploring/investigating
- **Interpersonal skills:** communication/teamwork

Key vocabulary

- 2D shape, straight, curved, side, vertex, vertices, circle, square, pentagon, hexagon, triangle, rectangle, quadrilateral, properties

Resources

- equal lengths of string, elastic, exercises bands or strip of fabric (ideally at least 30 cm and up to 1 m long, 8 per group)
- selection of plastic or wooden 2D shapes; at least six pieces of scrap paper per student

Language support

Make a large paper copy of each 2D shape and display it with its name. Refer to the display shapes when you talk about 2D shapes, modelling correct pronunciation.

 Introductory activity

Give small groups at least eight lengths of string or similar and set them the investigation in a large, open space, if possible. *How many different shapes can you make? Can you name them?* Working together in their groups, students should arrange the lengths of string to create 2D shapes.

Once groups have had a chance to make a number of shapes, ask other groups to circulate and look at all the shapes that have been made. Ask students if they know the names of some of the shapes. Select one of the shapes made (e.g. a square) and draw it on the board. *Do you know the name of this shape? Can you describe it? Can anyone else add to the description?* Introduce vocabulary such as **curved, straight** and **vertex** as appropriate.

 Main activity

Draw a **triangle** on the board (preferably not equilateral so that students become used to seeing a variety of triangles) and say, *This is a triangle.* Explain that 'tri-' means three, and

a triangle is any shape with three **sides** and three **vertices** (corners). Point to these on your triangle. Ask students whether they can think of another word that starts with 'tri-'. If students are unable to name anything, ask them how many wheels a tricycle has.

Ask each student to draw two different triangles on two pieces of scrap paper and then one shape that is not a triangle on another piece. Collect all the shapes in and then distribute them between small groups. Ask the groups to sort the shapes in some way and be able to explain their reasoning. They will be thinking about what is not a triangle, as well as what is. As groups work, circulate and ask questions.

Introduce 'quad-' and **quadrilateral**, explaining that **squares** and **rectangles** are just two sorts of quadrilaterals and that quadrilaterals are shapes with four straight sides and four vertices. Now repeat the previous activity, asking students to draw two four-sided shapes and one that isn't four-sided. Collect all the shapes and then share out the shapes for the groups to sort and explain their reasoning.

Ask students to continue to work in their small groups naming each shape pictured on Student Book page 135 and completing the information about each shape's properties. Before they begin, read the speech bubble text together. Explain that a **regular** shape has all its sides equal in length. Provide 2D shapes for students to pick up and work with. You can also refer students to the classroom display of shapes and their names.

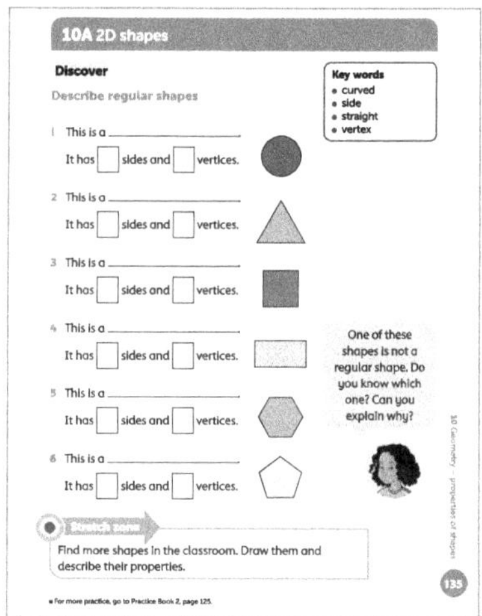

Differentiation

Supporting: Help students to describe how the shapes have been sorted, supporting them to use key vocabulary.

Consolidating: Ask students to explain why they sorted shapes in a particular way, referring to their properties.

Extending: Ask students to give examples of three shapes that have a given property (e.g. equal-length sides) and a shape that does not have that property.

Stretch zone: *Find more shapes in the classroom. Draw them and describe their properties.*

Encourage students to look beyond regular examples of shapes. Listen to students' descriptions and check for accuracy using correct vocabulary.

Reflection time

Play a game of 'Guess my shape' with the class. Students should ask questions to which you can only answer 'Yes' or 'No'. Draw a shape on a piece of paper, then hide it and reveal it when students have guessed it correctly. You should include a **circle**, a **square**, a rectangle and equilateral, isosceles and scalene triangles. Explain that **2D shapes** are flat shapes with straight sides. 2D shapes have no thickness.

Practice Book: Students can now complete the activity on page of 125 in the Practice Book. They should do this directly after the main activity, as homework, or as the focus of a separate mathematics session to help students consolidate their learning and build fluency. Suggest students use a ruler or a straight edge to help them draw their shapes accurately.

Differentiated outcomes	
All students	should recognise some regular 2D shapes.
Most students	will recognise some common 2D shapes and describe their properties.
Some students	will recognise and name regular and irregular 2D shapes and describe their properties.

Answers

Student Book page 135

1   This is a circle. It has 1 side and 0 vertices.

2   This is a triangle. It has 3 sides and 3 vertices.

3   This is a square. It has 4 sides and 4 vertices.

4   This is a rectangle. It has 4 sides and 4 vertices.

5   This is a hexagon. It has 6 sides and 6 vertices.

6   This is a pentagon. It has 5 sides and 5 vertices.

Practice Book page 125

Check that students have drawn the correct shapes against the shape names in the table.

Stretch zone: Check that students have drawn their own image made of more than one shape and have written a sentence about it.

10A 2D shapes

Explore   Student Book page 136 • Practice Book page 126

Specific learning focus

- Sort, name, describe, visualise and draw 2D shapes referring to their properties; recognise common 2D shapes in different positions and orientations.

Global skills

- **Creative skills:** exploring/investigating
- **Interpersonal skills:** communication

Key vocabulary

- properties, face, edge, curved, vertex, vertices, 2D shape, straight, side, pentagon, hexagon, triangle, quadrilateral, square, rectangle, pentagon, regular, irregular

Resources

- examples of 2D shapes: square, rectangle, pentagon, hexagon, triangle, pentagon
- strips of card in a standard length (e.g. 8 cm) and split pins (at least six strips and pins per student)
- extra strips of card in different lengths

Language support

Continue to refer to shapes by their correct names and link to other objects that are the same shape, or with faces that are the same shape, where possible, throughout the school day.

Introductory activity

Give each student a 2D shape. When you call out the name of a particular shape, all students with that shape should come to the front of the class. Ask the group how many sides and vertices their shape has and what its name is if they know it (e.g. 6 sides, 6 vertices: a **hexagon**). Explain that the **properties** of a hexagon are that it has 6 straight sides and 6 vertices. Repeat with other shapes.

After going through several shapes individually, call out both 'square' and 'rectangle', asking students with either of these shapse to come to the front of the class *What name can we give all these shapes?* (quadrilateral) *How do you know?* (4 sides, 4 vertices) You could also call out other statements such as: *Come to the front of the class if you have the shape often found on the surface of a football. Come to the front if you have the shape that looks like the full moon in the sky.* Use other links to real life that students may have noticed.

Main activity

Give students several strips of card (all the same length) and split pins. Ask them to use three strips and three pins to make a triangle. Students can work in pairs for discussion but each make the shapes.

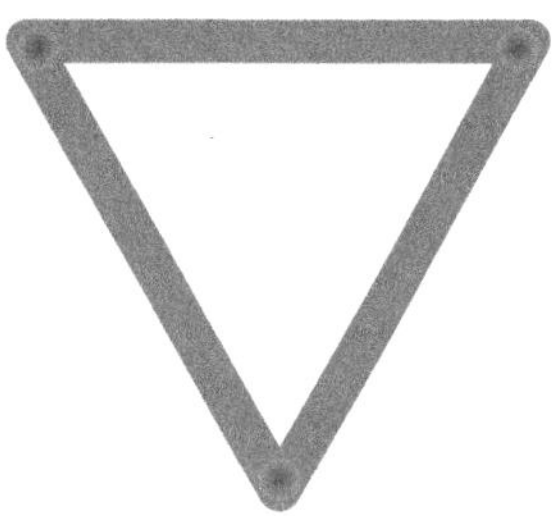

Ask students to undo a split pin and add another side. *What shape do you have now?* Explain that, although all the sides are the same length, the shape is only a square when students set it out that way. The vertices are flexible, allowing the sides to move, making other quadrilaterals.

Ask students to add another side and to arrange the sides to make a shape similar to a usual **pentagon**. Ask students what shape they have made. Say that this is called a **regular** pentagon, which means that all its sides are the same length, although many people miss out the word 'regular' and just call it a pentagon. Explain that when a five-sided shape is not in the usual pentagon shape, because it has sides of different lengths, it is called an **irregular** (not regular) pentagon. Ask students to add another side to their pentagon and explain that this six-sided shape is a regular hexagon.

Give students some extra strips of card of different lengths. Encourage students to explore and make new irregular shapes including triangles, quadrilaterals and pentagons.

Show students page 136 in the Student Book and talk through the colours needed. Write the colour names in that colour on the board as a reminder. Ask students to complete the page individually. While they are working, walk around the classroom and ask them to explain how they know what colour to use for each shape. Ask questions such as: *How do you know that that shape is a triangle, not a square? How do you know this is a rectangle and not a square? Could I colour this square red? Can you explain why? Could I colour this rectangle orange? Why not?*

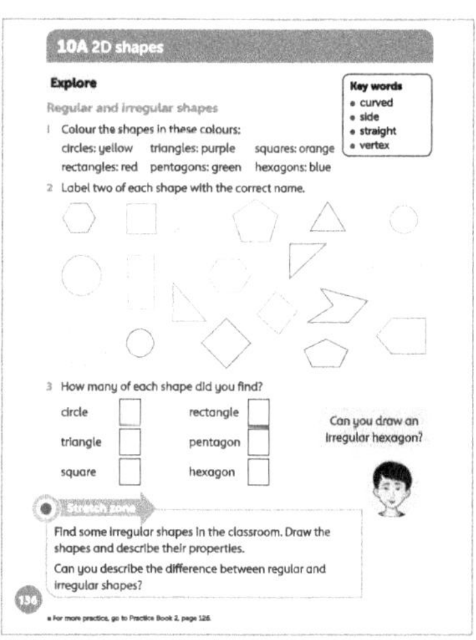

Differentiation

Supporting: Help students to identify regular 2D shapes.

Consolidating: Ask students to explain what makes a shape regular or irregular.

Extending: Ask students to name a range of 2D shapes and describe what is the same and what is different about regular and irregular shapes.

Stretch zone: *Find some irregular shapes in the classroom. Draw the shapes and describe their properties. Can you describe the difference between regular and irregular shapes?*

Ask students to share their work with another student and describe each shape's properties aloud. Does the other student agree? Can they add anything to the description?

 **Reflection time**

Ask students whether they noticed that the ending of the words 'pentagon' and 'hexagon' are the same. (Both end in '-gon'.) Explain that the '-gon' part tells us that the shape has vertices and the parts 'pent' and 'hex' (like 'tri') tell us how many vertices. and they will come across it again when they look at 2D shapes with even more sides in the future.

Practice Book: Students now complete the activity on page of 126 in the Practice Book. They should do this directly after the main activity, as homework, or as the focus of a separate mathematics session to help students consolidate their learning and build fluency. Suggest that students use a ruler or a straight edge to help them draw their shapes accurately.

Differentiated outcomes	
All students	should recognise some common 2D shapes.
Most students	will recognise some common 2D shapes and describe their properties.
Some students	will recognise and name all the common shapes and describe their properties.

Answers

Student Book page 136

1 Check that students have coloured all the circles yellow, the triangles purple, the squares orange, the rectangles red, the pentagons green and the hexagons blue.

2 Shape   How many?

circle   3

triangle   3

square   4

rectangle   1

pentagon   4

hexagon   4

Practice Book page 126

Check that students have written two properties for each chosen shape.

Stretch zone: Check that students have written two properties of the shape they drew on page 125.

10B 3D shapes

Discover
Student Book page 137 • Practice Book page 127

Specific learning focus

- Sort, name, describe and make 3D shapes referring to their properties; recognise 2D drawings of 3D shapes.

Global skills

- **Creative skills:** exploring/investigating
- **Real-world skills:** presenting information
- **Interpersonal skills:** communication/teamwork

Key vocabulary

- 3D shape, properties, face, edge, curved surface, vertex, cube, cuboid, cone, cylinder, square-based pyramid, sphere

Resources

- a selection of 3D shapes and a bag to put them in
- sets of 3D shapes to match those shown on page 137 of the Student Book
- modelling clay

Language support

Some of the words we use to describe shapes are likely to be unfamiliar to the students. Focus on modelling 'vertex' and 'vertices'. Emphasise the word 'face' by drawing a smiley face on each shape face. Continually name the shapes and ask students to name them too. Help them to remember the names by linking them with some of the everyday objects on the poster created in the Engage activity. For example, a sphere is similar to a ball.

 **Introductory activity**

Put a selection of 3D shapes in a bag. Ask one student to come to the front of the class and feel one of the shapes in the bag. They should describe it without taking it out of the bag or looking at it. The rest of the class guess what the shape is from the properties that are being described. Pull the shape out of the bag to check. Introduce its correct name, if necessary. Some students may be able to identify **cubes** and **cuboids**, and perhaps **cones**, but may find other shapes more challenging. Repeat with a different student for each shape. Model the use of vocabulary such as **face**, **edge**, 'vertex' and 'vertices' as appropriate, pointing to the parts of the shapes.

 **Main activity**

Show students a square-based pyramid, point to one of the triangular faces and say *What shape is this face?* (a triangle) Then do the same with the square face and say *Three triangle faces and one square face: this is a* **square-based pyramid**. *The pyramid is a* **3D shape** *and each of its faces is a 2D shape. 3D shapes have thickness as well as width and length.*

Arrange students in small groups and give each group a set of 3D shapes like the ones on page 137 of the Student Book. Ask students to explore each shape in their groups to find out how many edges, faces and vertices it has, and then to complete the table on page 137.

Once they have completed the table, invite groups to share their findings and describe how they found the number of edges, faces and vertices. *Do we all agree? Can anyone show me on a shape how you know?* Discuss why the **sphere** has no faces and say that we describe it as having a **curved surface**. *Which other shape has a curved surface?* (**cylinder, cone**)

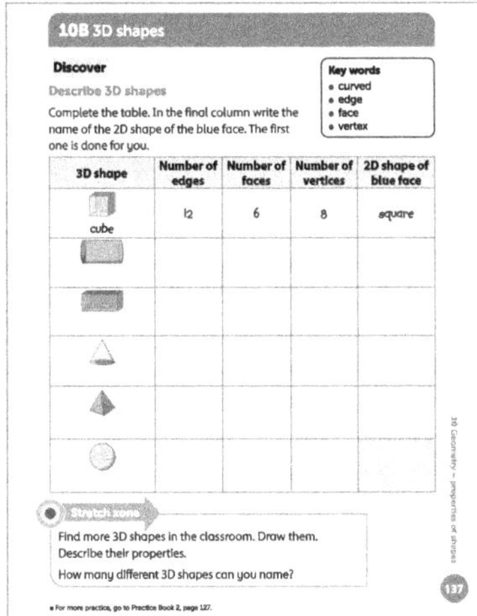

Differentiation

Supporting: Help students to count the faces, edges or vertices by marking them as they count, perhaps with a small blob of modelling clay.

Consolidating: Ask *Which shape has the most/fewest faces? How many of these shapes have a circular face?*

Extending: Ask students *Can you explain how a cube and cuboid are the same and how they are different?*

Stretch zone: *Find more 3D shapes in the classroom. Draw them. Describe their properties.*
How many different 3D shapes can you name?

Check that students are naming and describing shapes correctly. Encourage them to look for shapes beyond the ones they have seen in the lesson and use a mathematics dictionary or search online to find their names.

 **Reflection time**

Play the game from the introduction again. See whether the class have got better at recognising the shapes as a result of the main activity.

Practice Book: Students now complete the activity on page of 127 in the Practice Book. They should do this directly after the main activity, as homework, or as the focus of a separate mathematics session to help students consolidate their learning and build fluency. Drawing 3D shapes will be challenging. Encourage students to try their best, but emphasise that accuracy is not essential at this stage. Alternatively, they can take photos and stick them on the page.

Differentiated outcomes	
All students	should recognise and name some common 3D shapes.
Most students	will recognise some common 3D shapes and describe their properties.
Some students	may be able to compare the properties of 3D shapes as well as name and describe their properties.

Student Book page 137

3D shape	Edges	Faces	Vertices	2D shape of face
Cube	12	6	8	square
Cylinder	2	2	0	circle
Cuboid	12	6	8	rectangle
Cone	1	1	1	circle
Square-based pyramid	8	5	5	square or triangle
Sphere	0	0	0	none

Practice Book page 127

Check that students have stuck pictures in or have drawn shapes that match the shape names in the table.

Stretch zone: Check that students have drawn their own image made of more than one 3D shape and have written a sentence about it.

10B 3D shapes

Explore Student Book page 138 • Practice Book page 128

Specific learning focus

- Sort, name, describe and make 3D shapes referring to their properties, recognise 2D drawings of 3D shapes.

Global skills

- **Creative skills:** exploring/investigating
- **Real-world skills:** presenting information
- **Interpersonal skills:** communication/teamwork

Key vocabulary

- 3D shape, properties, face, edge, flat, curved surface, vertex, cube, cuboid, cone, cylinder, square-based pyramid, sphere

Resources

- selection of common 3D shapes (8–10 shapes per pair)

Language support

Many students will find it hard to pronounce 'sphere' correctly. Ask them to hiss the 's' sound, then say 'fear'. Encourage them to run the two together, with no pause in between.

 **Introductory activity**

Play 3D shape I-spy. Model a starting statement and question such as: *I spy a cuboid over by the door. What is it? I spy a sphere in the corner at the back of the room. What am I looking at?* Ask the student who correctly identifies the item to take a turn at describing a new 3D shape. You may need to support students with the name of the 3D shapes at first.

 **Main activity**

Ask students to work in pairs to make a simple model with some of the 3D shapes. It may be useful to limit them to no more than eight shapes. Ask students to tell you which 3D shapes they used to make their model. When they name a shape, ask how many of that shape were used in the model.

Show students a different model using 3D shapes (or display one on the IWB, if possible). Ask students which 3D shapes have been used and how many of each shape have been used.

Explain that you would now like students to work with their partner to make the models pictured in the Student Book page 138 and then create a table in their notebooks to show which shapes, and how many of each shape, they used for each model.

Model number	Shape		How many?

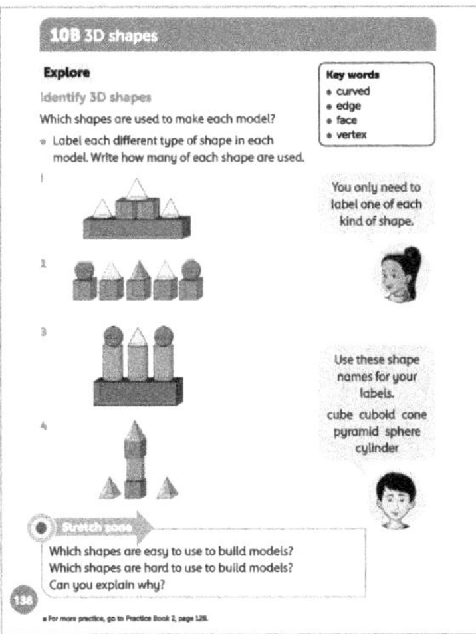

Differentiation

Supporting: Ask students questions to help them recall the shape names (e.g. *Can you show me three cones?*)

Consolidating: Ask students to describe the properties of the 3D shapes that they name.

Extending: Ask students to name a wide range of 3D shapes and describe their properties. *What do some of the shapes in the model have in common? How are they different?*

Stretch zone: *Which shapes are easy to use to build models? Which shapes are hard to use to build models? Can you explain why?*

Check that students can explain their answers using correct vocabulary, referring to curved surfaces. They are likely to say that spheres are hard to build with as they have no flat faces. Pyramids and cones can sit on top of other shapes but cannot be built on.

 Reflection time

Choose pairs to describe one the models from page 138 of the Student Book. They should describe the shapes used but also the correct positional vocabulary. For example (model 4), 'This model is made of three pyramids, two cubes and one cylinder. On top of one of the cubes there is a cylinder and then another cube, and then a pyramid on top of that. On either side of this tower is a pyramid.' Discuss why some shapes are more challenging to build with. *Why? What do you need to do with a cylinder if you include it in your model?*

Practice Book: Students now complete the activity on page of 128 in the Practice Book. They should do this directly after the main activity, as homework, or as the focus of a separate mathematics session to help students consolidate their learning and build fluency. Revise the name of each shape before they begin.

Differentiated outcomes	
All students	should make models and name some of the 3D shapes that they used.
Most students	will name the 3D shapes that they used and describe the properties of the shapes.
Some students	may know the names of all the 3D shapes, describe their properties and describe their position.

Answers

Student Book page 138

1 cube, 2; cuboid, 1; cone, 3

2 sphere, 2; cube, 5; cone, 2; pyramid, 1

3 sphere, 2; cylinder, 3; cuboid, 1; cone, 1

4 cylinder, 1; cube 2, pyramid, 3

Practice Book page 128

1 cylinder; 2 faces, 1 curved surface

2 cube; 8 vertices, 12 edges, 6 faces

3 cuboid; 8 vertices, 12 edges, 6 faces

4 square-based pyramid; 5 vertices, 5 faces, 8 edges

Stretch zone: Check that students have correctly written two different properties of the shape they drew on page 127 made from more than one 3D shape. For example, it may have all flat faces.

10C Sorting shapes using a Venn diagram

Discover
Student Book page 139 • Practice Book page 129

Specific learning foci

- Sort, name, describe and make 3D shapes referring to their properties; recognise 2D drawings of 3D shapes.
- Use Venn diagrams with one intersection to sort 2D and 3D shapes using two criteria.

Global skills

- **Creative skills:** investigating
- **Real-world skills:** presenting information
- **Interpersonal skills:** communication/teamwork

Key vocabulary

- 3D shape, properties, face, edge, flat, curved surface, vertex, cube, cuboid, cone, cylinder, square-based pyramid, sphere, Venn diagram, criteria, intersection

Resources

- selection of common 2D and 3D shapes, cloth bag or envelope

Language support

Continue to model correct pronunciation of all shape vocabulary. Display some 3D shapes, or pictures of them, with face, vertex and edge labelled.

 **Introductory activity**

Place a 2D or 3D shape in a cloth bag or padded envelope without students seeing what it is. Challenge them to identify the shape after asking you three questions, but tell them that you can only answer 'Yes' or 'No'. Have a trial run, modelling the questions students could ask, for example, *Does the shape have any curved surfaces? Is the shape 2D?* Give students a point each time they identify the shape after three questions, and yourself a point when they don't. Keep a running score over a few sessions.

 **Main activity**

Remind students of how a Venn diagram works by demonstrating a simple example on the board, for example, using 'triangle' and 'red' as your sorting criteria. Draw two separate circles and label them 'triangle' and 'red'. Explain that 'triangle' and 'red' are the sorting rules, or **criteria**. Hold up different 2D shapes and ask students where you would place them on the diagram. Continue

to sort shapes into one of the two circles, or outside the circles, until you find a shape that belongs in both circles (i.e. a red triangle). Ask students what they can suggest to solve this. Some may remember previous examples where the circles were moved together so that they overlap, and the item that is in both being placed in the overlap. Change the diagram to have an overlap and remind students that a shape that belongs in both circles goes in the **intersection**. Remind students that the circles need a box around them and a title (e.g. '2D shapes') so that we know what we are sorting, otherwise anything red, from a jumper to a tomato, could go in the red circle. Remind students that this type of diagram is called a **Venn diagram.**

Ask students to look at the Venn diagram in the Student Book page 139. Ask a student to read the title (3D shapes). *What does this tell us about what we are going to sort in this Venn diagram?* (3D shapes) *What are the sorting rules, or criteria?* (flat faces and curved surfaces). *Which shape can you see that fits the sorting rule 'flat faces'?* (a cube) *What shape can you see that fits the sorting rule 'curved surface'?* (a sphere) *Where would you put a shape that fits both criteria?* (in the overlap, the intersection)

Ask students to complete the diagram and answer the questions in the Student Book individually. Students are likely to find drawing 3D shapes difficult, so they may record the name of the shape instead. Give them a list of names and matching shapes to support them.

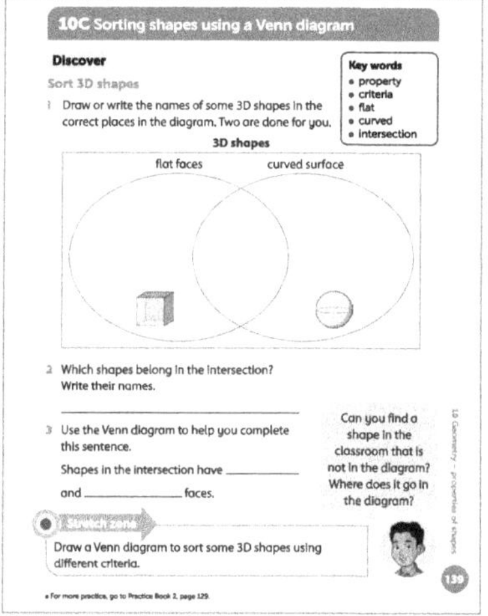

Differentiation

Supporting: Ask students to name the shapes they are sorting. Model the correct pronunciation of mathematical terms.

Consolidating: Ask students to describe the properties of the shapes they are sorting.

Extending: Ask students for alternative ways of sorting the shapes.

Stretch zone: *Draw a Venn diagram to sort some 3D shapes using different criteria.*

Check that students sort the shapes correctly according to the sorting criteria they choose. Support students to come up with ideas for criteria, if necessary.

 ## Reflection time

Ask students to share ideas about which shapes they placed where on the diagram. Check which shapes they listed as belonging in the intersection (cone and cylinder). Do all students agree? Ask individuals to justify their choices.

Practice Book: Students now complete the activity on page of 129 in the Practice Book. They should do this directly after the main activity, as homework, or as the focus of a separate mathematics session to help students consolidate their learning and build fluency. Encourage students to explain to an adult how they know which shapes have been sorted incorrectly.

Differentiated outcomes	
All students	should recognise and name the 3D shapes and sort some shapes using two criteria.
Most students	will sort 3D shapes into a Venn diagram, including the intersection.
Some students	may sort 3D shapes using a range of criteria.

10C Sorting shapes using a Venn diagram

Explore Student Book page 140 • Practice Book page 130

Specific learning foci

- Sort, name, describe, visualise and draw 2D shapes referring to their properties; recognise common 2D shapes in different positions and orientations.
- Sort, name, describe and make 3D shapes referring to their properties, recognise 2D drawings of 3D shapes.
- Use Venn diagrams with one intersection to sort 2D and 3D shapes using two criteria.

Global skills

- **Creative skills:** exploring/investigating
- **Real-world skills:** presenting information
- **Interpersonal skills:** communication/teamwork/ leadership

Key vocabulary

- 3D shape, properties, face, edge, flat, curved, vertex, cube, cuboid, cone, cylinder, square-based pyramid, sphere, Venn diagram, criteria, intersection

Answers

Student Book page 139

1 Flat faces: cube, cuboid, square-base pyramid.

Intersection: cylinder and cone.

Curved surface: sphere.

2 cylinder and cone

3 Shapes in the intersection have flat faces and a curved surface.

Practice Book page 129

1 Rectangle should not be in the intersection (does not have equal sides).

2 Cuboid does not have a square face. Cone does not have a triangular face.

Stretch zone: Left-hand circle 'Has at least one square face'; right-hand circle 'Has no irregular-shaped faces'.

Students' ideas for sorting the shapes in a different way could include 'Has at least one triangular face' and 'Has at least one rectangular face' (with sphere outside the circles) or 'Has no triangular faces' and 'Has at least one regular face'.

Resources

- selection of 2D and 3D shapes (one set per student)

Language support

Photocopy some examples of the students' work to display with the shape work. Continue to model correct pronunciation of all the shape words. Display some 3D shapes or pictures of them with 'face', 'vertex' and 'edge' labelled.

 ### Introductory activity

Play the three-question shape game from the last lesson again. As students ask questions, build up a list of yes/no questions for students to ask in future rounds. After two or three turns, invite a student to choose a shape and challenge the class to guess the shape after three questions.

 ### Main activity

Draw or show a blank Venn diagram with one intersection on the board. Revise with students how the Venn diagram works. *If I labelled one circle 'yellow' and the other circle 'squares' what would that tell us? Where can I find the intersection? What belongs in the intersection? What could we label the outer box? If an item is in the box but outside both circles what does that tell us?* Invite students to ask you, or their classmates, questions.

Ask students to complete page 140 of the Student Book. Explain that they can use the Venn diagram to sort shapes, either 2D or 3D shapes, or real-life examples from around the classroom, in any way they wish. Provide them with a selection of shapes to sort. Ask students to discuss possible sorting criteria in pairs before they begin but to complete their diagrams individually. Once they have sorted their shapes using their chosen criteria, they can swap with a partner to check and ask questions about each other's work.

While students sort their shapes, walk around the classroom and ask them to explain why they chose to place each shape in its position on the Venn diagram. Where students are sorting 2D and 3D shapes on the same diagram, check that they haven't mixed up 'faces' (of 3D shapes) with 'sides' (of 2D shapes).

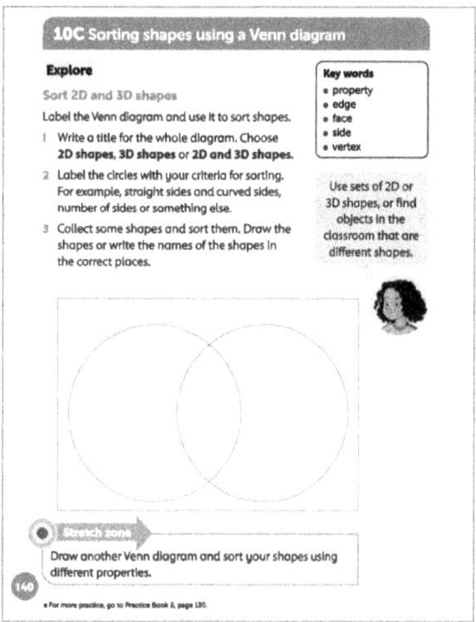

Differentiation

Supporting: Ask students to name the shapes they are sorting. Model the correct pronunciation of mathematical terms.

Consolidating: Ask students to describe the properties of the shapes they are sorting.

Extending: Ask students for alternative ways of sorting the same shapes. *How will that change the way you sort the shapes?*

Stretch zone: *Draw another Venn diagram and sort your shapes using different properties.*

Check that students draw the diagram correctly according to the sorting criteria they choose.

 ## Reflection time

Invite pairs of students to come to the front and draw one of their Venn diagrams on the board, explaining what they did. Invite the rest of the students to ask questions or suggest where the student could include a particular shape.

Practice Book: Students now complete the activity on page of 130 in the Practice Book. They should do this directly after the main activity, as homework, or as the focus of a separate mathematics session to help students consolidate their learning and build fluency. Encourage students to use a ruler to draw the shapes with increasing accuracy. Can they name each of these shapes?

Differentiated outcomes	
All students	should sort shapes using a Venn diagram with support in choosing sorting criteria and shapes to sort.
Most students	will choose suitable criteria to sort shapes, label a Venn diagram with reasonable accuracy and sort shapes correctly.
Some students	may create and label a Venn diagram accurately and use it to sort shapes of their choice confidently using a range of criteria.

Answers

Student Book page 140

Check that students have chosen reasonable sorting criteria and have sorted their shapes correctly.

Practice Book page 130

Straight sides – square, hexagon, quadrilaterals.

Curved sides – circle.

Intersection – semi-circle.

Stretch zone: Check that students have drawn four correct shapes in the Venn diagram.

10D Sorting shapes using a Carroll diagram

Specific learning foci

- Sort, name, describe, visualise and draw 2D shapes referring to their properties; recognise common 2D shapes in different positions and orientations.
- Sort, name, describe and make 3D shapes referring to their properties, recognise 2D drawings of 3D shapes.
- Use Carroll diagrams to sort shapes using two criteria; explain choices using appropriate language, including 'not'.

Global skills

- **Creative skills:** exploring/investigating
- **Real-world skills:** presenting information
- **Interpersonal skills:** communication/teamwork

Key vocabulary

- 3D shape, properties, face, edge, flat, curved, vertex, cube, cuboid, cone, cylinder, square-based pyramid, sphere, Carroll diagram, criteria

Resources

- selection of 2D and 3D shapes

Language support

Display a Carroll diagram with different parts labelled.

 Introductory activity

Give each pair of students some 3D shapes from a selection of cube, cuboid, pyramid, cone, cylinder and sphere. Allow them time to recall in pairs the properties of the shapes – faces, edges, vertices and curved surfaces. Then ask students to focus on how many edges each shape has. Ask students to describe what an edge is, and then for each shape you can elicit from them how many edges it has.

 Main activity

Draw an empty Carroll diagram on the board. *Can you recall what this diagram is a called and what it is used for?* Say that it is a different kind of sorting diagram, called a **Carroll diagram**. Remind students that in this diagram there is no intersection, so the sections are usually labelled with opposites, such as 'red' and 'not red'. Label the columns at the top of the diagram with these criteria.

Add the labels 'more than 4 sides' and 'not more than 4 sides' to the rows at the side of the diagram. Then use the Carroll diagram to sort some 2D shapes. Ask, for example, *Which box should this shape go in? How do you know?*

Ask students to look at the Carroll diagram in the Student Book page 141. Read the title and labels together. Explain that Carroll diagrams sometimes do not have a title because it is often clear from the labels what we are sorting but here we have one. *What are we sorting?* (2D shapes)

Give each student a selection of 2D shapes to work with. Ask students to choose some 2D shapes, decide in which section they should be sorted and then draw them in the Carroll diagram. While students carry out the activity, walk around the classroom and check that students have sorted the shapes into the correct sections on the Carroll diagram. Ask them to justify why they have placed a shape in a particular position.

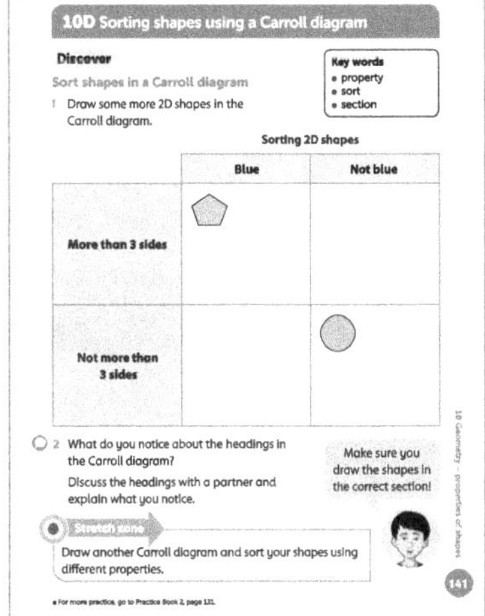

Differentiation

Supporting: Ask students to name the shapes they have chosen to sort, and name for them any they are unsure of. Describe their properties together.

Consolidating: Ask students to describe the properties of the shapes they are sorting.

Extending: Ask students for alternative ways of sorting the same shapes. *Are there shapes that cannot be sorted into any sections?*

Stretch zone: *Draw another Carroll diagram and sort your shapes using different properties.*

Check that students draw the diagram correctly according to the sorting criteria they choose.

 Reflection time

Ask students to share ideas about which shapes they placed where on the diagram in their Student Books. Do all students agree? Ask students to justify their choices. Draw some additional shapes on the board. *Into what section should we sort these shapes? Why?*

Practice Book: Students now complete the activity on page of 131 in the Practice Book. They should do this directly after the main activity, as homework, or as the focus of a separate mathematics session to help students consolidate their learning and build fluency. Ask students to look at the criteria. *What type of shape will you need to sort?* (3D) Tell students that they may write the names of the shapes rather than draw them if they prefer.

Differentiated outcomes	
All students	should name common concrete 2D shapes and sort them using a Carroll diagram.
Most students	will draw, name and sort 2D shapes using a Carroll diagram.
Some students	may sort 2D shapes using a Carroll diagram, choosing their own criteria.

Student Book page 141

1 Check that students have sorted shapes correctly into the different sections of the Carroll diagram.

2 Students should notice that one label is the negative of the other on each side of the Carroll diagram, for example, 'has 3 sides' and 'does not have 3 sides'.

Practice Book page 131

1 Check that students have put a suitable shape for each criterion in the table.

2 cylinder

3 cuboid

4 sphere

Stretch zone: On a cube and cuboid, all the faces are rectangular, but on a cube, all the faces are square.

10D Sorting shapes using a Carroll Diagram

Explore   Student Book page 142 • Practice Book page 132

Specific learning foci

- Sort, name, describe, visualise and draw 2D shapes referring to their properties; recognise common 2D shapes in different positions and orientations.
- Sort, name, describe and make 3D shapes referring to their properties, recognise 2D drawings of 3D shapes.
- Draw and use a Carroll diagram to sort shapes using two criteria; explain choices using appropriate language, including 'not'.

Global skills

- **Creative skills:** exploring
- **Real-world skills:** presenting information
- **Interpersonal skills:** communication

Key vocabulary

- 3D shape, properties, face, edge, flat, curved, vertex, cube, cuboid, cone, cylinder, square-based pyramid, sphere, Carroll diagram, criteria

Resources

- selection of 2D and 3D shapes

Language support

Display some 3D shapes or pictures of them with 'face', 'vertex' and 'edge' labelled. Continue to model correct pronunciation of all the shape words.

 ## Introductory activity

Play several rounds of the game 3D shape 'I-spy'. Remind students how to play, modelling a starting statement and writing it on the board, for example, *I spy a sphere on the table. What am I looking at?* Ask the student who correctly identifies the item (e.g. blue ball) to take a turn at describing a new 3D shape somewhere in the room.

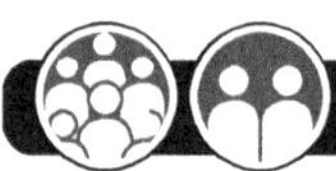 ## Main activity

Ask students, *How do I make a Carroll diagram?* Together, work through how to make one, including deciding on what to sort and which criteria to use. Choose a student to draw the diagram as other students describe what should be included. Ask for examples for possible things to sort and which criteria they could use. Ask students to sort the objects, encouraging them to explain how they know they have sorted each object correctly.

Ask students to look at page 142 in the Student Book. Explain that they can use the blank Carroll diagram to sort shapes any way they wish. Talk through some possible criteria for sorting and then ask the students to work in pairs to create two Carroll diagrams (e.g. one investigation could involve 2D shapes and the other 3D shapes). They could use one student's book for their first investigation and the other for their second investigation. Encourage students to use real 2D and 3D shapes for support.

While pairs carry out the activity, walk around the classroom and check that they have sorted the shapes into the correct places on their Carroll diagrams. Ask them to explain what should go into each part of their Carroll diagram and ask them to justify why they have placed a shape in a particular position.

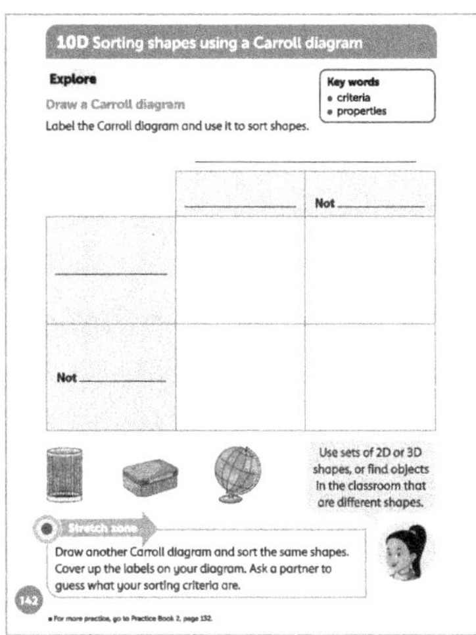

Differentiation

Supporting: Ask students to name the shapes they are sorting. Model the correct pronunciation of mathematical terms.

Consolidating: Ask students to describe the properties of the shapes they are sorting.

Extending: Encourage students to think of alternative ways of sorting the shapes.

Stretch zone: *Draw another Carroll diagram and sort the same shapes. Cover up the labels on your diagram. Ask your partner to guess what your sorting criteria are.*

Check that students are using the criteria correctly when sorting.

 **Reflection time**

Invite pairs of students to come to the front and draw one of their Carroll diagrams on the board and explain to everyone what they did. Invite the rest of the class to ask questions or suggest where the students could include a particular shape.

Practice Book: Students now complete the activity on page of 132 in the Practice Book. They should do this directly after the main activity, as homework, or as the focus of a separate mathematics session to help students consolidate their learning and build fluency. Review 2D shape names and ask, *What does 'vertices' mean? What is a word we could use in place of not regular?* (irregular)

Differentiated outcomes	
All students	should recognise one criterion for sorting shapes and will choose some shapes that match the criterion.
Most students	will choose criteria to sort shapes using the Carroll diagram with some assistance and use it to sort shapes effectively.
Some students	may make a Carroll diagram, labelling it clearly, choosing appropriate criteria and sorting shapes accurately.

Answers

Student Book page 142

Check that they have sorted the shapes into the correct boxes on their Carroll diagrams using their chosen criteria.

Practice Book page 132

Check that the students have placed the shapes in the correct boxes on the Carroll diagram, according to the sorting criteria.

Stretch zone: Check that students have sorted the shapes into the correct places on their Carroll diagrams using their chosen criteria.

Discover
Student Book page 143 • Practice Book page 133

Specific learning foci

- Draw lines of symmetry in simple shapes.
- Identify reflective symmetry in patterns and 2D shapes.

Global skills

- **Creative skills:** exploring/investigating
- **Interpersonal skills:** communication/teamwork

Key vocabulary

- symmetry, line of symmetry, line symmetry, symmetrical, horizontal, vertical

Resources

- paper and scissors for each student
- paper rectangle and small mirror for each student
- A4 sheet containing all the letters of the alphabet in capitals, one for each student

Language support

Display vocabulary associated with symmetry, for example, 'line of symmetry', 'symmetrical', 'vertical', 'horizontal', alongside labelled examples. Refer to the display when students are engaged in activities exploring symmetry.

 **Introductory activity**

Give students some paper and scissors. Ask them to draw a shape of their choice on the paper, then cut it out. Once they have cut it out, ask them to fold their piece of paper through the middle of the shape, then open it and see whether it looks the same on both sides of the fold line. If it does look the same on both sides, ask them to say what their shape was. Now ask them to try to draw a shape that they think will look the same on each side of the fold line (they should cut it out and fold to check). Explain to students that the shapes that looked the same on each side of the fold line are said to be **symmetrical**. The fold line is called a **line of symmetry**.

 **Main activity**

Give each student a paper rectangle and ask them to fold it in half. Ask *Do the two halves of the rectangle line up with each other?*

Give each student a small mirror and show students how to place the mirror on the lines of symmetry of their rectangles. They should see the same complete shape with the mirror and without it. Ask them to try placing the mirror on the diagonal. This will show that it is not a line of symmetry.

Students can use the mirrors to help them find the lines of symmetry on the letters and shapes in the Student Book page 143. When they have found these lines of symmetry, they could go on to check the rest of capital letters of the alphabet.

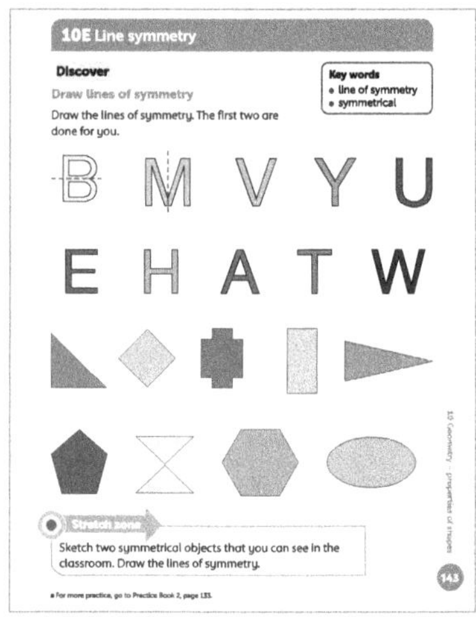

Differentiation

Supporting: Support students in using mirrors to check lines of symmetry on the letters of the alphabet.

Consolidating: Students can check a variety of shapes to see whether they are symmetrical.

Extending: Ask students to write a word that has a line of symmetry, for example, HIDE.

Stretch zone: *Sketch two symmetrical objects that you can see in the classroom. Draw the lines of symmetry.*

Check that students have identified the line of symmetry correctly. *Can they explain how they know it was symmetrical?*

 **Reflection time**

Ask students to show where they drew the line of symmetry on each letter or shape. If students have drawn different lines on the same shape, check that they are all correct. Students may notice that there is a **horizontal** and **vertical** line of symmetry for some shapes. Explain that many shapes have more than one line of symmetry. Revisit all the definitions from the opening of the lesson by asking questions such as *What is a symmetrical shape? What is a line of symmetry? Can a shape have more than one line of symmetry?* Agree one definition that can be written on a poster which clearly explains what **line symmetry** or reflective symmetry is (e.g. when a shape can be divided in half by a line).

Practice Book: Students now complete the activity on page of 133 in the Practice Book. They should do this directly after the main activity, as homework, or as the focus of a separate mathematics session to help students consolidate their learning and build fluency. Students will need a ruler, scissors and glue for this activity. They can take photos of symmetrical shapes and print them or find pictures of symmetrical shapes in magazines and cut them out. Once students have chosen their pictures, ask them to talk to an adult about how they knew the shape had a line of symmetry.

<table>
<tr><td colspan="2">Differentiated outcomes</td></tr>
<tr><td>All students</td><td>should identify lines of symmetry in some letters and shapes.</td></tr>
<tr><td>Most students</td><td>will correctly identify one or more lines of symmetry in all letters and shapes.</td></tr>
<tr><td>Some students</td><td>may correctly draw all lines of symmetry in letters and shapes.</td></tr>
</table>

Answers

Student Book page 143

Letters

One horizontal line of symmetry: B, E

One vertical line of symmetry: M, V, Y, U, A, T, W

Two lines of symmetry (vertical and horizontal): H

Shapes

One horizontal line of symmetry: pink triangle

One vertical line of symmetry: brown irregular pentagon

One slanting line of symmetry: red triangle (from vertex at the right angle to middle of opposite side)

Two lines of symmetry (vertical and horizontal): purple dodecagon, blue rectangle, yellow hexagon, turquoise oval

Four lines of symmetry (vertical, horizontal and 2 diagonal): green square

Six lines of symmetry (vertical, horizontal and 4 diagonal): orange regular hexagon

Practice Book page 133

Check that students have chosen symmetrical shapes and have correctly drawn a line of symmetry on each.

Stretch zone: Students should say, for example, the shape has a line of symmetry and that means the object looks the same on both sides of the line.

10E Line symmetry

Explore Student Book page 144 · Practice Book page 134

Specific learning focus

- Identify reflective symmetry in patterns and 2D shapes; draw lines of symmetry.

Global skills

- **Creative skills:** exploring/investigating
- **Real-world skills:** research
- **Interpersonal skills:** communication

Key vocabulary

- symmetry, line of symmetry, reflective symmetry, line symmetry, symmetrical

Resources

- paper square for each student
- small mirror for each student
- squared paper

Language support

Display any sketches from the previous lesson's stretch zone activity, with the other symmetrical pictures. Add speech bubbles with statements such as, 'Can you spot the line of symmetry?' 'This object is symmetrical.'

 Introductory activity

Display or draw a square on the board. Ask students, *Is it symmetrical?* (yes) Ask a student to explain why. They should be able to say that it has a line of symmetry that divides it in half. Now give each student a square of paper. Ask them to fold it in half vertically (demonstrate this) and then fold it in half again, this time horizontally. Now open it up – ask *What can you tell me about the folds on your square?* Students should notice that both fold lines are lines of symmetry.

Ask students to fold the square in half again (diagonally or vertically/horzontally) and then to cut a shape or design from one or more sides of their square (leaving at least some of the fold line intact). Now ask them to unfold the square and look at the pattern they have made. Ask *What do you notice about your pattern? Is it symmetrical? How do you know?* (because both sides of the fold line are the same) They should notice that the shape is symmetrical on each of the fold lines.

 Main activity

Give each pair of students a mirror and check that they recall how to use it to check if an object or picture is symmetrical. Allow them time to find a symmetrical object and a non-symmetrical object in the classroom – these must be things that can be moved and put on a table (e.g. studded blocks, leaves, rulers, plastic digits, counters, pattern blocks). When the time is up, set aside one table for symmetrical objects and one for non-symmetrical objects. Give students some more time to check some of the objects on the tables. As students check, ask, in relation to a specific object, *Could you tell whether this was symmetrical or not just by looking at it or did you need to use a mirror to check?*

Students then work as individuals to decide whether the pictures on page 144 the Student Book are symmetrical or not. Ask students to predict whether a picture is symmetrical or not and then use their mirror to check. They can then check one another's answers as a pair.

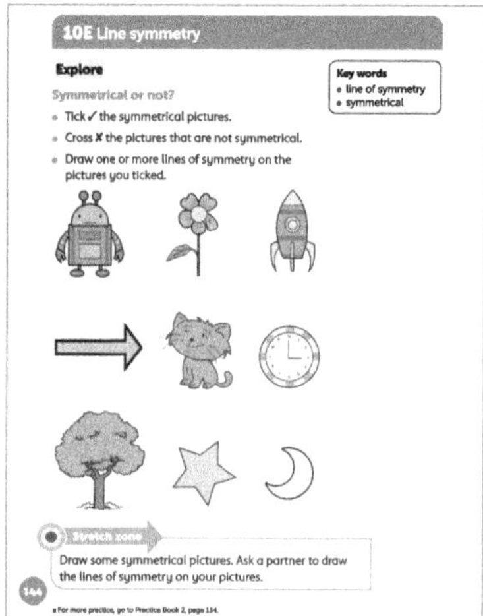

Differentiation

Supporting: Help students to place their mirror correctly to see the symmetry.

Consolidating: Ask students to explain their strategy for predicting whether a picture has a line of symmetry or not.

Extending: Ask students to write a description of a line of symmetry.

Stretch zone: *Draw some symmetrical pictures. Ask a partner to draw the lines of symmetry on your pictures.*

Suggest students draw their pictures on squared paper; this will make it easier to draw a symmetrical picture.

Check that students have drawn a symmetrical picture and that their partner has identified the lines of symmetry correctly.

 Reflection time

Look together at the pictures on page 144 of the Student Book. Display them on the IWB, if possible. Agree which ones are symmetrical and which ones are not. Ask students to explain their strategies for drawing the lines of symmetry. Ask whether anyone drew more than one line of symmetry on any picture and, if so, ask them to explain or show everyone else what they did. Focus on some errors and use the definition from the previous lesson to explain these errors.

Practice Book: Students can complete page 133 of the Practice Book. They should do this directly after the main activity, as homework, or as the focus of a separate mathematics session to help students consolidate their learning and build fluency. Tell students that if they are struggling to find shapes with three or four lines of symmetry, they could do some internet research with an adult to find some examples and then copy them into their Practice Books.

Differentiated outcomes	
All students	should identify a vertical line of symmetry in some pictures.
Most students	will correctly identify both vertical and horizontal lines of symmetry in some pictures.
Some students	may find more than two lines of symmetry in some symmetrical shapes.

Answers

Student Book page 144

Symmetrical pictures:

Robot: 1 vertical line of symmetry

Rocket: 1 vertical line of symmetry

Arrow: 1 horizontal line of symmetry

Star: 5 lines of symmetry (from each point of the star to the opposite vertex)

Moon: 1 line of symmetry (slanting through the middle)

Practice Book page 134

Check that students have correctly drawn correct shapes with the appropriate lines of symmetry.

Stretch zone:

square: true

rectangle: false

triangle: false

10 Geometry – properties of shapes

Big idea

- I can describe the properties of 2D and 3D shapes. I can name some 2D and 3D shapes.

Global skills

- **Creative skills:** exploring/investigating
- **Real-world skills:** presenting information
- **Interpersonal skills:** communication/teamwork
- **Self-development skills:** reflecting on learning

Key vocabulary

- triangle, pentagon, circle, sphere, cylinder

Resources

- selection of 2D and 3D shapes

Language support

Encourage students to look at the shape display to remind themselves of the names of each shape and how to spell them.

 **Introductory activity**

Hand out one shape to each student from a selection of common 2D and 3D shapes. Call out a property of shape, for example, *I have square faces.* Students who have a shape with this property stand up and hold up their shape. *Does everyone agree? Why or why not?* Ask whether anyone else would like to stand up now. This will give students who were previously unsure a chance to answer the question correctly. Repeat with other shape properties.

 **Main activity**

Ask students to look at page 145 of the Student Book. Read the question text together to check understanding. Then ask *What type of shapes are on each planet? Are they 3D or 2D shapes? How do you know?* Agree that the descriptions in the questions give them clues: shapes with sides are 2D, shapes with faces are 3D. Shapes with a curved edge or face are 3D.

Ask students to work in groups to investigate and identify which shapes they could find on each planet. They can use a selection of 2D and 3D shapes to help

them. As groups work, ask them to explain why they have decided a specific shapes could not be found on one of the two planets.

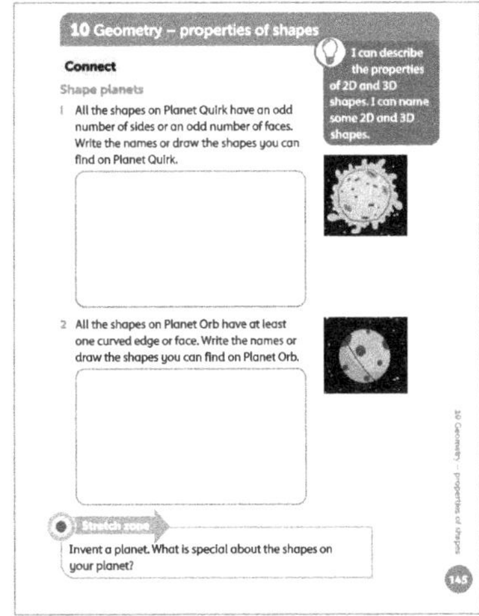

Differentiation

Supporting: Help students to work through shapes systematically, asking one question at time, eliminating shapes as you go along.

Consolidating: Ask students to describe why they chose and why they did not choose certain shapes for each planet.

Extending: Ask students to try to picture and describe a shape that could be found on each planet that they have not learnt yet. Tell them the name of the shape they describe.

Stretch zone: *Invent a planet. What is special about the shapes on your planet?*

This stretch zone will be accessible to students of a wide range of abilities. If possible, leave time for all or most groups to complete this activity. Check that students have written accurate descriptions of the shape properties.

 **Reflection time**

Choose a group to describe one of their shapes. Can the other groups work out which planet it is from and which type of shape it is? Invite a student from the group to draw the shape on the board or hold up an example of it. Did the other groups guess correctly? Ask, *Was the description accurate? Could it be improved upon? How?* Repeat with other groups.

Differentiated outcomes	
All students	should sort shapes according to defined properties with support.
Most students	will find most shapes that fit the set criteria.
Some students	may identify all possible shapes that fit the criteria by working systematically.

10 Geometry – properties of shapes

Global skills

- **Creative skills:** exploring
- **Real-world skills:** presenting information
- **Interpersonal skills:** communication
- **Self-development skills:** reflecting on learning

Student Book

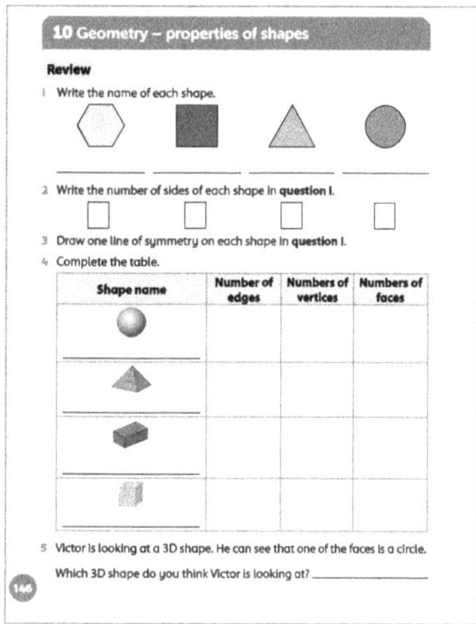

With young children, assessment activities are most effective when carried out as an everyday classroom activity. Students could have access to 2D and 3D shapes during the review to support them, if necessary.

Watch as students identify the properties for each shape. Listen to check that they understand the names of shapes and their properties, for example, faces, edge and vertices. Ask questions to check understanding: for example, *What type of shape is this? Are these shapes regular or irregular? How do you know? What is the difference between a cube and cuboid? Does this shape have any lines of symmetry?*

Answers

Student Book page 146

1 hexagon, square, triangle, circle

2 6, 4, 3, 1

3 Check that students have drawn a line of symmetry correctly for each shape.

Shape name	Edges	Vertices	Faces
Sphere	0	0	0
Square-based pyramid	8	5	5
Cuboid	12	8	6
Cube	12	8	6

Practice Book

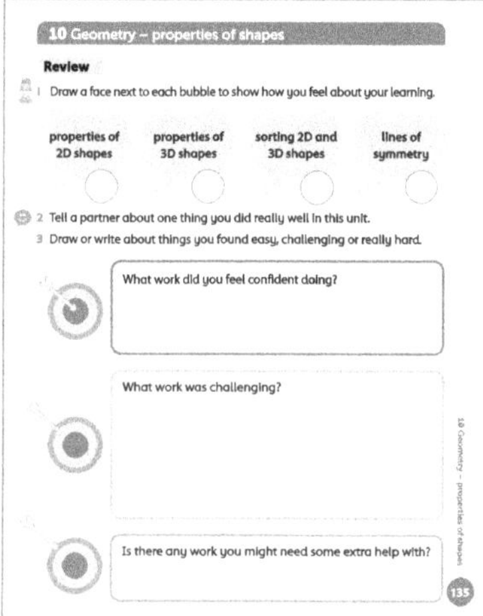

It is appropriate to complete this Practice Book review as a whole-class discussion. You may choose to keep a record of the class discussion or a copy of the review page for your own records. The review provides an opportunity for students to reflect on their learning from the unit, to discuss any areas of mathematics that they feel went particularly well, and any areas that they feel less confident about. Ensure all students have a copy of the Student Book as a reminder of the areas of mathematics that they have worked on in this unit.

Allow students plenty of time for discussion before asking them to complete the Practice Book page individually, and then, if appropriate, to share their responses with the rest of the class. If students complete this self-assessment at home, encourage them to discuss this with adults. Make a note of areas that students still feel unsure about.

Additional material

There are additional end-of-unit assessments available on the *Oxford Owl* website.

11 Geometry – position and direction

Big idea

The Big idea in this unit is that geometry enables us to understand our physical world, through describing it and representing it. Students need to develop a sense of space, which is also called a spatial awareness. This plays an important role in other areas of mathematics as well across the curriculum, for example in geography, science and art.

Students need to be able to locate and describe the position of objects and their movements, including their own movements. They need the relevant mathematical language to describe position and movement so that they can discuss and visualise.

Early positional language usually focuses on fixed positions. Words and phrases such as before, after, behind, in front of, on top of and below are good starting points but need to be developed to include measures and turns. Position and movement games will help students to develop their understanding. These skills play an important role in representing and solving problems in all areas of mathematics and in the real world.

Students will become more aware of angles in relation to position and direction, for example turning to face different directions.

Look out for

- **Students who understand angles as a fixed property rather than as a movement (rotation), a measure of turn.** Build on their real-life experience. Young students will have discovered through experience that objects, including themselves, can be moved and turned. They will have played with toys, jigsaw puzzles, building blocks and much more, often having to turn the object to place it where they want. You can use the movement of a door or window to show that angle is a measurement of the amount of turn.

- **Students who confuse left and right.** Show students how the left-hand thumb and fingers make an L-shape when held out face down.

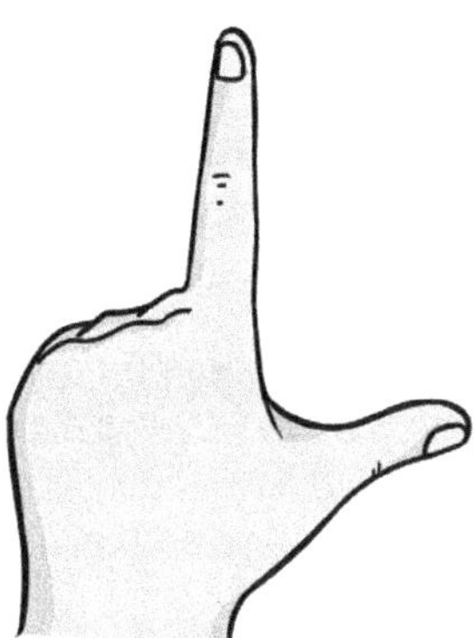

Possible misconceptions

- **Students may only see angles in shapes and not in turns.** Provide lots of opportunities to practise making quarter turns, both clockwise and anti-clockwise, and describing them as right angles.

Key vocabulary

- backward, forward, left, right
- clockwise, anti-clockwise, right angle, turn
- whole turn, half turn, quarter turn, three-quarter turn
- pattern, grid, diagonal, route
- landmark

Coverage in lessons

Learning focus	Learning outcomes (the ENC objectives)
Turns and right angles	Use mathematical vocabulary to describe position, direction and movement, including distinguishing between rotation as a turn and in terms of right angles for quarter, half and three-quarter turns (clockwise and anti-clockwise).
Travelling	Use mathematical vocabulary to describe position, direction and movement, including movement in a straight line.

11 Geometry – position and direction

Big question

- How can I describe turns and give accurate directions?

Global skills

- **Creative skills:** exploring
- **Interpersonal skills:** communication/teamwork

Key vocabulary

- clockwise, anti-clockwise, turn, half turn, quarter turn, three-quarter turn, backward, forward, left, right

Resources

- maps of the local area

Language support

Display a simple map of the local area. Use different-coloured pens to draw in the route from one place of interest to another. Write instructions for each route using the matching coloured pen and display the directions to the route. You could also display a map of the school and the local area with the students' maps of their routes arranged around it.

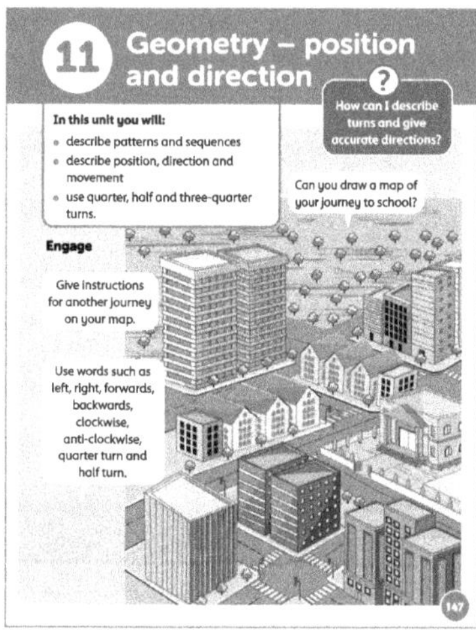

Introductory activity

Ask students to look at page 147 of the Student Book. Display it on the IWB, if possible. *What can you see?* Students should describe the different buildings. Encourage them to use accurate vocabulary of shape (e.g. the tall yellow cuboid). Encourage them also to describe their position in relation to other buildings (e.g. the row of yellow buildings is next to the red building

with 20 black windows). Next, ask questions such as, *How do I get from the tall yellow building to the brown wooden building?* Encourage students to use phrases such as '**turn left/right**' and 'go straight/**forward**' and any other language they are familiar with, drawing up language in the speech bubble.

Main activity

Ask students if they can describe the route from school to a place they know nearby. Share ideas. Show students a map of the area (display on the IWB, if possible). Challenge students to find their own home on it, and other local places of interest.

Draw a very simple route map from school to, for example, the supermarket or somewhere else that will be familiar to students. Talk about it as you draw, for example: *I go all the way along High Street, then I turn left onto School Road.*

Ask pairs to draw a map, in a similar way to yours, of one of their journeys to school. Ask: *Would someone visiting us be able to find their way to the school using your map?* Agree that it must have enough information so they do not get lost but not too much that it gets confusing. Remind students that they do not travel in a straight line, they have to **turn** corners and you want to see those in their map. Once they have drawn their map, ask them to write instructions for the journey. Direct them to the speech bubble on page 147 of the Student Book for a list of words they can use in their instructions.

Differentiation

Supporting: Ask students to show you their route. Modelling clear instructions, describe the route as you follow it with your finger on the map. Ask students to repeat the instructions after you.

Consolidating: Ask students to share their route with you.

Extending: Ask students to give instructions to their partner. Can they work out where the route starts and follow it to its end point?

Reflection time

Ask students to look at each other's maps. If one student knows where another student lives, ask them to look at the student's map and imagine travelling there. *Does the map make sense? Would you go a different way?*

11A Turns and right angles

Specific learning focus

- Recognise whole, half and quarter turns, both clockwise and anti-clockwise.

Global skills

- **Creative skills:** exploring
- **Interpersonal skills:** communication/teamwork

Key vocabulary

- clockwise, anti-clockwise, turn, half turn, quarter turn, whole turn, right angle

Resources

- circles of thin card (one per student)
- sets of wooden or plastic 2D shapes (for extension only)
- set of 'turn cards' (quarter turn, half turn, whole turn only), several of each
- mini whiteboards and markers

Language support

Take photographs of a student at a starting point and after a quarter turn, one right-angle turn, half turn, two right-angle turns, whole turns and four right-angle turns in both clockwise and anti-clockwise directions. Display these with turn card labels attached, showing that a quarter turn clockwise and turning one right angle clockwise are identical.

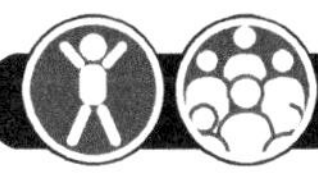

Introductory activity

Ask all students to stand up and arrange themselves in an array. If you can do this in an open space such as the hall or playground, you can be more adventurous.

Ask the whole class to move through a quarter turn. Explain that a **quarter turn** is also called a **right angle**. Model turning a quarter turn for students to copy.

Repeat giving instructions for a **half-turn** (two right-angle turns), a **whole turn** (four right-angle turns) in both **clockwise** and **anti-clockwise** directions. After a series of movements, the class can check they are all facing the same way.

What happens when you turn through two right angles? (You face in the opposite direction.)

How many right angles do you turn through to make a whole turn? (4) *Which way will you be facing?* (the same direction as originally)

Main activity

Model how to make a right-angle checker by folding a circle of thin card into four quarters and using the folded quarter or cutting out one of the quarters. Ask students to look at the image on page 148 of the Student Book for an additional visual reference. Say, *This corner shows a right angle.* Choose an object with a right angle (e.g. a book). *Does this book have a right angle?* Show students how to check the corner of a book using the right-angle checker. Continue to check all four angles and say that the book has four right angles. Choose another object (e.g. the end of the table). *Do you think this tabletop has any right angles? How many?* Check each angle and confirm how many. Choose an item with no right angles and check one of its angles using the right-angle checker. Agree that it is not a right angle and ask students whether it is larger or smaller than a right angle.

Ask students to look at page 148 in the Student Book. *What can you see?* Students should describe different elements of the picture. Encourage them to look for angles in the picture. Introduce them to relevant new vocabulary if necessary. Then ask students to work in pairs, first making their own right-angle checker, then checking all the angles in the picture to see whether they are right angles.

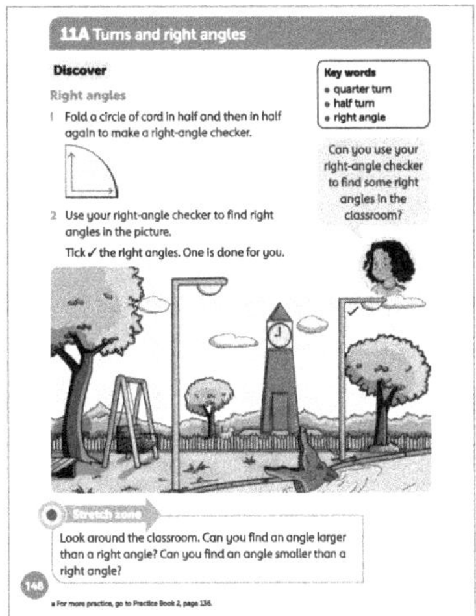

Differentiation

Supporting: Model how to use the right-angle checker to check for right angles.

Consolidating: Ask students to predict whether an angle will be a right angle before checking it.

Extending: Ask students to sort a selection of 2D shapes into those with right angles, those with angles less than a right angle and those with angles greater than a right angle.

Stretch zone: *Look around the classroom. Can you find an angle larger than a right angle? Can you find an angle smaller than a right angle?*

Check that students' vocabulary when talking about right angles is accurate. Have they identified angles larger and smaller than a right angle? How do they know?

Reflection time

Make some 'turn cards' with quarter turn, half turn and whole turn written on them (several of each card). Shuffle the cards. Ask a student to come to the front and choose four cards. They move according to the cards, taking a pause between each turn. Ask the other students to record on mini whiteboards each turn that they saw. Reveal the cards the student used and check that the class saw the matching movements. Repeat several times with different students.

Practice Book: Students can now complete the activity on page of 136 in the Practice Book. They can do this directly after the main activity, as homework, or as the focus of a separate mathematics session to help students consolidate their learning and build fluency. Students may make a new right-angler checker or use the one from the main activity. They can find real-life examples or pictures in magazines or newspapers.

Differentiated outcomes	
All students	should use the right-angle checker to find right angles.
Most students	will recognise right angles without using an angle checker.
Some students	may understand the equivalence between a right angle and a quarter turn and understand clockwise and anti-clockwise.

11A Turns and right angles

Explore 1 Student Book page 149 • Practice Book page 137

Specific learning focus

- Recognise whole, half, quarter and three-quarter turns, both clockwise and anti-clockwise.

Global skills

- **Creative skills:** problem solving/ investigating
- **Interpersonal skills:** communication/teamwork

Key vocabulary

- clockwise, anti-clockwise, turn, half turn, quarter turn, three-quarter turn, right angle

Resources

- simple shapes such as isosceles triangles, semi-circles or rectangles, circles of thin card
- set of 'turn cards' (including some step cards – see language support)
- mini whiteboards and markers

Student Book page 148

Examples of right angles in the picture include:

- hands on the clock
- jaws of the crocodile
- four angles at the corners of the door on the clock tower
- angle between back and seat of the swing
- angle between each fence post and base of the fence.

Practice Book page 136

Check that the angles drawn for the chosen examples do show right angles.

Stretch zone: Answers will vary but students should say something like 'I open the door wide so that it makes a quarter turn from its closed position.'

Language support

Add more cards to the set of turn cards. Include some step cards, saying, '1 step forward' (or 2, 3, …, 5 steps).

 Introductory activity

Repeat the introductory activity from Lesson 11A Discover but this time also include a **three-quarter turn**. Practise quarter, half, three-quarter and whole turns in both clockwise and anti-clockwise directions and then add instructions such as 'move forward/backward 3 steps'. Check that students recognise the effect of a half turn and a whole turn.

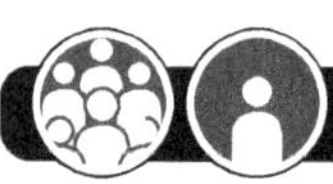 Main activity

Give each pair of students a simple shape such as a triangle (not equilateral), semi-circle or rectangle (not square) to draw round. After students have completed the first drawing, give instructions to turn their shape a quarter, half or whole turn clockwise or anti-clockwise for the next drawing. *What is the same? What is different?* Ask further questions to encourage students to predict what the shape will look like after a certain number of turns. For example, *How many more turns until your shape is in its start position? Will this vertex be pointing left or right after a half turn clockwise?*

With students referring to their drawings, ask:

How many right angles do you turn to turn one quarter?

I make a half turn. How many right-angle turns did I make?

How many more right-angle turns do I need to make after a three-quarter turn to make a whole turn?

Students then complete the activity in the Student Book page 149. Encourage them to look at the speech bubbles for support. Emphasise that they do not need to copy the shapes perfectly, but it needs to be clear how the shape has turned.

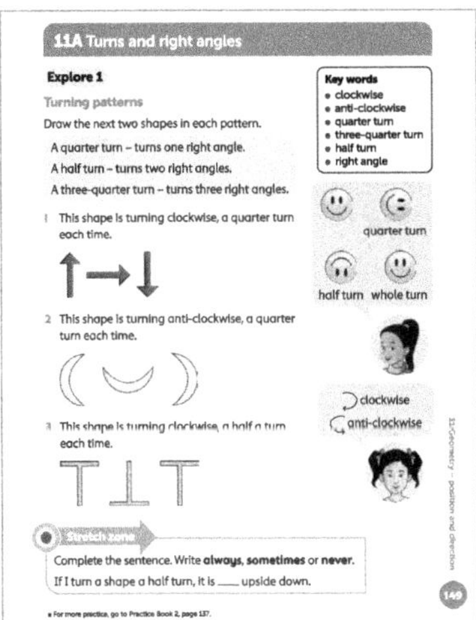

Differentiation

Supporting: Give students a sheet of paper with 'left' and 'right' written in the appropriate corners and curved arrows labelled 'clockwise' and 'anti-clockwise' as a guide.

Consolidating: Ask students to give you instructions for rotating their images. Make deliberate mistakes so that they have to correct you.

Extending: Ask students to identify right angles and angles greater and less than a right angle in their images.

Stretch zone: *Complete the sentence. Write 'always', 'sometimes' or 'never'.*
If I turn a shape a half turn, it is _______ upside down.

Check that students have written 'always'. Ask them to explain why, using a drawing or actions to support their thinking.

 **Reflection time**

Ask students to work together using a set of 'turn cards' to direct a student from the front of the class to the door, for example. Can directions for a shorter journey between the two places be given? Encourage students to include half, quarter and three-quarter turns, if possible, in their instructions.

Practice Book: Students can now complete the activity on page of 137 in the Practice Book. They can do this directly after the main activity, as homework, or as the focus of a separate mathematics session to help students consolidate their learning and build fluency. Students may find it easier to trace over the shapes, cut them out, turn them and then stick them in their books.

Differentiated outcomes	
All students	should rotate images clockwise and anti-clockwise and recognise half and quarter turns.
Most students	will recognise right angles in the shapes that they have drawn.
Some students	will recognise angles that are greater and less than right angles in the shapes they have drawn.

Answers

Student Book page 149

Check that students have drawn the correctly turned shapes to make each pattern.

Practice Book page 137

Check that students have drawn the correctly turned shapes to make each pattern.

Stretch zone: In the first line, each shape is rotated one right angle clockwise from the previous one. In the second line, each shape is rotated one right angle anti-clockwise from the previous one.

11A Turns and right angles

Explore 2
Student Book page 150 • Practice Book page 138

Specific learning focus

- Recognise whole, half, quarter and three-quarter turns, both clockwise and anti-clockwise.

Global skills

- **Creative skills:** exploring/investigating
- **Interpersonal skills:** communication

Key vocabulary

- pattern, clockwise, anti-clockwise, turn, quarter turn, half turn, three-quarter turn

Resources

- set of simple 2D shapes or other shapes that can be easily rotated (e.g. plastic letters of the alphabet)

Language support

Continue to use the set of turn cards. Include different numbers of right-angle turns, which say clockwise or anti-clockwise and some step cards.

 **Introductory activity**

Hold up an asymmetric shape for the students to see (e.g. a 'P' shape). Ask them to draw the shape as they see it. Now tell them that you are going to turn the shape a quarter turn clockwise. Ask students to draw what they think it will look like when you have turned it. Once everyone has drawn their prediction, turn the shape and ask students to check whether they were correct.

1 quarter turn clockwise

Repeat this a few times, starting from one position and turning the shape either a quarter turn, half turn or three-quarter turn, either clockwise or anti-clockwise.

 **Main activity**

Draw a simple asymmetric shape on the board. Ask a student to suggest how much of a turn to give the shape and ask another to choose either clockwise or anti-clockwise. Now ask another student to come to the front of the class and draw the position of the shape after that turn. *Ask the rest of the class to check and confirm the result of the turn.* Tell students they are going to keep using the same turn and make a **pattern** of shapes. *So what turn do we need to do again?* (e.g. three-quarters turn anti-clockwise)

Ask students to turn the shape another three-quarters turn anti-clockwise and draw the shape in its new position. Repeat until the shape is back to the same position it started from. Then repeat with a different shape and a new turn instruction.

Students then complete the activities on page 150 of the Student Book individually. Encourage students to use a ruler to make their shapes more consistent in size and shape. Suggest that they use coloured pencils to define, for example, one vertex of their shape so that the turn is more obvious to see. After they have completed their pattern, they show it to a partner and describe it.

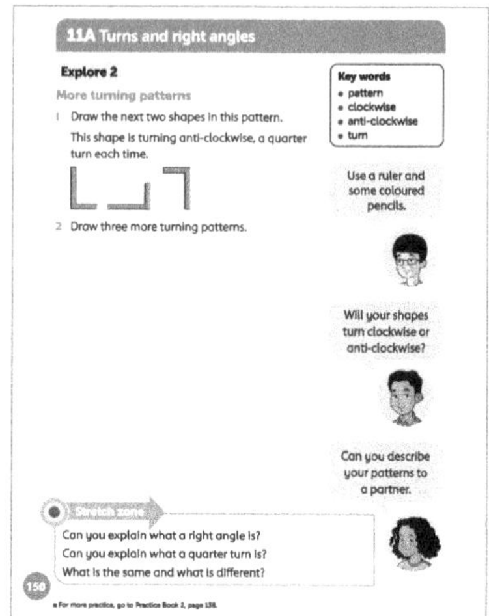

Differentiation

Supporting: Cut a small version of the shape for students to move to see the positions. A shape stencil may also be useful.

Consolidating: Ask students to predict what position the shape will be in two moves' time.

Extending: Show students a shape and turn it without them seeing. *Can you describe the movement in more than one way?* (e.g. a quarter turn clockwise is the same as a three-quarter turn anti-clockwise)

Stretch zone: *Can you explain what a right angle is? Can you explain what a quarter turn is?*
What is the same and what is different?

Students may say that a right angle is like an L-shape, or refer to examples such as the corners of a book. A quarter turn might be described as turning to one side. If students answer these questions confidently, extend with further questions, for example, *How many right angles in a half/three-quarter/whole turn? How do you know?*

 **Reflection time**

Ask students to share the patterns they have made for page 150 in the Student Book. From showing the pattern of positions for their chosen shape, can another student work out which turn was used each time? Which patterns took longest to get back to the starting position?

Practice Book: Students can now complete the activity on page of 138 in the Practice Book. They can do this directly after the main activity, as homework, or as the focus of a separate mathematics session to help students consolidate their learning and build fluency. Ask students to think about possible shapes that would be easy to draw in different turns, drawing on their work in the Student Book activity.

Differentiated outcomes	
All students	should rotate images clockwise and anti-clockwise and recognise half and quarter turns.
Most students	will predict the outcome of a turn without needing to use a shape.
Some students	may be able to predict the next two shapes in a pattern of turns.

Answers

Student Book page 150

Check that students have drawn correct turning patterns.

Practice Book page 138

Check that students have drawn suitable shapes and drawn them after the turns indicated.

Stretch zone: Students may say that the two turns result in the shape being back where it started.

11B Travelling

Discover Student Book page 151 • Practice Book page 139

Specific learning focus

- Follow and give instructions involving position, direction and movement.

Global skills

- **Creative skills:** exploring
- **Real-World skills:** interpreting information
- **Interpersonal skills:** communication/teamwork

Key vocabulary

- clockwise, anti-clockwise, turn, half turn, quarter turn, grid, right-angle turn, diagonal, route

Resources

- squared paper
- mini whiteboards and markers
- large 100-square for front of class

Language support

Draw up a list of direction words with the students and discuss the meaning of each word: 'clockwise', 'anti-clockwise', 'right angle', 'right', 'left', 'turn', 'diagonal', 'route'.

Talk through how they know what direction to turn for 'clockwise' and 'anti-clockwise'. Do the same for 'left' and 'right'. Add the words to the display if they are not already included. Display a speech bubble next to each word with its meaning.

 **Introductory activity**

Set up some cones labelled with letters around the classroom or, preferably, a large open space. Choose a student to come forward, then direct them from one letter to another. For example, *Walk forward 5 steps and stop. Turn right and walk 5 more steps. Make a quarter turn clockwise. You have arrived.* Repeat with another student, this time including a poor instruction. Do the students catch the mistake? Can they tell you what the instruction should be? Students should then take it in turns to direct their partner from one letter to a second and third letter.

 **Main activity**

Model drawing a simple journey with some turns on a squared-paper background on the board, moving along vertical or horizontal lines of grid squares. Write the instructions for the journey, using full wording such as 'Forward 2 squares, quarter turn anti-clockwise, forward 5 squares, one right-angle turn anti-clockwise.' Ask students to check that the instructions match the journey. Students can use these written instructions as a guide later in the activity. Give each pair of students some squared paper and ask them to draw a simple journey on it (three or four stages only). They should then record instructions for the journey on mini whiteboards. Ask pairs to swap journey and instructions with another pair to check.

Ask students to look at the map on page 151 of the Student Book (display on the IWB, if possible). Check students' understanding of the map by asking questions. *What can you see? What do you think the blue shows? What is a swamp?* Read through the 'travel advice' as a class. Do you remember what the word **diagonal** means? *If I made a diagonal move from the boat where might I end up?* (a swamp or Skull Hill)

Ask students to complete the activity on Student Book page 151 in pairs. Observe students as they talk about the different journeys and check that they are using the vocabulary accurately. For example, from the boat to Pointed Hill could be: quarter turn clockwise, forward 2 squares, quarter turn clockwise, forward 1 square.

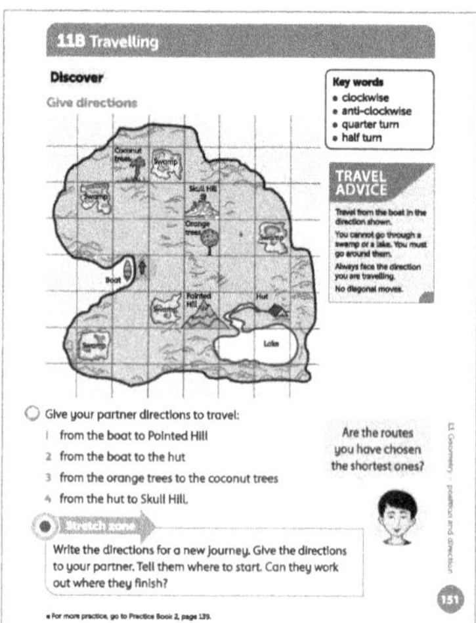

Differentiation

Supporting: Help students to keep their place with a finger as they move on the map.

Consolidating: Ask students to give you directions to move on the map. Make deliberate mistakes so that they have to correct you.

Extending: Ask students to create more complex journeys on the map.

Stretch zone: *Write the directions for a new journey. Give the directions to your partner. Tell them where to start. Can they work out where they finish?*

Listen to students' reasoning and check that they can use the correct vocabulary in describing their journey.

 Reflection time

Display a 100-square. Ask students to imagine a car on 23, facing 22. Say that the car needs to get to 76, but there is a forest covering some squares, say 51 to 57. Invite different students to describe a route for the car to take. Repeat with different numbers and scenarios.

Practice Book: Students can now complete the activity on page of 139 in the Practice Book. They can do this directly after the main activity, as homework, or as the focus of a separate mathematics session to help students consolidate their learning and build fluency. Students will now be writing their directions. Encourage them to refer to the vocabulary listed at the top of the page to help them write them or ask an adult for help. Explain the word 'grid' before they start.

Differentiated outcomes	
All students	should give simple directions orally.
Most students	will give accurate multi-step directions orally.
Some students	will give accurate directions orally and write them down.

Answers

Student Book page 151

Answers will vary. Observe and listen to students' descriptions of the journeys as they work.

Practice Book page 139

Answers will vary. Check that directions match the map.

11B Travelling

Explore 1 Student Book page 152 · Practice Book page 140

Specific learning foci

- Plan a journey.
- Follow and give instructions involving position, direction and movement.

Global skills

- **Creative skills:** exploring
- **Interpersonal skills:** communication/teamwork

Key vocabulary

- clockwise, anti-clockwise, turn, half turn, quarter turn, grid, right-angle turn, route

Resources

- large 100-square for front of class
- blank 10-by-10 grid or squared paper (for each pair)
- dice

Language support

Display a 100-square with labels showing directions from one number to another. Add students' own maps to the classroom display.

Introductory activity

Give each pair of students a blank 10-by-10 **grid** and explain that they are going to plan a route to move around the grid. Ask students to imagine that this is a 100-square. They should write 'Start' where 1 would be and 'Finish' where 100 would be. Ask them to write in four numbers (e.g. 25, 47, 62 and 84) in their correct positions.

Next, ask them to colour in any 30 squares on the grid, but explain that no more than six connected squares can be coloured.

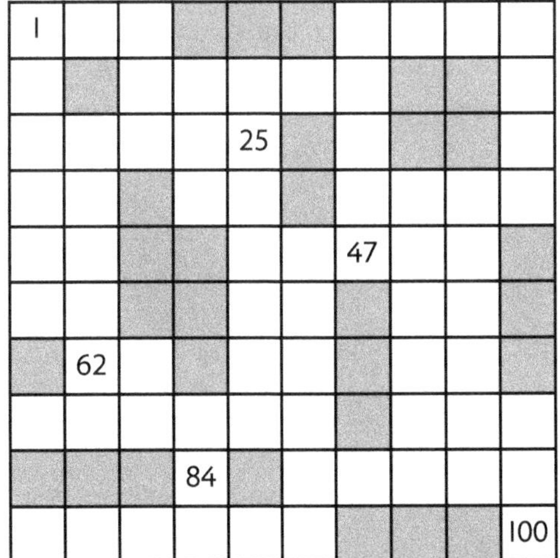

Explain that each pair should plan a route from square 1 to 100 and describe each of their moves in writing. (Discuss what is meant by 'a move'.) Explain that their coloured squares are blocks that have to be travelled around, not through. Say that the **route** must pass through at least two of their numbered squares. Challenge pairs of students to write a route from start to finish in a particular number of moves. If turns and movements forward count as separate moves, you may need to allow a larger number of moves. Now suggest to students that they find a route between the same two squares in the smallest number of moves. Pairs can swap grids with each other and check routes.

Main activity

Ask students to look at page 152 of the Student Book. Display it on the IWB, if possible. As with the previous main activity, check that they are familiar with all the places and vocabulary. Review the travel advice. Ask a few questions to get students thinking about moving around the map. *I start at the big wheel and move up one square. I make a quarter turn anti-clockwise. What am I facing?* (ice cream stand)

Ask students to complete the activity in the Student Book page 152 individually. Once students have completed their route, they pair up with another student so that they can check each other's directions and compare journeys.

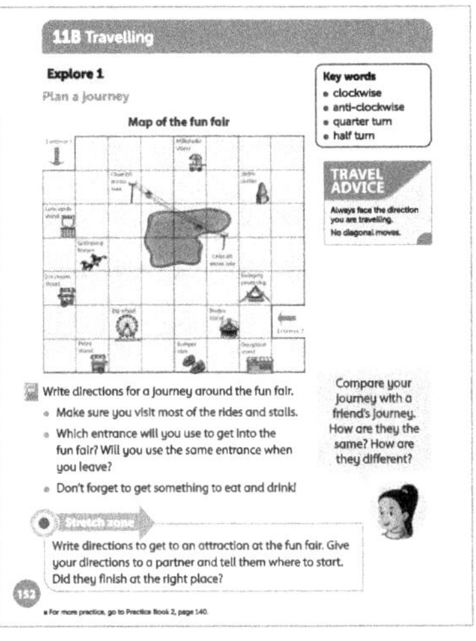

Differentiation

Supporting: Ask students to give you verbal directions for their journey to move on the map rather than written.

Consolidating: Ask students to read you the directions for their journey for you to trace with your finger on the map. Make deliberate mistakes so that they have to correct you.

Extending: Ask students to create more complex journeys on the map.

Stretch zone: *Write directions to get to an attraction at the fun fair. Give your directions to a partner and tell them where to start. Did they finish at the right place?*

Encourage the partner to feedback on how any directions could be clearer. Can they plot another route?

 ## Reflection time

Ask some students to talk through their set of directions. See if other students can follow the route on their own maps. Ask for example, *How did you decide on the order for your route? Did you miss any of the rides or stalls? Why?*

Practice Book: Students now complete the activity on page of 140 in the Practice Book. They can do this directly after the main activity, as homework, or as the focus of a separate mathematics session to help students consolidate their learning and build fluency. Suggest that if students are finding drawing the objects challenging, they can mark them on the grid with an 'X' and then label them with a word. Explain that they can ask an adult to help them with the writing, if necessary.

Differentiated outcomes	
All students	should give accurate directions orally.
Most students	will give accurate directions orally and write them down.
Some students	will give more complex directions orally and write them down.

Answers

Student Book page 152

Check students' instructions for their journeys around the fun fair. Note who is confident in identifying quarter turns clockwise and anti-clockwise, and who requires more practice in using clockwise and anti-clockwise correctly.

Practice Book page 140

Check that students' instructions are correct for the journeys they have chosen. Note who is still having difficulties telling clockwise from anti-clockwise and provide further practice and support.

Explore 2 Student Book page 153 • Practice Book page 141

Specific learning foci

- Draw a reasonably accurate map of a classroom.
- Follow and give instructions involving position, direction and movement.

Global skills

- **Creative skills:** exploring/investigating
- **Real-world skills:** research/presenting information
- **Interpersonal skills:** communication/teamwork

Key vocabulary

- clockwise, anti-clockwise, turn, half turn, quarter turn, three-quarter turn, grid, right-angle turn, landmark, route

Resources

- 10-by-10 grid or squared paper

Language support

Encourage the use of correct mathematical language throughout the activities. Students will learn from talking, and listening to others talking, and show their understanding to you as they describe routes using positional language as displayed in the previous lessons.

Introductory activity

Draw or display on the board a simple plan of a bedroom on a square grid, showing basic furniture, door, window (you could search for one on the internet). *What things are shown on the map?* (e.g. bed, chair, window) *Can you think of some things that might be in the room that are not shown?* (e.g. clothes, shoes, books) *Can you give me an instruction to get from the bed to the chair on this map?* Repeat with other instructions.

Main activity

Explain to students that they are going to draw a map of the classroom on page 153 of their Student Book. *What things might be important to include?* (e.g. desks or tables, doors, cupboards) Explain that we can call these **landmarks** as they are important features of the classroom. *What things should you not include?* (e.g. pencils on the desks, books on a shelf, chairs that fit under desks/tables) *Which things are going to take up most of the space on your map?* (student desks/tables)

Students may want to do a rough map first on a separate sheet of grid paper before drawing in their Student Book.

Once they have finished their maps, they complete question 2 on page 153 of the Student Book and record their directions in their notebooks. They can pair up with another student to work together to find and describe different routes on their maps (question 3).

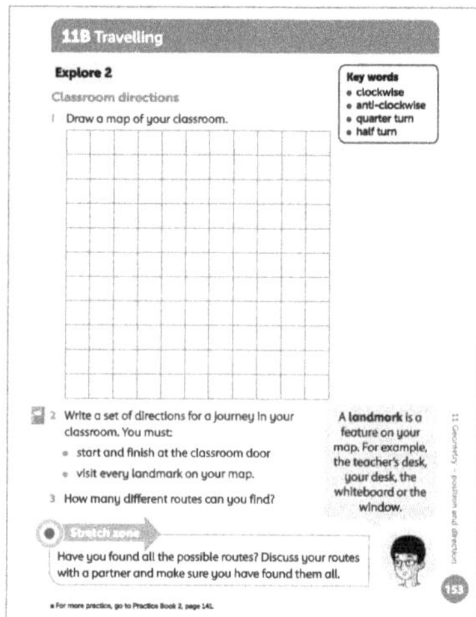

Differentiation

Supporting: Help students to compare the size of different items of furniture in the room and assign them a square or part of a square.

Consolidating: Ask students to create an alternative set of directions, then compare them to see which they think is more efficient in getting around the classroom.

Extending: Ask students to try measuring some items in the room to improve the accuracy of the sizing of objects on the map.

Stretch zone: *Have you found all the possible routes? Discuss your routes with a partner and make sure you have found them all.*

Check that students have found all the routes and can explain how they know they haven't missed any.

 Reflection time

Ask some students to share their map of the classroom and explain how they made it. *What did you include? What did you leave off? How did you decide which way up to draw the map? How did you decide on the sizes of the things on the map? How did you decide on the directions you chose to move from one part of the classroom to another?*

Practice Book: Students now complete the activity on page of 141 in the Practice Book. They can do this directly after the main activity, as homework, or as the focus of a separate mathematics session to help students consolidate their learning and build fluency. Before they begin, ask them to think about the maps they saw as part of the main activity to help them create the map of a room in their homes. What worked well? What did not? What did they like in other students' maps?

Differentiated outcomes	
All students	should draw some features of the classroom on a map in a fairly accurate position.
Most students	will draw a map of the classroom including the main fixed objects.
Some students	may use some measurements to make relative sizes of items on their map mroe realistic.

11 Geometry – position and direction

Connect Student Book pages 154–155

Big idea

I can use angles to describe turns and to give accurate directions.

Global skills

- **Creative skills:** exploring/investigating
- **Real-world skills:** presenting information
- **Interpersonal skills:** communication/teamwork
- **Self-development skills:** reflecting on learning

Key vocabulary

- clockwise, anti-clockwise, turn, half turn, quarter turn, grid, right-angle turn, right angle

Resources

- rulers, tape measures, metre sticks, trundle wheels

Language support

Encourage students to use the correct language for the location of objects rather than pointing or saying, 'there.'

 **Introductory activity**

Revise turns with the students by asking them to play a game. They all stand up and face the front of the class. You call out instructions, for example, 'quarter turn left', 'half turn', 'whole turn' and so on. Students who turn the wrong way should sit down.

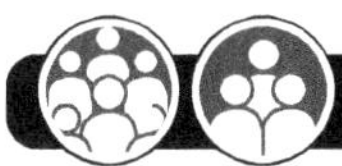 **Main activity**

Ask students to discuss in their pairs how they might produce a list of possible (generic) instructions to include as part of directions to move around the school.

Answers

Student Book page 153

Check that students have drawn a reasonably accurate map and described routes around the classroom.

Practice Book page 141

Check that students have drawn a reasonably accurate map and described routes around their chosen room at home.

Get students thinking about distance as part of their directions by asking, for example, *If you were giving directions from your desk to the door, how could you measure the distance?* Agree that students could use strides (paces) or metres and take suggestions of metre sticks, measuring tape or, ideally, a trundle wheel.

Arrange students in small mixed-attainment groups to work on writing the directions for moving on one journey around the school. This could include actual distances In whole metres if appropriate for your students. Each group should work on a single journey. Ask students to look at page 155 of the Student Book. Although some possible routes are suggested in the Student Book, you may want to use others that are more suited to your school. These should be agreed before students start work.

Once each group has completed the directions for their route, they should swap with another group and follow their directions to see if they work.

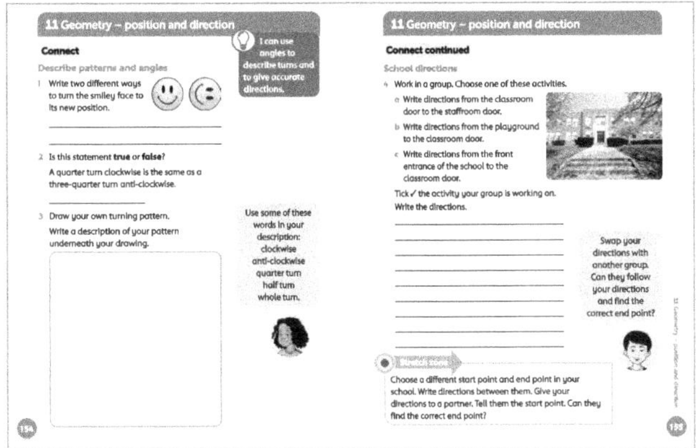

Students can then complete page 154 of the Student Book individually.

Differentiation

Supporting: Ask students to describe the directions for their turning pattern, for you to write them down.

Consolidating: Ask students to share turning patterns in pairs and describe each other's patterns. Do their descriptions match? Why or why not?

Extending: Encourage students to write more elaborate directions that include a variety of different types of instructions including accurate distances (e.g number of metres before a turn) if possible.

Stretch zone: *Choose a different start point and end point in your school. Write directions between them. Give your directions to a partner. Tell them the start point. Can they find the correct end point?*

Encourage the pair to discuss how to improve their directions so they are easier to follow.

Reflection time

Give each group a few minutes to make a presentation to the rest of the class about their directions. *Did any groups follow these directions? Were they helpful? Did you arrive at the final destination?* After all groups have presented, discuss what makes a good set of directions. Take suggestions such as clear instructions, specific number of steps or metres to walk, getting right and left correct, taking the simplest route.

Differentiated outcomes	
All students	should describe journeys using directions.
Most students	will give directions which are clear and accurate.
Some students	May give directions that include a variety of instructions including metres or centimetre distances.

Answers

Student Book page 154

1 quarter turn clockwise, three-quarter turn anti-clockwise

2 true

3 Check that students have drawn a turning pattern correctly.

Student Book page 155

Check that groups have written their directions in a sensible and clear way.

11 Geometry – position and direction

Review
Student Book page 156 • Practice Book page 142

Global skills

- **Creative skills:** exploring /investigating
- **Real-world skills:** presenting information/ interpreting information
- **Interpersonal skills:** communication
- **Self-development skills:** reflecting on learning

Student Book

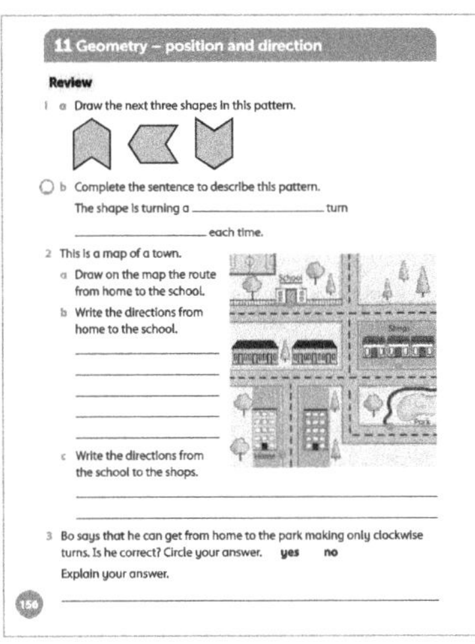

With young children, assessment activities are most effective when carried out as an everyday classroom activity. Students should be able to describe directions using accurate turns and continue patterns involving turns.

Watch as students plan routes using the map and write directions from home to school. Alternatively, you could ask a student to describe their route orally if writing the directions is proving very challenging.

As students work, ask them questions to further assess their understanding such as: *Imagine the first shape in the pattern was actually the second. What would the shape before look like? How many quarter turns to get back to your start position?*

Answers

Student Book page 156

1 a  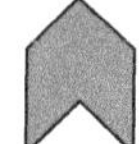 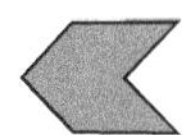

   b The shape is turning a quarter turn anti-clockwise each time.

2 Check that the routes are drawn correctly and that the directions are accurate.

3 No, Bo needs to make an anti-clockwise turn when he leaves home and to turn towards the park in his last move.

Practice Book

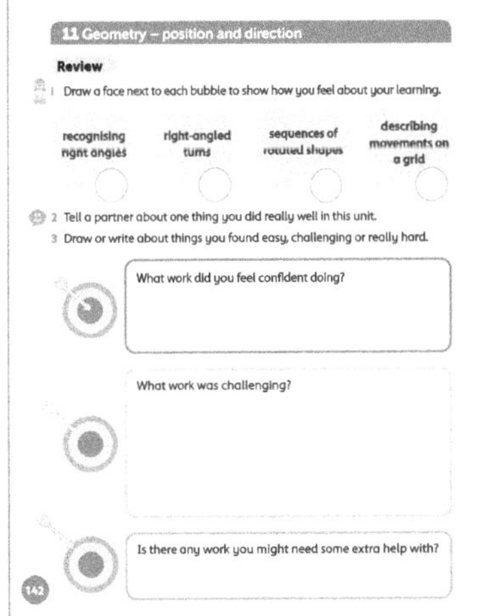

It is appropriate to complete this Practice Book review as a whole-class discussion. You may choose to keep a record of the class discussion or a copy of the review page for your own records. The review provides an opportunity for students to reflect on their learning from the unit, to discuss any areas of mathematics that they feel went particularly well, and any areas that they feel less confident about. Ensure all students have a copy of the Student Book as a reminder of the areas of mathematics that they have worked on in this unit.

Allow students plenty of time for discussion before asking them to complete the Practice Book page individually, and then, if appropriate, to share their responses with the rest of the class. If students complete this self-assessment at home, encourage them to discuss this with adults. Make a note of areas that students still feel unsure about.

Additional material

There are additional end-of-unit assessments available on the *Oxford Owl* website.

12 Statistics

Overview

Big Idea

The Big idea for this unit is that the same information can be presented in a variety of ways, in order to help us to read, describe, interpret and evaluate that information. Data can be presented visually using tables, charts and diagrams. Students can gather, organise and display data to answer a question. Simple surveys, with a limited number of responses, help students to develop their understanding of data presentation and interpretation.

Students need lots of experience of sorting and classifying objects, making choices and justifying their decisions. They need to see how objects can share a particular property and then represent their ideas using pictures, drawing and numbers.

Students can demonstrate their understanding of the data by telling a matching story, making comments about, and comparing different parts of, the data. It is important that they use accurate mathematical language to describe what they have noticed and to ask questions about what they see.

Students should justify their opinion or interpretation by referring to specific parts of the data. A useful way to encourage students to refer to the data is to ask: *What makes you think that? or What is it that tells you that?*

Various computer programs can be used to help students organise, sort and present the data they have collected.

Look out for

- **Students who use one-to-one correspondence when constructing tally charts, bar diagrams and pictograms.** A good understanding of number and place value is essential for them to be able to read and interpret scales, particularly when they move on to represent two or more objects or responses with one block, picture or something else. Students need to count in multiples rather than ones.

- **Students who become confused about what the numbers represent.** Often different sets of numbers are in use when we are handling data and students can become confused. They should refer to the title and labels when making statements about the data.

Possible misconceptions

- **Students may think that a symbol in a pictogram can only represent one of something.**
- **Students may think that it is not possible for an object to sit outside the circles of a Venn diagram because all objects can be sorted using a Carroll diagram.** Choose simple criteria that make it easy to see that this is possible (e.g. numbers to 15: criteria 'even numbers' and 'multiples of 3').

Key vocabulary

- survey, data, frequency table, tally mark, tally chart, pictogram, key, block diagram
- Carroll diagram, Venn diagram, sort, criteria, intersection
- odd, even, multiple

Coverage in lessons

Learning focus	Learning outcomes (the ENC objectives)
Pictograms	Interpret and construct simple pictograms, tally charts and tables.
	Ask and answer simple questions by counting the number of objects in each category and sorting the categories by quantity.
Block diagrams	Interpret and construct simple block diagrams, tally charts and tables.
	Ask and answer simple questions by counting the number of objects in each category and sorting the categories by quantity.
Sorting using Carroll diagrams	Record, organise and represent data using a Carroll diagram. [Optional non-ENC objective]
Sorting using Venn diagrams	Record, organise and represent data using a Venn diagram. [Optional non-ENC objective]

12 Statistics

Big question

- What is the best way to represent data?

Global skills

- **Creative skills:** exploring/investigating
- **Interpersonal skills:** communication

Key vocabulary

- data, pictogram, block diagram

Resources

- squared, lined and plain paper
- poster-sized paper
- sticky notes or small pieces of paper and sticky tack

Language support

Display the poster prepared in the introductory activity, showing the sticky notes, with its labels, as a complete block diagram. Add speech bubbles with the questions you asked. Include further speech bubbles containing the answers. Keep this on view for reference in future lessons.

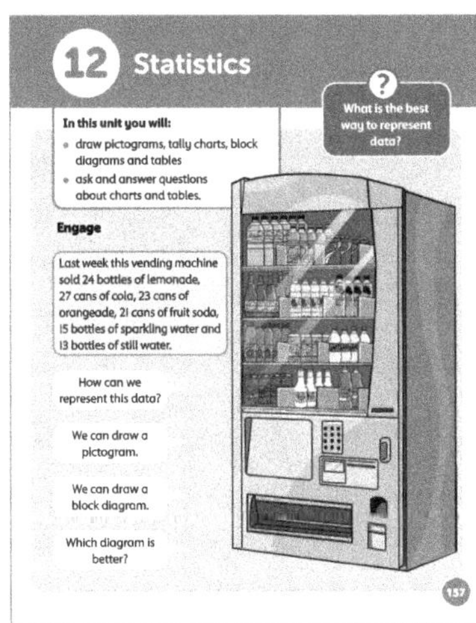

Introductory activity

Ask students what some popular drinks are among the class. Discuss and agree the top five drinks for students to choose from.

Give each student a sticky note or small square piece of paper (all the same size) to record their choice of favourite drink on. Prepare a large poster with the five drinks written as column descriptions at the foot of an empty chart. Ask each student in turn to stick their paper to the correct column on the chart. Students should add their squares of paper one above the other in each column.

Once complete, explain that what they have done is create a picture about their favourite drinks, called a **block diagram**. Add labels and a scale for the vertical axis to show the number of students.

Ask a range of questions such as: *Which is the most popular drink? Which is the least popular drink? How many more students liked X than liked Y?*

Main activity

Ask students to look at Student Book page 157. Display it on the IWB, if possible. Read together the description of what was sold through the vending machine. Ask questions about what was sold. For example, *Which drink did the machine sell most of?* (cola) *Which drink did it sell least of?* (water) *Was more water sold than cola?* (no) *Which did the machine sell more of, lemonade or orangeade?* (lemonade)

Then say, *All the information about the drinks sales is one sentence. Did this make it easy or difficult to answer the questions? How could we make it easier? How could we represent this information?* Agree that drawing a diagram would make it easier to understand.

What kind of diagram could we use? Some students may recall **pictograms** from the previous year. If not, suggest this and agree that pictograms or block diagrams would be good diagrams to use. Show students an example of a pictogram to remind them what they look like. As a class discuss their similarities and differences of the two representations.

Next, discuss what a pictogram and block diagram might look like using the vending machine data. *What title could we have for this block diagram? How many columns of blocks would there be? What symbol could we use for our pictogram?* Explain that you will be looking in more detail at how to present information in different ways in later lessons.

Differentiation

Supporting: Ask students simple questions about the vending-machine data.

Consolidating: Ask students specific questions about possible features of each representation. *If each symbol represents 1 drink, how many symbols would you draw to represent a can of fruit soda? Would that be more or less than you would draw for bottles of lemonade?*

Extending: *Which do you think would be better, a block diagram or pictogram? Why?*

Reflection time

Ask students to compare the two suggested representations, a block diagram and a pictogram. *Could both be used to answer the questions easily? Which do you think is easier to understand?*
What questions could you ask about the information in the diagram?

Ask students to imagine a classroom or playground snack sale. How could they find out what students would like to be able to buy from the snack sale?

12A Pictograms

Specific learning focus

- Answer a question by collecting and recording data in a frequency table and representing it as a pictogram to show results.

Global skills

- **Real-world skills:** research/presenting information/interpreting information
- **Interpersonal skills:** communication

Key vocabulary

- data, frequency table, pictogram, key

Resources

- squared paper, counters
- example of a completed pictogram with a symbol that represents two items

Language support

Add a pictogram to the classroom display with question speech bubbles and arrows to answers in other speech bubbles.

Introductory activity

Explain to students that they are going to collect some **data** and record it in a **frequency table**. Explain that data is another way of describing information that they have collected by observing, measuring or asking a question. The data will answer the question: 'Which is the most popular of these types of pet in our class?' Draw a frequency table on the board listing four possible pets: dog, cat, bird and fish. Explain to students that they should choose which one of these animals would be their favourite as a pet. Ask those who have chosen 'dog' to raise a hand – count how many hands are up and record the total in the table next to 'dog'. Repeat for the other three choices. When the table is complete, ask students *Which animal is the most popular choice? What else does this data tell us?* (e.g. which pet is the least popular, which is more popular than another)

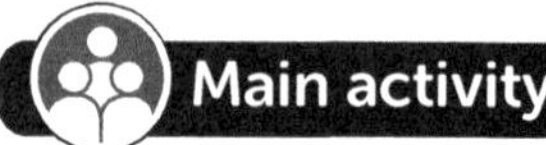

Main activity

Ask students to recall what a pictogram is. Remind them that it is a way of showing data about something on a diagram that uses pictures to represent the data.

Show students an example of a simple pictogram (e.g. showing how some students travel to school: by bus, by car or on foot) which has a **key** showing that one picture represents two responses (e.g. one smiley face represents two students). List choices in a column on the left, with the picture symbols arranged in rows, like the pictogram on page 158 of the Student Book. *What does the key tell us? What does each picture represent? What does the pictogram tell you about the data?* Help students to understand that the key on a pictogram is important because it tells us how many items each picture represents, for example each picture might represent one item or more.

Explain to students that they are going to create a pictogram to represent the favourite fruits of everyone in the class. Ask how they might collect the data. Some students may suggest asking everyone to just name their favourite fruit. Others may suggest they should they give a choice of, say, five fruits and ask which is the favourite among those. Agree that a fixed number of choices will work better. Discuss with the students why the pictogram would be a good way to represent the data (looks more interesting, easy to read) and why it might not (only useful for a small number of choices).

Ask students to work in small groups (or pairs) to collect data from all the students in the class about their favourite fruit and to complete the table in question 2 on Student Book page 158.

After groups have collected their data and completed their frequency table, ask students to draw their pictograms individually, completing the diagram on page 158 of the Student Book. Tell students that they will use one symbol to represent two students on the pictogram. Ask students how they could represent an odd number of responses on their pictogram (half a symbol).

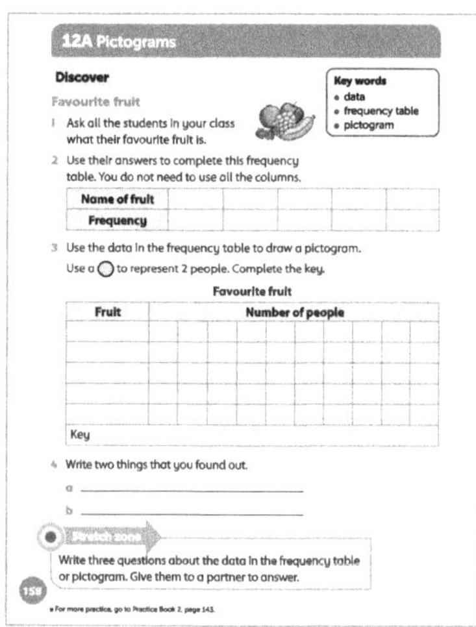

Differentiation

Supporting: Ask students to use counters to represent responses (1 counter representing 1 person) and then to replace every two counters with a symbol (or a different colour of counter) in the correct row on their pictogram.

Consolidating: Ask students questions about the data represented in their pictograms.

Extending: Ask students what additional questions they could ask as a follow-up to their investigation.

Stretch zone: *Write three questions about the data in the frequency table or pictogram. Give them to a partner to answer.*

Check that students can formulate sensible questions about the pictogram and that their partner's answers are clear.

 **Reflection time**

Ask individual students to feedback on what they did, how they collected and represented the data and what they found out. Ask students to think about why some of the pictograms might be different from others. For example, students may not have included the same responses, they may have omitted to ask some people or they may have included different fruits as options.

Practice Book: Students can now complete the activity on page of 143 in the Practice Book. They can do this directly after the main activity, as homework, or as the focus of a separate mathematics session to help students consolidate their learning and build fluency. Check that students recognise each type of vegetable. You may want to make substitutions if some are unfamiliar to your class. You could ask students to complete the first part of the practice activity (asking 10 people which vegetable they like best) in class and then complete the pictogram for homework.

Differentiated outcomes	
All students	should ask a question and record the responses in a frequency table. They should understand how a pictogram represents data.
Most students	will use data they have collected and recorded in a table to create a pictogram.
Some students	may make a range of statements about the data in displayed in their pictogram.

Answers

Student Book page 158

Assess the responses as you move around the classroom. Check that students have recorded their data correctly in the frequency table and used it correctly to complete the pictogram.

Practice Book page 143

Check that students have recorded their data correctly in the frequency table and used it correctly to complete the pictogram.

Stretch zone: Answers will vary. Check that students' statements reflect the data in their pictogram.

12A Pictograms

Explore Student Book page 159 • Practice Book page 144

Specific learning focus

- Use data presented in a block diagram to complete a pictogram to represent the same data.

Global skills

- **Creative skills:** exploring
- **Real-world skills:** presenting information/ interpreting information
- **Interpersonal skills:** communication/teamwork

Key vocabulary

- data, tally chart, tally mark, frequency table, pictogram, block diagram, survey

Resources

- squared paper

Language support

Continue to use the displayed diagram and pictogram to help students remember what these are called and their features. Encourage them to use the associated vocabulary (e.g. data, frequency table, survey, most, least) as they work on the activities.

 **Introductory activity**

Display on the board a list of five vegetables that students will be familiar with. Explain that you are going to do a survey and make a **tally chart** of their favourite two vegetables from the list. Ask whether any students have heard of a **survey** before, and confirm through discussion that a survey is when we collect information from people. After giving students a little time to choose their two vegetables from the list, go around the class asking for their responses. Record each student's two favourites by making a **tally mark** in the table. When any tally reaches 5, explain that the fifth mark is drawn across the existing four to make a group of five which makes the marks easier to count at the end. Model this for students.

Once you have recorded each student's choices, work out the total for each vegetable, modelling how you count up in fives and then ones to find the total (e.g. 5, 10, 11, 12).

Main activity

Ask students to look at the block diagram at the top of page 159 of the Student Book. Display it on the IWB, if possible. *What data is displayed? What question do you think was asked to collect this data? Who do you think was asked? Why?*

Students should discuss the block diagram further in pairs. Then ask each pair to share a piece of information from the block diagram, for example, 'The same number of people said coffee was their favourite drink as tea.' Ask them how they know that their statement is true. (Because the bars for coffee and tea are the same height.)

Explain that students are going to complete a pictogram to show the same data as in the block diagram. Ask pairs to compare the 'Tea' section of the block diagram with the completed row of the pictogram. Ask: *What do you notice?* They should notice that the block diagram represents the ten people who said tea was their favourite using ten blocks but the pictogram has only five smiley faces. *Can you explain why? Does the key help you?* Students should explain that each smiley face represents 2 votes for a drink so for 10 votes you use 5 smiley faces. Ask students to complete the pictogram and answer the questions in pairs.

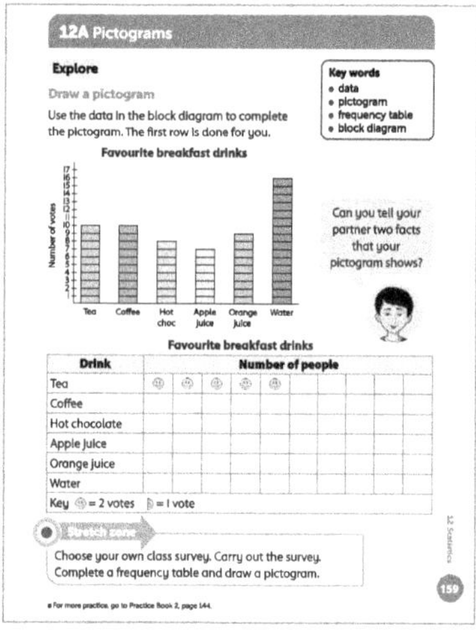

Differentiation

Supporting: Ask students to tell you how many votes for each category and then count up in twos from 0 to that number to help them work out how many smiley faces to draw for each drink.

Consolidating: Ask students to interpret the graphs and make statements about the most and least popular drink, for example.

Extending: Ask students to work out from the pictogram how many people were asked for their favourite drink.

Stretch zone: *Choose your own class survey. Carry out the survey. Complete a frequency table and draw a pictogram.*

Students may struggle to come up with a reasonable question. Support them to brainstorm some ideas, for example, *Could you ask something about school lunches? Do you know how many students have a brother or a sister? Do some have more than 1? More than 2?* Discuss what key is most appropriate, given how many people they plan to ask.

Reflection time

Ask questions about the breakfast drink data for students to answer. For example, *Were there more people who preferred tea or apple juice? How many more?* Ask which question they found easier to answer with the block diagram and which with the pictogram. Can they explain why one graph was easier to use than the other? For example, the block diagram may be easier to count up totals.

Practice Book: Students can now complete the activity on page of 144 in the Practice Book. They can do this directly after the main activity, as homework, or as the focus of a separate mathematics session to help students consolidate their learning and build fluency. You may prefer to have students complete the first part of the practice sheet (asking 20 people which of four foods they like best) in class and then complete the rest for homework.

Differentiated outcomes	
All students	should interpret simple graphs and charts.
Most students	will create a pictogram from the information.
Some students	may ask further questions after interpreting the data.

Answers

Student Book page 159

Check that each group has produced an accurate pictogram based on the block diagram.

Practice Book page 144

Check that students have tallied the foods and used the data to produce an accurate pictogram.

Stretch zone: The pictogram shows the most popular and least popular foods, for example.

12B Block diagrams

Discover Student Book page 160 • Practice Book page 145

Specific learning focus

- Answer a question by collecting and recording data in lists and tables, and representing it as a block diagram to show results.

Global skills

- **Real-world skills:** research/presenting information/interpreting information
- **Interpersonal skills:** communication/teamwork

Key vocabulary

- data, tally marks, frequency table, block diagram

Resources

- squared paper
- example of a complete block diagram

Language support

Add an example of a frequency table and tally chart to the class display to help students remember the names of the features on each. Include questions in speech bubbles such as 'What does this column tell us?', 'How many groups of 5 in this row?'

 **Introductory activity**

To help students recall what a block diagram is, ask them questions about the favourite drinks block diagram from the last lesson (page 159 of the Student Book). *What kind of data does the diagram show? How does it show how many? What do the numbers represent? What does each block represent?*

 **Main activity**

Explain to students that they are going to create their own block diagram to represent the favourite colours of everyone in the class. Ask how they might collect and record the data (e.g. asking a question and recording the answers in a tally chart or frequency table). *Should you ask everyone to just name their favourite colour? Or should you give a choice of four or five colours and ask which is the favourite among those?* Agree that a choice of four of five will work better. *What do you think the most popular colour will be? Is there a colour you think few people would choose?* Write the names of colours on the board as students suggest them so that students can refer to them later when labelling their tables and diagrams.

Tell students they should collect the data in pairs but they will each make their own frequency table and block diagram on page 160 of the Student Book. Each pair of students should first decide which four or five colours to use from the list on the board. Then they should ask each student in the class and record their responses in the frequency table. Check that students write their colour choices in the top row of the table first. Allow them to choose their own method of recording how many.

As they work, ask students questions about how they will display the data in their block diagram. For example: *How will you represent the total number of votes for a particular colour?* (with one block on the diagram) *What labels will you need to add to your block diagram?* (colour names) *Where will you write these?* (at the bottom of the diagram, under the column that shows how many votes for that colour)

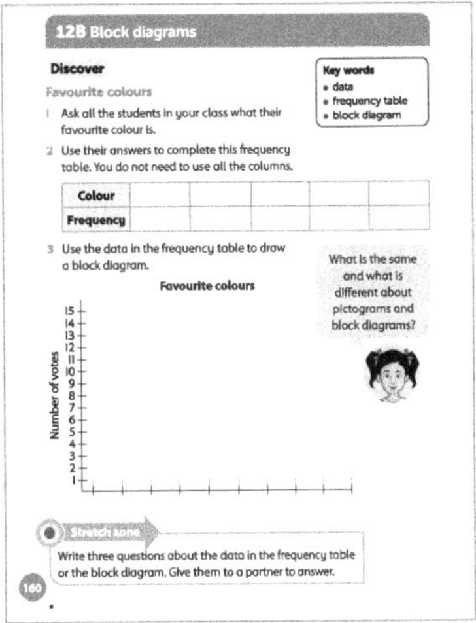

Differentiation

Supporting: Help students with recording the data by marking the tallies in the chart.

Consolidating: Ask students to compare their frequency table and block diagram to check the data matches.

Extending: Ask students to think of comparison statements they can make using their data (e.g. 7 more people prefer pink than brown).

Stretch zone: *Write three questions about the data in the frequency table or block diagram. Give them to a partner to answer.*

Check that students can formulate sensible questions and that their partner can answer them.

 **Reflection time**

Ask individual students to feedback on what they did, how they collected and represented the data and what they found out. *Did the data they collected support what they thought would be the most popular colour?*

Practice Book: Students now complete the activity on page of 145 in the Practice Book. They can do this directly after the main activity, as homework, or as the focus of a separate mathematics session to help students consolidate their learning and build fluency. Ask students to look at the empty answer lines. Discuss with them what information should be in each. Explain that they can find all the information in the pictogram.

Differentiated outcomes	
All students	should complete their own block diagram accurately.
Most students	will conduct a survey and use the data to complete the frequency table and block diagram accurately.
Some students	may devise questions to interpret the data in their frequency table and block diagram.

12B Block diagramss

Explore Student Book page 161 · Practice Book page 146

Specific learning focus

- Answer a question by collecting and recording data in lists and tables and representing it as a block diagram to show results.

Global skills

- **Creative skills:** exploring/investigating
- **Real-world skills:** presenting/interpreting information
- **Interpersonal skills:** communication

Key vocabulary

- data, tally marks, tally chart, frequency table, block diagram

Resources

- squared paper, coloured pencils

Language support

Add block diagrams and related frequency tables to the classroom display to help students remember the names of the features on each. Include further questions in speech bubbles such as 'How do the tallies make it easier to count?' and so on.

Student Book page 160

Assess the responses as you move around the classroom. Check that students have recorded their data correctly in the frequency table and used it correctly to complete the block diagram.

Practice Book page 145

Check that students have used the data correctly from the pictogram to complete the block diagram.

Stretch zone: The bar chart shows which day most apples were sold, which day least apples were sold and how many apples were sold in total for the week, for example.

 **Introductory activity**

Ask students what they like to eat and drink for breakfast. Explain that some schools set up a breakfast club for students. Students have breakfast at school with their schoolmates rather than at home with their families. Talk through the reasons why this is useful for some people – they can get to work earlier, they are not in such a rush at home or something else that students might suggest.

 **Main activity**

Explain that a school has decided to set up a breakfast club, and has to find out what students like to eat for breakfast. There are many different kinds of breakfast cereals, so the school has chosen five cereals to ask students about in a survey.

Ask students to look at page 161 in the Student Book and read the first speech bubble together. *What do the tally marks show? How could we count the number of votes for each type of cereal?* Students may suggest counting each tally mark. *Is there a more efficient way we could do this? What does each group of tally marks with a mark through it tell us?* (there are 5 votes) Together count the votes up in fives and ones for the first type of cereal. (9) *How many votes are there for Crispy rice?* (12)

Show students how the number of votes for Cornflakes and Crispy rice is represented as a block in the block diagram on page 161 of the Student Book. Point out the scale at the side of the diagram showing how many votes for each cereal. Ask students to tell you how tall each of the other bars in the diagram should be. Start with Fruity flakes as this is the easiest.

Ask students to work with their partner to draw the missing blocks for the other cereals to complete the block diagram. *The Cornflakes block is green and the Crispy rice block is red. Why do you think they are different colours?* Agree that choosing a different colour of block for each cereal will make it easier to 'read' the information and answer the questions.

After they have completed the chart and their diagrams, they should share with their partner two facts that the diagram shows. They can also ask their partner questions about the data shown in the diagram (e.g. 'Which cereal is more popular, Crispy rice or Fruity flakes?')

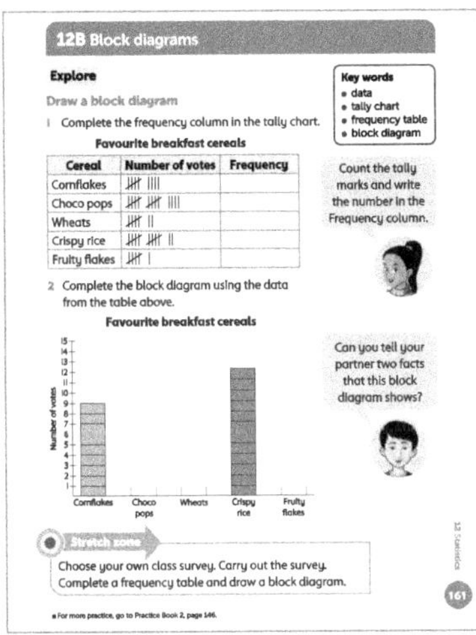

Differentiation

Supporting: Ask students to use a ruler to help guide them to draw their blocks.

Consolidating: Ask students to devise simple questions about their block diagrams.

Extending: Ask students to calculate how many people were surveyed altogether. *What is the difference in votes between the least and most popular cereal?*

Stretch zone: *Choose your own class survey. Carry out the survey. Complete a frequency table and draw a block diagram.*

Check that students have drawn the correct block diagram for the data they collected.

 ## Reflection time

Ask questions relating to the data such as: *If the makers changed the recipe for Choco pops and students did not like them anymore, which cereal do you think would then be the most popular?* Then ask each pair to form their own word problem relating to the data. Work on some of these problems as a class.

Practice Book: Students now complete the activity on page of 146 in the Practice Book. They can do this directly after the main activity, as homework, or as the focus of a separate mathematics session to help students consolidate their learning and build fluency. You may prefer students to complete the first part of the practice sheet (asking ten people in what month their birthday is) in class and then complete the rest for homework. Students may also want to keep a tally of their votes as well before they complete the frequency table.

Differentiated outcomes	
All students	should complete a frequency table and use the data to complete a block diagram.
Most students	will make at least two statements based on the data in a block diagram.
Some students	may ask and answer more complex questions relating to the data in a block diagram.

Answers

Student Book page 161

1 Cornflakes 9, Choco pops 14, Wheats 7, Crispy rice 12, Fruity flakes 6

2 Check that students have produced an accurate block diagram based on the data from the tally chart.

Practice Book page 146

Check that students have produced an accurate block diagram based on the birthday data from the frequency chart.

Stretch zone: The block diagram shows which three-month period had most birthdays and which three-month period had least birthdays, for example.

12C Sorting using Carroll diagrams

Discover Student Book page 162 • Practice Book page 147

Specific learning foci

- Sort numbers into categories such as odd/even, multiples of 2, 5 and 10.
- Use Carroll diagrams to sort numbers using one criterion and begin to sort numbers and objects using two criteria; explain choices using appropriate language, including 'not'.

Global skills

- **Creative skills:** investigating
- **Real-world skills:** research/presenting information
- **Interpersonal skills:** communication/teamwork

Key vocabulary

- Carroll diagram, criteria, even, multiple, sort

Resources

- none needed

Language support

Add a Carroll diagram to the classroom display with parts labelled for reference, for example: 'Title', 'Sorting criteria', 'Section'.

 **Introductory activity**

Ask all students to stand up. Now ask students who have a sister to stay standing and everyone else to sit down. *What can we say about those who are standing?* (They have at least one sister.) *What can we say about the students who are sitting down?* (They do not have a sister.) Now ask students to stay standing if they have a pet. Again, ask what we can say about those who are standing. (They have at least one sister *and* one or more pets.) Continue with a range of conditions (e.g. have brown hair, wearing something green) until one student is left standing. Ask what we can say about those who are standing at each stage.

 **Main activity**

Draw a single-criterion **Carroll diagram** on the board labelling it **'even'** and 'not even'. Ask students to suggest some numbers for each section. Suggest some more challenging numbers for students to **sort** (e.g. 84).

Draw another simple two-way Carroll diagram. Put the numbers 15, 30, 85 and 100 in the first section and 13, 36, 78 and 94 in the second section. *How could we label the diagram to explain how the numbers are sorted?* Give students some time to dsicuss in pairs how the numbers are similar and different. Ask questions to help, for example, *How many digits do the numbers have?* (2) *So it can't be that they are sorted into 2-digit numbers or not 2-digit numbers. What do you notice about the digits in the ones place?* (In the first group, they all end in 0 or 5.) *What type of number ends in 5 or 0?* (**multiples** of 5) Agree that the labels could be 'multiple of 5' and 'not a multiple of 5'. Say, *These are our sorting **criteria**.*

Ask students to look at the simple Carroll diagram on page 162 of the Student Book. Students should decide in pairs how to label the diagram and write at least ten numbers in each section of the diagram.

As an extension, some pairs could swap with another pair, cover or hide their labels at the top of their diagram and ask the other pair to try to work out what the sorting criteria are.

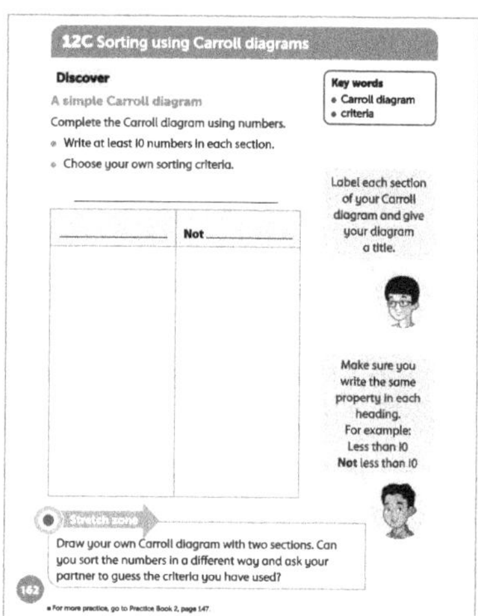

Differentiation

Supporting: You can help write the answers for students. They should tell you which section the numbers should go in. Keep referring to the sorting criteria.

Consolidating: Ask students to justify the choices they are making.

Extending: Ask for alternative sorting criteria and encourage students to create their own Carroll diagrams using these new criteria.

Stretch zone: *Draw your own Carroll diagram with two sections. Can you sort the numbers in a different way and ask your partner to guess the criteria you have used?*

Check that students have placed their numbers correctly.

 **Reflection time**

A single-criterion Carroll diagram can be used to sort anything into a category and its matching negative, such as green foods and foods that are not green. Choose something like this to explore with students. Encourage them to make suggestions for sorting criteria.

Practice Book: Students now complete the activity on page of 147 in the Practice Book. They can do this directly after the main activity, as homework, or as the focus of a separate mathematics session to help students consolidate their learning and build fluency. Students have seen Carroll diagrams with two different sorting criteria (a four-way table) in Year 1 so this can be used as an opportunity to recall that learning; it will be revisited in the next lesson.

Differentiated outcomes	
All students	should sort using one set of criteria for a simple Carroll diagram.
Most students	will use properties of numbers to sort using a Carroll diagram.
Some students	may suggest a range of ways of sorting using a Carroll diagram.

Student Book page 162

Observe students while they complete the activity in the Student Book, and check that they are completing the Carroll diagram according to the criteria that they have chosen.

Practice Book page 147

1 Check that students have placed their choice of objects correctly and chosen suitable criteria for the numbers. However, this will be developed further in the next lesson.

2 Title: Numbers from 1 to 30
Row headings: More than 14; Not more than 14
Column headings: Even, Not even

Stretch zone: Students may say, for example, that they must remember to label each part of their diagram and use is/is not to sort.

12C Sorting using Carroll diagrams

Explore Student Book page 163 • Practice Book page 148

Specific learning foci

- Sort numbers into categories such as odd/even, multiples of 2, 5 and 10.
- Use Carroll diagrams to sort numbers or objects using two criteria; explain choices using appropriate language, including 'not'.

Global skills

- **Creative skills:** investigating
- **Real-world skills:** presenting/interpreting information
- **Interpersonal skills:** communication

Key vocabulary

- Carroll diagram, criteria, odd, even

Resources

- large 100-square for front of class, individual 100-squares

Language support

Photocopy and display the students' Carroll diagrams from the previous lesson. This will be a useful reminder of what they did and why. Add labels to some of the numbers or items. For example, in a Carroll diagram with the sorting criteria of 'multiple of 10' and 'not multiple of 10', put a label next to 25 saying, '25 is not a multiple of 10 so it goes in this section'.

 **Introductory activity**

Display a large 100-square. Count in tens with students from any single-digit number to the bottom row of the 100-square and back again. Ask students to tell you what steps they were counting in. Count in multiples of 2, 3, 4 and 5 forward and back, depending on which count you think students need to practise.

 **Main activity**

Display an empty Carrol diagram on the board which has two criteria for sorting a set of numbers, for example, for the numbers 1–20, sort according to whether they are even or not and whether they are in the 5 times table or not. Work through the numbers from 1 to 20 one at a time asking students to say which box each number should go in, and why.

Talk to students about the kind of criteria that can be used on a two-criteria (four-way) Carroll diagram. These might include multiples of two different numbers, perhaps 3 and 4, or **odd** numbers and multiples of 3. Remind students that there is nowhere on the diagram for numbers that belong in two places at the same time because there is no overlap in a Carroll diagram. There is also nowhere to put the numbers that do not belong in either box. As the criteria include the opposite in some way, there is usually nothing that does not belong in one box or the other.

Ask students to look at the empty Carroll diagram on page 163 of the Student Book and explain that they are going to choose two sorting criteria for numbers that will go in this diagram. They can work in pairs to agree criteria and then complete the Carroll diagram in their books individually. Alternatively, after completing the Carroll diagram in one student's book, the pair could choose a different set of criteria and sort numbers for this diagram in the other student's book.

Observe students while they complete the Carroll diagram in the Student Book, and check that they write the numbers in the correct places according to the criteria they have chosen.

Ensure that their criteria are correctly worded. For example, the headings along the top could be 'multiple of 5' and 'not a multiple of 5', and the headings down the side could be 'odd numbers' and 'not odd numbers'.

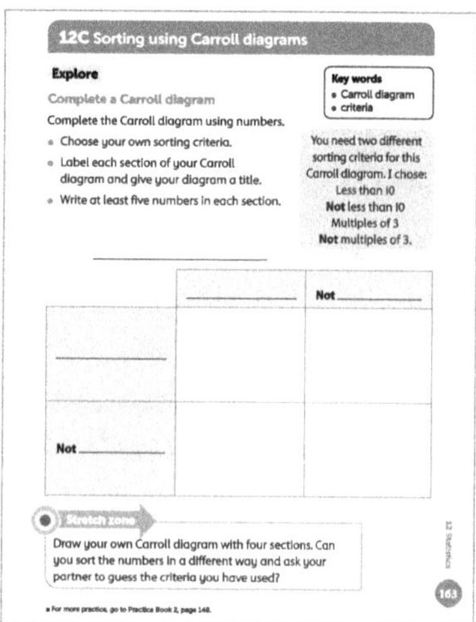

Differentiation

Supporting: You can help write the answers for students. They should tell you which section the numbers should go in. Keep referring to the sorting criteria.

Consolidating: Ask students to justify the choices they are making.

Extending: Ask for alternative sorting criteria and encourage students to create their own Carroll diagrams using these new criteria.

Stretch zone: *Draw your own Carroll diagram with four sections. Can you sort the numbers in a different way and ask your partner to guess the criteria you have used?*

Check that students have placed their numbers correctly according to their criteria before giving it to their partner to guess.

 **Reflection time**

Ask one pair of students to draw their Carroll diagram on the board, without labelling the criteria. Invite other students to try to work out what the sorting criteria were. Repeat for other pairs. Remind students that the criteria should be recorded in the style, 'criterion' and 'not criterion'.

Practice Book: Students now complete the activity on page of 148 in the Practice Book. They can do this directly after the main activity, as homework, or as the focus of a separate mathematics session to help students consolidate their learning and build fluency. Discuss the sorting criteria on the diagrams to make sure students understand them. Ask, *How many sides does a triangle have?* (3) *Can you give me some examples of 2D shapes that are not triangles?* (e.g. square, rectangle, pentagon) If students suggest a circle or semi-circle, explain that for this activity they should only use shapes with straight sides. Discuss what equal length means and demonstrate if necessary. You may want to give students shape stencils or 2D shapes they can trace around so that they can make equal-sided shapes more easily.

Differentiated outcomes	
All students	should understand how to sort by one criterion using a Carroll diagram.
Most students	will understand how to sort by two criteria using a Carroll diagram.
Some students	may suggest a range of ways of sorting using a Carroll diagram.

Answers

Student Book page 163

Check that students' criteria were reasonable and correctly worded and that numbers are written in the correct places according to the criteria chosen.

Practice Book page 148

1

	Even	Not even
In the 3 times table	12 30	15 27
Not in the 3 times table	2 26 14	7 29 11

2 Check that students have drawn the correct shapes in each box.

Stretch zone: Students may add an irregular hexagon, for example.

12D Sorting using Venn diagrams

Discover Student Book page 164 • Practice Book page 149

Specific learning focus

- Use Venn diagrams to sort numbers, objects and preferences using two criteria.

Global skills

- **Real-world skills:** presenting/interpreting information

Key vocabulary

- Venn diagram, criteria, odd, even, intersection

Resources

- none needed

Language support

Photocopy the Venn diagrams about students and their families to add to the classroom display with the Carroll diagrams. Make sure that the two different types of sorting diagrams are clearly labelled, and that each section of the diagrams includes an explanatory label to help students remember which diagram is which and what information goes in each section.

 Introductory activity

Draw two large overlapping circles on the board. Label one circle as 'Has an older brother or sister' and the other as 'Has a cousin'. Talk through how you would decide which circle to place each student's name in. Ask students, for example, where someone would go if they have a cousin but no older siblings and so on.

Ask, *Who has an older brother or sister?* Ask these students to move to one side of the class. *Who has a cousin?* Ask these students to move to the other side of the class. Stop any students moving from one group to another. *You can't be in two groups at the same time!* Explain that in a Venn diagram students meeting both criteria (having an older sibling and a cousin) can go in the overlap which is called the **intersection**. Make an intersecting group and then record all the names in the correct part of the diagram on the board. Agree that students having neither an older sibling nor a cousin will be outside the circles. Draw a box to surround the diagram (including any names outside the circles) and write a title above the box 'Students in our class'.

 Main activity

Ask students to look at the empty Venn diagram on page 164 of the Student Book. Explain that they are going to add information about themselves and a partner to the diagram. Model this for a few examples on the board with one student and yourself, showing how to record them in the diagram, for example, 'has brown hair', 'has a pet', 'likes to swim' and so on. Discuss each fact idea with the student before you record it on the diagram. if there is fact that you share, ask students where it should be recorded. Draw out that the overlap (intersection) is the correct place to write this. Ask students to think what the title could be for the Venn diagram. Agree that something like 'Facts about _____ and _____ ' would work.

Ask students to work with a partner to add information about each other to the diagram in their Student Book. Remind them to agree each fact choice first and that anything that is true for both of them must go in the intersection. Students could both record the same diagram. Alternatively, after completing the Venn diagram in one student's book, students could swap partners and complete another diagram so that each Student Book has a different Venn diagram.

As students work, ask questions to help consolidate their thinking, for example, *How have you labelled your diagram? Can you explain why the circles intersect? What is in the intersection in your diagram?*

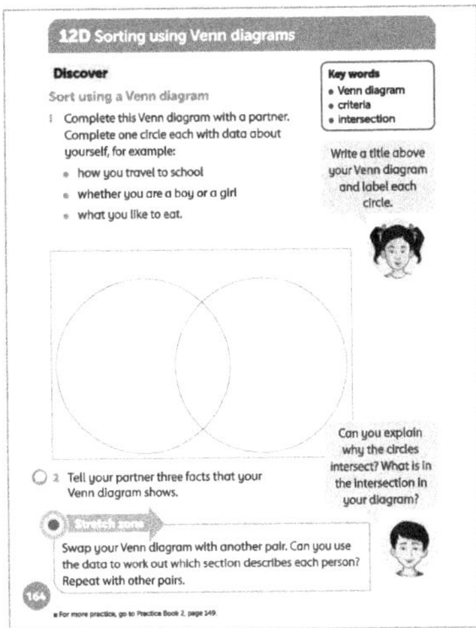

Differentiation

Supporting: Help pairs to label their Venn diagram. Ask questions to help them with sorting.

Consolidating: Ask students to justify how they have sorted different preferences.

Extending: Ask students whether they and their partner can decide on some information they could put outside their circles?

Stretch zone: *Swap your Venn diagram with another pair. Can you use the data to work out which section describes each person? Repeat with other pairs.*

Tell students they should cover the labels on their diagrams before swapping.

Ask pairs of students to share their Venn diagram with the rest of the class. Alternatively, complete a Venn diagram together, focusing on something of interest to students. Discuss what information belongs in each part of the Venn diagram, including the box surrounding it.

Practice Book: Students now complete the activity on page of 149 in the Practice Book. They can do this directly after the main activity, as homework, or as the focus of a separate mathematics session to help students consolidate their learning and build fluency. Discuss what criteria they might use to sort the shapes. They may want to use similar criteria to those used with the previous Carroll diagram, for example, '3 sides or less' and 'all sides equal'. You may want to give students shape stencils or 2D shapes they can trace around so that they can draw shapes more easily.

Differentiated outcomes	
All students	should sort using two given criteria for a Venn diagram.
Most students	will choose and use criteria to sort using a Venn diagram.
Some students	may suggest a range of ways of sorting using a Venn diagram.

Answers

Student Book page 164

Observe students while they complete the activity in the Student Book, and check that they are completing the Venn diagram according to the criteria they have chosen.

Practice Book page 149

Check that students have placed their shapes correctly in the Venn diagram.

Stretch zone: Students should add an appropriate shape to the intersection depending on the criteria they have chosen for the Venn diagram.

12D Sorting using Venn diagrams

Explore   Student Book page 165 • Practice Book page 150

Specific learning focus

- Use Venn diagrams to sort numbers using two criteria, e.g. odd numbers and multiples of 3.

Global skills

- **Creative skills:** exploring
- **Real-world skills:** presenting information
- **Interpersonal skills:** communication/teamwork

Key vocabulary

- Venn diagram, criteria, intersection, odd, even, multiple

Resources

- none needed

Language support

Display the list of things that are the same and things that are different about Venn and Carroll diagrams (from Reflection time activity). This will help students to remember which diagram is which.

 **Introductory activity**

Play a game of 'fizz-buzz' with the class. Start counting around the class in order, and if a student says a multiple of 3, they say 'fizz', and if they say a multiple of 4, they say 'buzz'. If their number is a multiple of both numbers, they say 'fizz-buzz'. Anyone saying the wrong number or the wrong 'fizz' or 'buzz' sits down and is out of the game. Keep counting until one student is left standing.

The counting will start like this: 1, 2, fizz, buzz, five, fizz, 7, buzz, fizz, 10, 11, fizz-buzz, 13, 14, fizz, buzz, 17, fizz, 19, buzz, … and so on.

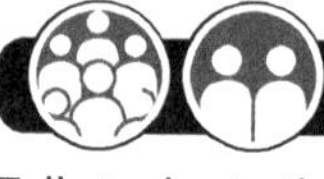 **Main activity**

Tell students that today they are going to use a Venn diagram to sort numbers. Draw a large empty Venn diagram on the board with two overlapping circles. Discuss with students how they might label the Venn diagram box and the two circles inside it. Choose an example such as 'numbers 0 to 100' for the diagram title with the two circles labelled 'even numbers' and 'numbers over 50'. Ask students to suggest some numbers for each section of the diagram, explaining why they belong there.

Ask students to look at the empty Venn diagram on Student Book page 165 and to work in pairs to choose a title and criteria for this diagram. They should then sort numbers to complete it. Students should both record the same diagram. Explain that they should then individually draw a second Venn diagram in their notebooks and use it to sort numbers, but not to label it. Can their partner work out which criteria they used? Alternatively, they can create the second Venn diagram in pairs and then swap with another pair.

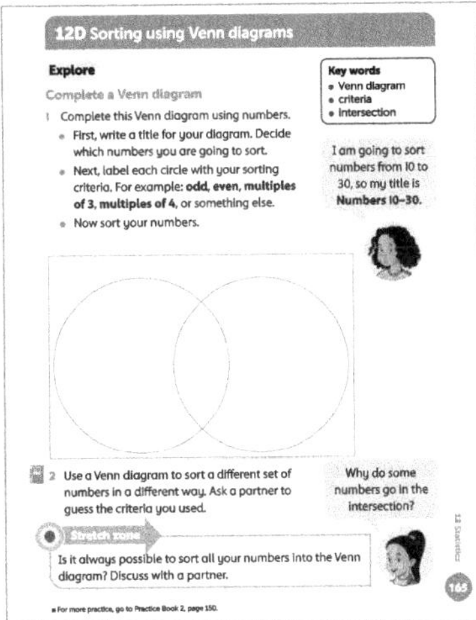

Differentiation

Supporting: Help students decide on their sorting criteria and label the diagram together.

Consolidating: Ask students to justify the choices they are making.

Extending: Ask students to sort a larger set of items, for example, numbers to 50 sorted as multiples of 3 and even numbers.

Stretch zone: *Is it always possible to sort all your numbers into the Venn diagram? Discuss with a partner.*

Check that students can reason why not all numbers can be placed. For example, if sorting into multiples of 5 and even numbers, then odd non-multiples of 5 will not fit, students might suggest that 7 is not a multiple of 5 or even.

 **Reflection time**

Ask pairs to swap books and check each other's Venn diagrams. *What is the same and different about Venn diagrams and Carroll diagrams?* Use their ideas to draw up a list of differences and similarities.

Ask students to take the information from a Venn diagram and see whether it can be placed on a Carroll diagram (or the other way around). This will help reinforce the differences between the two types of diagrams. Ask students to say what is the same and what is different about the diagrams.

Practice Book: Students can now complete the activity on page of 150 in the Practice Book. They can do this directly after the main activity, as homework, or as the focus of a separate mathematics session to help students consolidate their learning and build fluency. Show students examples of symmetrical and non-symmetrical shapes to remind them of what 'symmetrical' means before they begin the activity.

Differentiated outcomes	
All students	should sort numbers using Venn diagrams with given criteria.
Most students	will choose two criteria to sort numbers using a Venn diagram.
Some students	may suggest a range of ways of sorting, with a variety of criteria, using a Venn diagram.

Answers

Student Book page 165

1 Observe students while they complete the Venn diagram in the Student Book, and check that they write the numbers in the correct places according to the criteria they have chosen.

2 Students choose their own set of numbers and sorting criteria for the Venn diagram.

Practice Book page 150

1 Multiples of 3: 3, 6, 9, 12, 18, 21, 24, 27

Multiples of 5: 5, 10, 20, 25

Multiples of 3 and 5: 15, 30

3 Check that the shapes are drawn and sorted correctly.

Stretch zone: Check that students have drawn a 2D shape outside the two ovals that is not a triangle and not symmetrical, e.g. an irregular quadrilateral, an irregular pentagon, an L shape.

12 Statistics

Big idea

- I can choose a pictogram or a block diagram to represent data. I can use a Venn diagram or a Carroll diagram to sort data.

Global skills

- **Creative skills:** investigating
- **Real-world skills:** research/presenting information/interpreting information
- **Interpersonal skills:** communication/teamwork
- **Self-development skills:** reflecting on learning

Key vocabulary

- block diagram, pictogram, tally chart, frequency table, Venn diagram, Carroll diagram, criteria, odd, even

Resources

- class list, squared paper

Language support

Display a range of questions that students might use when conducting their survey on after-school activities, for example 'What activity do you like to do most?', 'How many after-school activities do you do each week?'. You can add others as they arise during the activity.

 **Introductory activity**

Ask students how you could quickly find out what activities students like to do after school or how many brothers and sisters each student has. Share ideas: for example, conduct a survey using a tally chart to record responses, make a list of choices and put a tick next to each choice when it is chosen. If no one suggests it, show students a class list and explain how this could be quickly passed around for students to tick one of several options or write a number as an option for each student.

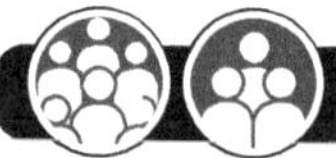 **Main activity**

Read the Connect activities on page 166 of the Student Book together. Discuss with the class how to carry out the survey. Ask students what question they are trying to answer. (What are the favourite after-school activities?) Talk through what information students need to collect, and how they will collect it. Agree together a list of five possible after-school activities the students can choose from for the survey. Discuss the question in the speech bubble: *What do you need to remember when you draw a pictogram or a block diagram?* Use their ideas to record a 'checklist' of things they will need to do.

As students are working in their groups on the survey, you can move around the room and ask questions to support them, or to assess their progress, for example, *Have you asked everyone in the class? Did you use a tally chart before you completed your frequency table? How will you number the diagram? What is your key?*

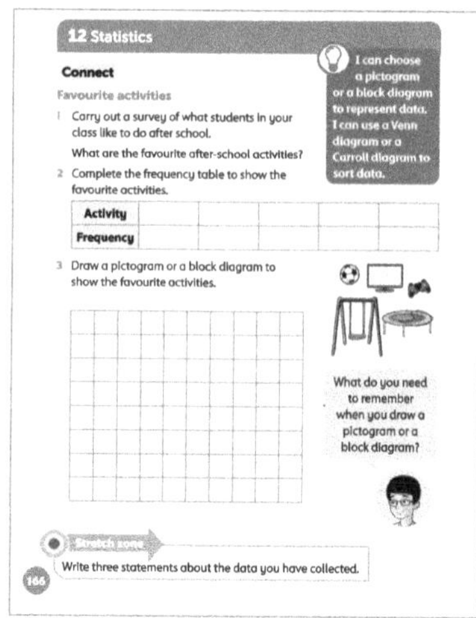

Differentiation

Supporting: Help students transfer their data. They can tell you how many students chose each activity and you can complete the frequency table for them.

Consolidating: Ask students to check their data to make sure that they have included it all on the pictogram or block diagram.

Extending: Ask students to think of questions they can ask to interpret their data, for example, 'Which activity is the least popular?'.

Stretch zone: *Write three statements about the data you have collected.*

Check that students' statements are correct. Ask further questions to extend their thinking, for example, If 'football' became more popular and double the number of students chose it as their favourite, how many students would that be?

 **Reflection time**

Invite some groups to share their diagram or pictogram. Can other students in the class make statements using this data? Once a couple of groups have presented, ask students to think about how the statements were the same and how they were different. Can they explain why that is?

Differentiated outcomes	
All students	should collect data about favourite after-school activities.
Most students	will use the data to complete a frequency table and a pictogram or block diagram.
Some students	may interpret their diagram and ask questions about the data.

12 Statistics

Global skills

- **Creative skills:** investigating
- **Real-world skills:** presenting information/ interpreting information
- **Interpersonal skills:** communication
- **Self-development skills:** reflecting on learning

Student Book

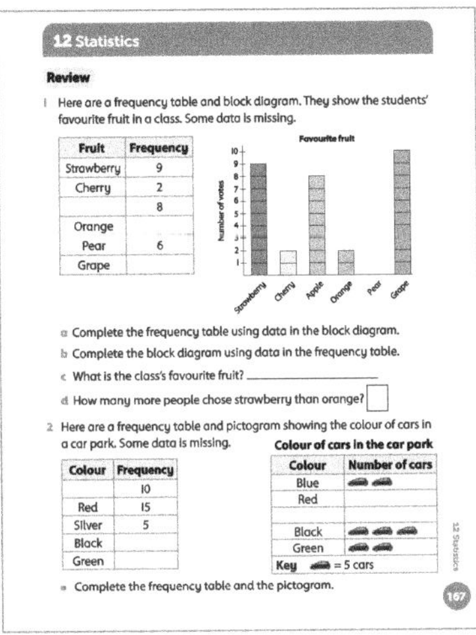

With young children, assessment activities are most effective when carried out as an everyday classroom activity. Students should be able to interpret frequency tables, pictograms and block diagrams and use them to complete missing data or answer questions about the data.

It may help to remind students of the features of the different diagrams, for example, the differences between pictograms and block diagrams, before they begin.

Answers

Student Book page 167

1 a Apple, 2, 10   **b** students draw 6 blocks for the pear

  c grape   **d** 7

1 a Blue, 15, 10   **b** 3 cars, Silver, 1 car

Practice Book

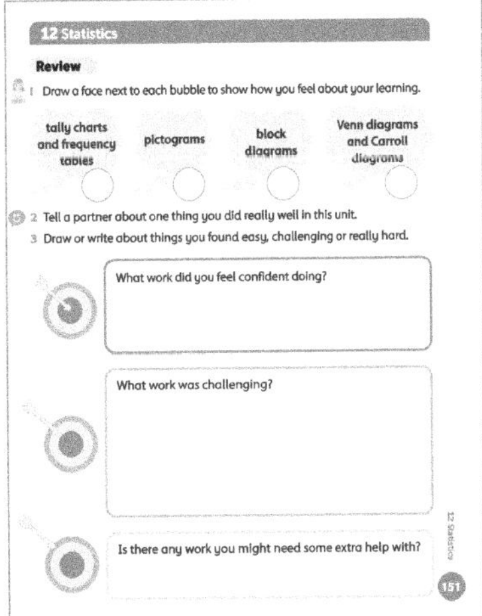

It is appropriate to complete this Practice Book review as a whole-class discussion. You may choose to keep a record of the class discussion or a copy of the review page for your own records. The review provides an opportunity for students to reflect on their learning from the unit, to discuss any areas of mathematics that they feel went particularly well, and any areas that they feel less confident about. Ensure all students have a copy of the Student Book as a reminder of the areas of mathematics that they have worked on in this unit.

Allow students plenty of time for discussion before asking them to complete the Practice Book page individually, and then, if appropriate, to share their responses with the rest of the class. If students complete this self-assessment at home, encourage them to discuss this with adults. Make a note of areas that students still feel unsure about.

Additional material

There are additional end-of-unit assessments available on the *Oxford Owl* website.

Glossary

addition	adding together to make a new number or amount. The sign for addition is +
analogue clock	a clock with hands that show the time
anti-clockwise	turning the opposite way to the hands of a clock
array	an arrangement of objects or numbers in columns and rows
block diagram	a type of graph that shows different numbers as rectangular blocks of different lengths
calculate	work out an answer
capacity	how much something holds
Carroll diagram	a table or diagram that sorts things into different categories, using 'X' or 'Not X' sorting rules
centimetre	a standard length shown on a ruler: 100 centimetres = 1 metre
circular	in the shape of a circle
clockwise	turning the same way as the hands of a clock
coin	an item of money made from metal
column	numbers arranged one below the other in a line going from top to bottom in a table
difference	the result of subtracting one number from another
digital	showing the time in displayed digits rather than with a hand or pointer
divide	share out equally
double	multiply a number by 2 or make a number twice as many
edge	where two faces of a 3D shape meet. An edge can be curved or straight
estimate	make a reasonable guess at an amount, number or value without calculating
even	describes a number that can be divided exactly by 2
exactly	neither more nor less than
face	the flat surface of a 3D shape
fewer	a smaller number of things
fraction	a part of a whole shape or a whole number
gram	**a** measurement used to weigh light things; 1000 grams = 1 kilogram
graph	a diagram showing the relation between two quantities, usually along two axes
group	a set of objects or numbers
half	one of two equal parts. You write a half as $\frac{1}{2}$
height	how tall something is or how far it is from the ground
hexagon	a flat shape with six straight sides
hundred-square	the numbers 1–100 arranged in order on a 10-by-10 grid
inverse	the opposite or reverse of something
investigate	explore to find out more about a number, shape or problem
kilogram	measurement used to weight heavy things; 1 kilogram = 1000 grams
least	the smallest
length	how long something (or a time) is. You can find the length by measuring the distance or time between two points.
line of symmetry	divides a shape in half so that one side is the reflection of the other
litre	a measurement of how much something holds. 1 litre = 1000 millilitres
mass	how much matter is in an object, measured using scales, usually in kilograms
measuring jug	a jug with a numbered scale used to show us how much of something we have
metre	a standard length used to measure long lengths or distances; 1 metre = 100 centimetres

millilitre	a small measurement of how much something holds; 1000 millilitres = 1 litre
minute	a short time: there are 60 minutes in 1 hour
most	the largest amount
multiple	the result of multiplying a number by another number: the multiples of 4 are 4, 8, 12, 16 and so on
multiply	adding the same number again and again
number bond	number fact, for example the number bonds for 10 are: $0 + 10 = 10$, $1 + 9 = 10$, $2 + 8 = 10$, $3 + 7 = 10$, $4 + 6 = 10$, $5 + 5 = 10$
number grid	a collection of numbers arranged in a square or rectangular gird
number line	a line with numbers marked in regular intervals
numeral	the symbol or symbols that represent a number
octagon	a flat shape with eight sides
odd	describes a number that cannot be divided into two equal parts
1-digit number, 2-digit number, 3-digit number	the digits 0 to 9 are used to make numbers. A single-digit number such as 4, 6 or 8 has only one digit. A 2-digit number such as 25 has two digits. A 3-digit number such as 240 has three digits.
part	a section or amount of something that, when combined with others, makes up a whole
partition	split numbers into smaller parts so that they are easier to work with
pentagon	a flat shape with five sides
pictogram	a graph that uses pictures to show amounts of something
place value	what each digit is worth in a number: for example the 5 in 54 represents 5 tens
predict	using what you know to work out the possible answer to a question or problem
quadrilateral	a flat shape with four sides
quarter	one of four equal parts of a quantity, shape or number. You can write a quarter as $\frac{1}{4}$

rectangular	in the shape of a rectangle
repeated addition	adding the same number many times: for example, $6 + 6 + 6 + 6$
right angle	a quarter of a whole turn
round	to change a number to a less exact number that is easier to work with: rounding 42 to the nearest 10 gives 40
row	numbers arranged one next to the other in a line going across a table from left to right
rule	something that tells you what to do
ruler	an instrument used for measuring the length of lines or objects and for drawing straight lines
scales	an instrument that measures mass or weight
second	a very short time: there are 60 seconds in 1 minute
sequence	numbers arranged in order following a rule
shape	the form of an object, its outer boundary
share	divide a larger number or amount into smaller numbers or amounts
side	one of the lines that make up a 2D shape
solve	find an answer to a number sentence, question or problem
straight line	the shortest line between two points
subtraction	the result of taking one number away from another (or the difference between one number and another). The sign for subtraction is –
sum	the total amount resulting from an addition of two or more numbers or amounts
surface	the outside part of a shape: it has length and width but no thickness. It can be flat or curved
symbol	a sign that is used instead of words, to represent something
symmetrical	describes a shape for which one half is a reflection of the other half

tally	a small mark. Tallies are used to show how often something has happened
temperature	the amount of heat in something, often measured in degrees Centigrade
three-quarters	three of four equal parts of a shape, quantity or number. You can write three-quarters as $\frac{3}{4}$
triangular	in the shape of a triangle
turn	move in a circular direction around a point
Venn diagram	a sorting diagram that uses circles to show sets of items with different properties. An overlap between two circles shows items that have both properties
vertex (plural vertices)	a point at which two or more lines meet in an object or shape
whole	not broken into parts
whole number	a positive counting number, such as 2, 16 or 84, that is not a fraction
zero	a word for the digit 0. It means no amount, nothing or nought